ANIMAL MOTIVATION

EXPERIMENTAL STUDIES ON THE ALBINO RAT

—

Prepared under the auspices of
COLUMBIA UNIVERSITY COUNCIL FOR RESEARCH IN THE SOCIAL SCIENCES

ANIMAL MOTIVATION

EXPERIMENTAL STUDIES ON THE ALBINO RAT

BY

C. J. WARDEN

THE ANIMAL LABORATORY, DEPARTMENT OF PSYCHOLOGY
COLUMBIA UNIVERSITY

WITH THE COLLABORATION OF

T. N. JENKINS
MARION JENKINS
L. H. WARNER
E. L. HAMILTON
H. W. NISSEN

NEW YORK
COLUMBIA UNIVERSITY PRESS
1931

Published March, 1931

Printed by The Torch Press
Cedar Rapids, Iowa

ACKNOWLEDGMENTS

To the Council for Research in the Social Sciences of Columbia University, generous thanks are due for their continued support of this project, extending to the publication of the final report in its present form. The project could not have been brought to successful completion without the hearty support, from year to year, of Dean F. J. E. Woodbridge, of the Graduate Faculties, and Professor A. T. Poffenberger of the Department of Psychology.

As a fitting recognition of their valuable contributions, the names of the five student co-workers who labored faithfully in carrying out the several major assignments have been placed upon the title page as collaborators, in the order in which they became connected with the project. The project owes its success in no small way to the willing coöperation of this group, and to the earlier work of Frances Holden, whose monograph has been included in the appendix. Miss Mercy Aylesworth deserves mention in connection with the study of motivation in discrimination work, which will also be found in the appendix.

Many valuable suggestions were offered by Professor R. S. Woodworth and Professor A. T. Poffenberger, of this department, relative to the preparation of the material for publication in the present volume.

C. J. Warden

PREFACE

In the spring of 1925, the writer submitted to the newly organized Council for Research in the Social Sciences of Columbia University the tentative plan of a comprehensive experimental investigation of animal motivation to extend over a period of several years. The plan called for the measurement and comparison of the reaction tendencies or drives associated with hunger, thirst, maternity, various sex conditions, etc., by means of the Columbia Obstruction Method, which had been devised by T. N. Jenkins and C. J. Warden in 1924, and was at the time being used by Holden in connection with her doctorate research. It was proposed to use the white rat, at first, in working the field over systematically and then, if possible, to duplicate "with the monkey and other higher types of animal," such parts of the work as offered the most promise.

The plan covering the contemplated work on the white rat was approved by the Council and an initial grant of $1000.00 for 1925-26 became available on July 1, 1925. A like amount was provided for the continuation of the work through the following year; the sum of $2000.00 was voted for further experimentation as planned for 1927-28, and a final grant of $1000.00 made to cover the 1828-29 budget. Of the allotment for 1928-29, one-half was provided to cover the cost of such clerical and statistical work as might be necessary to bring the results of the various studies together (Part VII) and to prepare the final report for publication in the present volume. The Council also generously assumed financial responsibility for the printing of this report. A summary of expenditures for the entire period shows the disbursement of the $5000.00 grant to be as follows: (1) for animals, $3,173.14; for apparatus and equipment, $752.13; (3) for food, bedding, etc., $647.71; (4) for clerical and statistical help, $427.02.

As will be seen, the above fund covered only the cost of the necessary animals and materials, and made no provision for stipends for those who should care for the animals and do the actual experimental work. From the beginning, it had been planned to make this project on animal motivation the major line of research in the departmental animal laboratory and to assign specific topics to approved graduate students, who should work under the supervision of the writer. Under this arrangement the animals were to be cared for, in the main, by the regular laboratory assistant and the data taken by approved students who should be given assignments. This plan has proven to be a wise economy and has worked out satisfactorily in every way. In most cases the first topics assigned were doctorate problems, and by encouraging each to take additional assignments it has been possible to limit the experimental work to a few mature and carefully selected students. By training each worker thoroughly in connection with his first assignment, the high level of uniformity in procedure necessary to warrant the widest interpretation of the results, taken as a whole, was maintained.

Much of the material of this volume has already appeared from time to time in the form of separate papers and monographs. In reprinting these various publications it was found desirable, in some cases, to modify the introductory portions so as to avoid needless repetition. Since the general method had been given detailed treatment in Part I, all diagrams and descriptions of the apparatus were deleted from papers included in the other parts. Aside from such modifications, the papers and monographs have been reprinted in full.

The author is under special obligations to the editor and publishers (The Williams and Wilkins Company, Baltimore, Md.) of The Journal of Comparative Psychology and Comparative Psychology Monographs for permission to reprint Part I, 1 and 2; Part II, 1; Part IV, 1; and Appendices 1, 2 and 4; and to the editor and publishers (Clark University Press) of The Journal of Genetic Psychology and the Genetic Psychology Monographs for permission to reprint chapters Part II, 2; Part III, 1; Part IV, 2 and 3; Part V, 1; Part VI, 1; and Appendix 3. The short introductory sections are, of course, new, and are intended to orient the reader with respect to the general topic of each main part. The opening section as well as Part VII, 1, in which the results are analyzed and interpreted, is also new.

C. J. Warden

CONTENTS

PART I

PART II

PART III

PART IV

PART I

THE COLUMBIA OBSTRUCTION METHOD

Part I

THE COLUMBIA OBSTRUCTION METHOD

C. J. Warden

I. Previous Methods of Studying Animal Motivation

The measurement and analysis of the motivation factor in infra-human behavior has been more or less neglected in the modern experimental movement in comparative psychology. The reason for this can hardly be due to lack of interest in the dynamic aspect of behavior, since theoretical discussions on this topic have not been wanting. Doubtless one important cause of this neglect has arisen from the obvious difficulties of applying experimental methods in this field. It is simple enough to get rough measures of learning performance under various conditions which may be compared and may thus lead to an understanding of the learning process. But precisely how can one measure the strength of tendencies in the organism aroused by hunger, thirst, sex deprivation, etc., and thus come to know these primary determiners of animal activity as they operate together in the daily life of various types of organism? The problem is made difficult by reason of the fact that one is here dealing directly with fluctuating physiological conditions that are difficult to control even roughly. Then too, those internal states which underlie various types of motivation carry with them the likelihood of a highly excitable organism, which predisposes to emotional disturbances during the testing process. When these and other factors (Part I, 1, 2) are taken into account, it is small wonder that experimentalists, interested in the quantitative analysis of infra-human behavior, should choose to work in the field of learning rather than in that of motivation.

Nevertheless several attempts have been made to investigate some of the problems of animal motivation in the laboratory, although usually these have been more or less preliminary and sporadic ventures. A brief account of the methods employed in this work will be given in the present section. This survey will be followed by a more detailed account of a new and thoroughly standardized method — the Columbia Obstruction Method — devised by T. N. Jenkins and C. J. Warden of this laboratory, in 1924, and, as later modified in certain indicated details, employed throughout the comprehensive project reported in this volume.

The Revolving Wheel Method seems to have been among the first used in the study of animal activity. As early as 1898, Stewart (15) applied the method to small mammals. The apparatus consists essentially of a revolving wheel attached to the living cage, after the manner of the old-fashioned squirrel cage, to be set in rotation by the animal inside whenever it is predisposed to be active. A measure of the state of activity of the animals at different times and under various conditions, can be secured by means of a kymograph record, a revolution counter, or some other such mechanical device. Slonaker (12) from about 1907 and onward, and Richter (10) and his pupils from about 1922 and onward have made no little contribution to our knowledge of motivation in small mammals by their extensive use of this method. Many problems intimately associated with the more analytical aspects of motivation can be studied by this method, although as we have before noted (Part I, 1) the indices obtained are of general activity which can seldom be related to specific incentives or incentive-seeking tendencies. Thus nocturnal-diurnal and other rhythms of activity can be measured, and the influence of gland extirpation, injections of various secretion extracts and of drugs, etc., can be determined by comparison of the indices secured with those of the normal animal. While the method does seem to give a fairly trustworthy index of the general tendency of the animal to be active or inactive under various conditions, it does not appear to lend itself readily to the isolation of many of the more important animal drives. Unfortunately for our present interest, the activity is always the same — running the wheel — regardless of the internal state induced in the organism; there is nothing characteristic in the behavior from one condition to another to give us the cue as to what may be the objective of the activity in the aroused organism. In the study of specific animal drives, it is desirable to set the conditions so that the amount of activity is registered, but also in such wise that this activity is exhibited as seeking behavior directed toward the specific type of incentive that the experimentally induced physiological state may call for. Certainly, the knowledge that a hungry rat is excited by food in the offing and is bending its energies to secure the food, to the exclusion of other types of activity, presents a more adequate biological and psychological picture of its behavior than the knowledge that, when equally hungry the same rat will turn a wheel round and round so many times per hour. The fact that the former statement indicates the relation of the behavior to a specific group of stimuli (food) is a vital point in the understanding of the nature of motivation in the situation.

The same general criticisms have been urged (Part I, 1) against the Activity Cage Method which was devised by Szymanski (17) and applied to a large number of animal types, and later used on small mammals by Richter and his pupils in America. This method has so far yielded much valuable information regarding some of the more fundamental activity rhythms in these organisms, and also concerning the manner in which these rhythms come to be overlaid by habit patterns. References to studies of activity of this type — "spontaneous" activity — are given in the bibliography, and in connection with a recent article by Richter (11). In this later article Richter has, indeed, applied the term "internal drives" to the behavior revealed by this method as well as the method discussed in the preceding paragraph, but it would seem wise to restrict the use of the term "drives" to activities depending upon the arousal of internal physiological states which can be definitely related to specific incentives, such as food, sex objects, etc., in the manner of approaching or avoiding responses.

The Choice Method of determining the relative potency of different incentive-drive situations may be considered an extension of the Preference Method, as widely used by Lubbock (5), Graber (2) and others in the study of sensory problems, during the latter half of the nineteenth century, to the field of motivation. The animal is stimulated by a pair of incentive objects (food-female, for example) and, with equally easy access to both, supposedly goes to the one having the highest incentive value at the moment. It is difficult to say who first applied this type of test, but certainly it has been used extensively by various workers on lower forms in approaching the problem of the relative potency of tropisms, and of instincts under different conditions. The literature of the last several decades contains so many studies of this sort that it is hardly necessary to cite specific investigations. Mast has perhaps used the method more widely than any other. More recently (1924) Moss (8) compared the food-sex drives in the white rat by this method, using a simple technique, which was greatly improved by Tsai (18) the following year. Jenkins (Part I, 1, last 4 pages) developed a standardized apparatus for testing animals by this method in 1925, but we have made no use of this device in our research work on motivation at Columbia, because we have felt that the method, even as improved by Jenkins, is of slight value in this field. In describing the latter apparatus, the writer (4) (Part I, 1, last 4 pages) pointed out the more obvious and fundamental defects of any form of Choice Method, as a means of analyzing dynamic behavior. It is quite impossible to isolate a single drive, since the method itself requires that two physiological states, corresponding to the two incentives used (food-sex, for example) be simultaneously aroused in the organism. But the two states induced are likely to be interdependent and the corresponding drives, unless they are genuinely antagonistic, thus influence one another in various manners and degrees instead of operating as distinct internal tendencies. Furthermore, it is impossible to say in our present state of knowledge precisely how much starvation should be balanced against a given amount of sex deprivation in such a case. The Method of Choice appears to be ill suited to quantitative, analytical work on motivation.

The Learning Method (or Training Method) of studying motivation has been employed with some success within a fairly limited field. This method is an adaptation of the usual learning experiment in that different incentives are used in the setting up of the same habit, and the various incentive-drive situations ranked in terms of some common index of learning efficiency (speed, accuracy, etc.). This method has been used in a score or more of studies beginning with the early work of Yerkes (1907) on punishment vs. reward; examples of the application of the method to certain special problems will be found in the last three sections of the appendix of the present volume. While the method must always be considered an important one in connection with the analysis of the incentive-drive factor in the learning process, it is doubtful whether it furnishes reliable indices of drives as such. For, it is clearly possible that a given internal physiological state may arouse an organism to activity without increasing thereby its efficiency in the adjustment-fixation process involved in habit formation. Indeed, too great internal arousal may lead to decreased learning efficiency, as when an animal is rendered so excited by a prolonged period of acute starvation as to be unable to run a maze efficiently. Conditions of motivation influence profoundly the stimulability of the animal and the entire repertory of movemnts, as well as the process of fixating new movements. Any comprehensive account of animal motivation must, of course, include the influence of incentive-drive factors on learning, but the more fundamental problem of the relation of incentives and drives to activity and movement must be obtained by some method that tests rather than trains. Aside from the narrow limitations of the learning method, it has also the very grave defect of introducing into the motivation index obtained, the various sources of error from chance factors that decrease the reliability of all training scores. It is common knowledge that learning scores in both the animal and human field involve gross inaccuracies because of the inability to control the experimental conditions adequately over the usual long training period. Then the very nature of the typical learning experiment in so far as it requires any considerable amount of repetition of the act to be learned, tends to lower rather than raise the value of the motivation factor in the process. All these criticisms suggest that the general problem of motivation should first be approached by means of some direct test in which the training factor is eliminated as far as possible. The learning method, however, may very well take care of the special problems as to the relation of motivation to different types of habit formation.

Several attempts have been made to develop a test method, as distinguished from a training method of measuring motivation. The basic principle employed in all these attempts is the general one involved in the usual learning experiment, which has come down to us from Lubbock. As early as 1882, Lubbock (6) laid down the principle that the best method of testing an animal as to its intelligence, etc., is to "interpose some obstacle" between the animals to be tested and some definite incentive object, thus requiring the animal to surmount the obstacle in order to secure the incentive. Nearly all our laboratory techniques rely upon the application of this principle in some form to arouse the animal to undergo the test or to work upon the problem. This direct stimulation by incentives of various sorts is, indeed, the sine qua non of experimental work on animals, whatever the main objective of the investigation may be. That the principle can be utilized in a test technique — that is, under conditions in which the training element is reduced to a minimum — as well as in the more usual learning technique is evident enough.

The use of this "obstacle" principle in a motivation test is fairly simple. By using a standard obstacle and varying the incentive from time to time, or from group to group, motivation indices that are comparable for various incentive-drive conditions may be secured. This procedure may be clearly distinguished from the training method for here the animal is required to learn nothing whatsoever; it must merely react to the incentive in accordance with the degree of arousal effected by its physiological condition under the circumstances. It is unnecessary to discuss in this connection the various applications of this general methodology to the problem at hand. As examples of its use on lower forms we may cite the work of Szymanski (16) and of Turner (19) on the cockroach, and that of Wodsedalek (23) on the May fly nymph. In 1922, Morgan (7) reported some tests made on the white rat which made use of this principle. He argued that "the amount of inhibition necessary to overcome any tendency may be used as a measure of the strength of that tendency" and suggests the possibility of thus measuring all tendencies including those classified as instincts. In 1924, Moss (8) applied this principle to the white rat more extensively than Morgan had done and stated his thesis in language similar to that of Morgan. Moss used as an inhibiting, or "opposing stimulus," as he speaks of it (the "obstacle" of Lubbock) a pair of electric grills. The more important defects in the apparatus and technique of Moss are indicated in Part I, 1, and require no further mention here. In no case had investigators carried their work far enough to develop anything that properly deserves to be called a general method of measuring animal motivation. The fundamental principle underlying the various techniques employed had been known for half a century and occasionally had been applied to motivation problems. The basic principle was the old one of Lubbock, and one that had been used by both biologists and comparative psychologists in the study of motivation. The matter of developing a standardized technique apparently received scant attention, even in the more recent work of Morgan and Moss. Nevertheless, it would seem to be desirable to refer to the technique of Moss as the Resistance Method in order to avoid confusing his apparatus and technique with that developed by Jenkins and Warden and called by them the Obstruction Method.

The term "obstruction method" was introduced by the writer in 1926 (4) to designate the general method of measuring animal drives which had been developed during the preceding two years in the Columbia laboratory. Several writers, evidently confused as to the facts in the case (I, 9), have spoken of the "obstruction method" of Moss, mentioning our own method as a modification of his technique. As before stated, our method was developed with direct reference to the "obstacle" principle of Lubbock, the term "obstruction" being used instead of obstacle because the latter seemed to us to have a much wider application than any single laboratory method. From the facts already cited it would seem that, so far as basic principle is concerned, neither ourselves, Moss, Morgan, Szymanski or other workers who adopted, knowingly or otherwise, the "obstacle" principle can lay any just claim to originality. It seems equally clear, on the other hand, that each of these workers, ourselves included, should be credited with the particular application of the principle which each has made in the way of apparatus and procedure. And especially so, when, as in the case of our own apparatus, the similarities with any other previously devised apparatus are of the most superficial sort. Naturally, we made use of the work of previous investigators in this field, including that of Moss, but only in a general way, and in the manner of rejecting outright rather than modifying the relatively simple apparatus and technique which others had used. It is hoped that this explanation will prevent any further confusion of the Columbia Obstruction Method with the Resistance Method of Moss, which have nothing important in common except that both apply the "obstacle" principle of Lubbock, as many earlier techniques have also done.

II. The Columbia Obstruction Method of Measuring Animal Drives

In 1924 the writer began the development of what was planned to be a general method of measuring motivation possessing a wide application in the animal field. It was the original intention to build a test method around the "obstacle" principle of Lubbock rather than to apply the principle as a learning technique as most previous workers in the field of motivation had done. It was considered not impossible that the experimental conditions could be so simplified that not only the different incentive-drive indices obtained for the same species, but also corresponding indices from species to species would be directly comparable.

The "obstacle" principle naturally calls for a three compartment test arrangement: an entrance or reaction compartment and an incentive chamber separated by an appropriate obstacle. The planning of this part of the apparatus thus presented no very great difficulties, except in the matter of devising automatic controls for the doors. Since we desired to carry out work on the white rat at first, it was necessary to construct the different compartments of this section of a suitable size for this animal; also to be careful not to introduce into it any device that could not be utilized in a corresponding testing section for animals of different species.

The matter of the general type of obstacle to be used was not so easily disposed of. To begin with it was felt that a general method of the sort here proposed should not be limited to a single type. Szymanski, Morgan, Moss and several others had used some form of electric shocking device as the obstacle, with its obvious advantages of ease of control in quantitative units of determinable stimulation value. After trying out a number of other types of obstacle, it was finally decided to use an electric grill, and to concentrate our efforts toward the development of a more adequate shocking system than had heretofore been used. However, it should be noted that, as a general method, the Obstruction Method is not limited to the shock type of obstacle, although it does seem that the electric shock, when properly applied, offers the greatest possibilities in the case of most types.

The problem remained as to how to make possible the selection of a suitable, standard shock by empirical tests to be used in as wide a range of conditions as possible. Any form of induction coil, such for example as that used by Szymanski and Moss would not do here, since in order to get the widest range of stimulation conditions from which to select a standard, amperage and voltage should be capable of independent change. Obviously the demand here was for a more or less elaborate system whereby a series of currents of graduated amounts could be had in any desired voltage.

Expert advice was needed at this point and, fortunately, T. N. Jenkins offered to devise and construct the electrical system needed. Too much credit can hardly be given to Dr. Jenkins for the months of experimental effort expended in perfecting this electrical system whereby a wide range of shocks of known stimulation value can be supplied to the obstruction grill. He found it necessary to construct resistances to be used, since ordinary commercial resistances were inadequate to carry the range of high voltages we wished to use at a constant value for long periods of time. The superiority of the Obstruction Method over previous techniques depends in no small degree upon the standardization of the "obstacle," or obstruction grill, by means of this system.

The apparatus as originally constructed by Jenkins and Warden in the fall of 1924, was first used in the doctorate research of Holden (Appendix 1), on the hunger drive in 1924-25. She employed three degrees of shock — low, medium and high — and from her results it was seen that a current of low amperage and of medium voltage would be best suited for use as a standard obstruction. Her study suggested also several other changes in apparatus and technique. The use of a tunnel over the grills to prevent the animals from attempting to jump across without getting a shock has proven to be more satisfactory than the arrangement of glass partitions employed in the original apparatus. The incentive compartment was found to be somewhat too short to contain the incentive object and allow room for the animal to react properly to it. This fact led the writer to devise a small fourth compartment, in direct line with the other three, in which the incentive object could be placed. When animals (male, female, litter, etc.) were to be used as incentive objects it would be necessary, of course, to close this small incentive chamber by a door. This called for some form of mechanical device to operate the door that should relieve the experimenter of the task, and also eliminate a possible source of error from human operation of the door. The automatic door release, illustrated in Figure 2, Part I, 1, was worked out by Jenkins and Warden and constructed by Warner in connection with his doctorate research in the spring of 1926. The grill used by Holden was also unsatisfactory on account of the posibility of shorting in case of micturition; this defect was remedied by a new type which allowed drops of urine to fall through and thus prevented shorting. The plan of this improved grill was suggested by Jenkins, and the grill itself constructed by Warner in the spring of 1926. Dr. Jenkins should also be credited with the device illustrated in Fig. 1, Part IV, 2, added in 1927 to facilitate the control of the door to the entrance compartment. The illumination hood, designed to eliminate any possible visual cue from the experimenter without the use of a curtain, was contributed by Warner in the spring of 1926. No further modifications were made in the apparatus, which, as thus standardized, was used throughout the project on animal drives reported in this volume. The first study of this series was that of Warner on sex behavior which was begun in the summer of 1926.

Several modifications were made in the technique originally planned for the method in connection with the work of Holden. Among the more important may be mentioned (1) the procedure whereby the preliminary and "shock" crossings precede immediately the test period proper ; (2) the lengthening of the test period from 10 to 20 minutes, and (3) the slow and careful manner of transferring the animal from the incentive compartment, after crossing, back to the entrance compartment, so as to prevent the possibility of emotional disturbances. The details of these changes can be had by comparing the technique used by Holden (Appendix I) with the new technique standardized for the project (Part I, 1, 2; Part IV, 1). Although the writer must be held responsible for the methodology finally adopted for use in connection with the obstruction apparatus, Holden, and later Warner deserve no little credit for carrying out extensive preliminary tests in the laboratory covering possible lines of procedure and these results were drawn upon in determining specific points of methodology. Both worked untiringly, especially in the matter of testing out group after group of animals, in the effort to determine the most suitable shock to be used in the project. Nissen also deserves special mention, not only for making the actual tests reported in Part I, 2, but also for his valuable suggestions in the planning of the analysis of the method there reported.

In view of the fact that the next two sections deal with the method in considerable detail, it seems unnecessary to discuss the matter further here. A complete description of the apparatus and general procedure will be found in Part I, 1, and additional points on procedure in Part II, 1. The function of the obstruction and of the incentive as related to this method are indicated in the experimental findings of Part I, 2, together with the effect of re-testing on individual and group scores. In both of these papers attention is called to the fundamental advantages of this method over other methods that had been previously used in the study of animal motivation.

In certain parts of the experimental work it was necessary to make use of special techniques lying outside the field of psychology proper. Each experimenter was required to master under competent workers in the appropriate field such special techniques as should be necessary to carry out his part of the project, such as the making of vaginal smears, both quick and permanent, and the technique of administering injections and of doing the necessary surgical work. Special credit has been given in connection with the individual papers to the several workers who took special interest in our project and gave advice and counsel on points lying outside our field.

III. A Note on Terminology

Unfortunately the recent attempts of comparative psychology to adjust itself to a natural science, or biological viewpoint has not as yet progressed to the point where a consistent objective terminology has come into general use. No field is more beset by difficulties in this regard than that of animal motivation, in which the phenomena have been described for ages past in anthropomorphic and teleological phraseology. Because of its historical antecedents, even the term *motivation* itself appears to most of us to involve a distinctly subjective connotation. Still, the present writer knows of no term less harmful in this respect which might be used instead.

As experimentalists, we are not especially interested in theoretical controversies on questions of terminology. Nevertheless, it is of some importance that we describe our findings and discuss the meaning of our results in language that will prove to be as accurate and as understandable as the present circumstances permit. In the absence of a generally accepted terminology in our field, we have found it necessary to select such terms as seemed best suited to our purpose, and have attempted to use these consistently throughout the various reports included in this volume. In order to avoid misunderstanding it seems desirable at this point to define as clearly as possible several of the more important terms employed.

By an *incentive* is meant an external object such as food, water, animal of the opposite sex, etc., that is capable of operating as a stimulus and arousing some fairly definite internal physiological state. While recognizing that any external object, in order to be classed as a stimulus at all, must be capable of arousing the organism in some manner and to some extent, we have purposely limited the term *incentive* in this report to those objects or classes of objects that satisfy, from the observer's point of view some fairly definite drive or seeking tendency.

By a *drive* we mean an aroused reaction tendency which is characterized primarily by the fact that the activity of the organism is directed toward or away from some *specific incentive*, such as food, water, animal of the opposite sex, etc. If the appropriate incentive object is not present, what appears to be seeking behavior to an outside observer is exhibited. While recognizing that all behavior represents in some manner and to some degree the dynamics of the organism, or the drive of the organism, still there appears to be some justification for limiting the term *drive* to reaction tendencies that are of sufficient biological importance to represent seeking behavior directed toward some fairly definite class of incentive objects, such as food, water, animal of the opposite sex, etc.

The term *drive* does not refer to the physiological state or system aroused either by an incentive or by deprivation of some sort, but to the behavior tendency resulting from the internal arousal. For example, *hunger drive* refers to the tendency of the animal to approach or to search for food and not to the physiological changes, such as stomach contractions, etc., that may be taking place at the time. In most cases we have not attempted to secure an independent index of the physiological state underlying the drive, so as to correlate various aspects of physiological change with corresponding changes in the reaction tendency, although something of the sort has been done in certain of the studies on the sex drive. Attention is called to this fact in order to emphasize the necessity of avoiding a commonly made confusion between the reaction tendency as a behavior phenomenon and the underlying physiological state.

By the term *incentive-drive* situation, and similar usage, we have meant to stress the fact that the internal and external factors may, and usually do operate together in behavior as induced in the usual motivation experiment. The methodology of the Obstruction Method actually requires that the incentive be either present or close at hand. In a sense, then, all our drive indices are in reality *incentive-drive* indices; but this must always be true if we define a drive as a reaction tendency directed toward an incentive, since this implies that the latter shall be present in some sense. The word *drive* has been used in most cases because the hyphenated term *incentive-drive* is more or less awkward; the latter could be substituted for the former throughout, however, without changing the essential meaning of the text.

Finally, the term *motivation* has been employed in the experimental sections to cover both the incentive and the drive aspects of the behavior investigated. In the discussion of The Columbia Obstruction Method (Warden) the meaning of the term has been extended to include studies of general or "spontaneous" activity by various methods as well as certain types of experimentation on "tropisms" and "instincts," which seem to deal primarily with the dynamics of behavior.

1. STANDARD APPARATUS FOR THE STUDY OF ANIMAL MOTIVATION [1]

T. N. Jenkins, L. H. Warner, and C. J. Warden

In the first section of the present report a description is given of an apparatus for the measurement of drives by the Method of Obstruction. The apparatus has been developed during the past two years in the Columbia laboratory in connection with a research project on motivation in the white rat. It can be readily adapted, however, to most common mammalian forms, and thus offers possibilities in comparative studies within this field. Great pains have been taken to standardize the essential features by the use of automatic devices when possible and by exactitude of detail in construction. The control procedure as developed will also be indicated.

The second section is a description of an improved apparatus of the choice method type, following the general pattern used by Tsai (18). The improvement consists, in the main, in the introduction of automatic electrical devices permitting a more adequate control of the animal and of the incentive stimuli. Standard problems and procedure are suggested.

I. The Obstruction Method Apparatus

As previously stated, this method involves the placing of an obstruction of some sort between the organism and the incentive stimulus. In order to obtain the latter and satisfy the dominant demand of the moment, the organism is required to surmount the obstruction. In the present apparatus the obstruction employed is an electric grill supplied with a constant current of quantitatively determined attributes. Various other types of obstruction might be used but the electric stimulation has the advantage of being more convenient to control and administer than most others would be.

[1] Reprinted with modifications from *The Journal of Comparative Psychology*, 1926, 6: 361-82.

For the sake of clarity in description the obstruction apparatus will be divided into two parts and each treated in order. First, the box for the control of the animal and the administering of the shock; and second, the device for the control of the electrical current supplied to the box in the obstruction section, permitting definite systematic variation of the same. This description of the apparatus will be followed by a discussion of methods of controlling the intra-organic state of the organism for this type of investigation, kinds of data that may be obtained by this general method, and standard problems to which the method offers an experimental approach.

1. The animal control box

The ground plan of this part of the apparatus is shown in figure 1. The box consists essentially of three compartments as follows: entrance compartment *A*, obstruction compartment, *B*, and incentive compartment, *C-D*, divided into two sections for reasons indicated later. The body of the box is built of whitewood and is finished both without and within in dull black. It rests upon a table 30 inches in height which also serves to support the elaborate electrical equipment necessary to furnish the shock to the grill in compartment *B*.

The entrance compartment *A* is 10 inches square and 10 inches deep, inside measurements. It is entirely separated from compartment *B* by a wooden partition except for a doorway 4 inches square

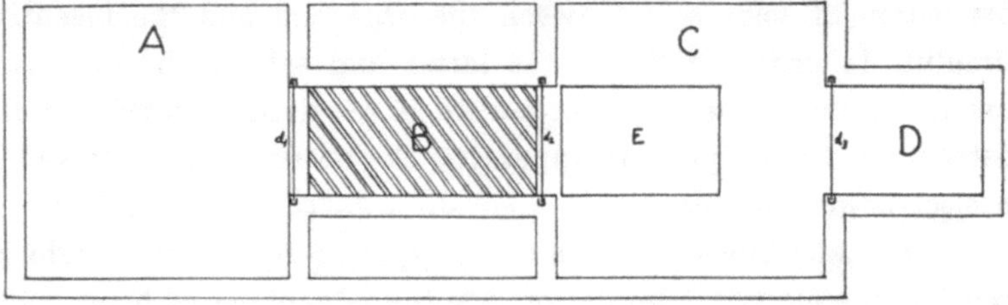

FIG. 1. DIAGRAM OF FLOOR PLAN OF THE OBSTRUCTION BOX

A, entrance compartment; *B*, obstruction compartment; *C*, *D*, divided incentive compartment; *E*, release plate; d_1, manually operated door of entrance compartment; d_2, automatic door (operated by release plate) between two divisions of incentive compartment.

leading into the tunnel that crosses the latter compartment. This doorway is fitted with a door of thin sheet steel, d_1, which slides in felt-lined grooves, and is therefore practically noiseless. It is operated by hand by means of a cord which passes through a bent glass tube (instead of the usual pulley device) to a point outside the box, so that its operation cannot be observed by the animal in the compartment. In closing, the door sinks into a narrow felt-lined groove in the floor. The purpose of the door is obviously to restrain the animal in the entrance compartment until such time as the experimenter may wish to test its reaction to the obstruction-incentive situation beyond. The compartment is covered by a panel of heavy plate glass, but may be opened by sliding the panel, which rests in a felt-lined groove, back over compartment *B*.

The dimensions of compartment *B* are exactly the same as those above enumerated for compartment *A*, except for the fact that the effective part of the compartment is restricted to a tunnel 4 inches wide and 4 inches high connecting the entrance and incentive compartments. The entire floor of compartment *B* consists of an electric grill set in the wooden base of the box so that it comes up even with the floor of the two adjacent compartments. The conducting elements in the grill are number 18 copper wires wound upon a slab of bakelite $\frac{1}{2}$ inch thick. The turns are $\frac{1}{4}$ inch apart, and are wound in such a way that the turns from the two terminals of the transformer alternate. A rat in crossing the grill will close the circuit, therefore, by stepping upon any two consecutive turns. This practically insures stimulation from the instant the animal steps upon the grill. Moss used, as a shocking device, two plates placed some distance apart along a passageway, simultaneous contact with both plates being necessary to produce the shock. The grill above described offers a much more dependable and uniform source of stimulation.

The presentation of the electrical stimulation to the animal offers certain difficulties. After trying out several other schemes the tunnel device was finally selected as being the most satisfactory. At first it was thought desirable to place the grill in the center of the box without any further separation into definite compartments. This would allow the animal to make contact at any point along the edge of the grill, and to have free passage from *A* to *C* across any portion of it. This arrangement was found to be unsatisfactory. Many of the animals would attempt, and sometimes with success, to jump from *A* to *C*, a distance of 10 inches, without touching the grill. To prevent this two pieces of plate glass, extending 6 inches from either wall at a point 3 inches from the edges of the grill, were so placed that the animal must pass through a winding pathway approximately 4 inches wide in crossing the grill. This arrangement eliminated the jumping behavior quite successfully and also aided in marking off the obstruction part of the box from the entrance part.

However, a further difficulty arose which led us to separate the box into 3 definite compartments, as above described, by partitions, and to install the tunnel device. It became clear from observing the behavior of the animals, especially when a high degree of shock was used, that they did not discriminate readily between the entrance and obstruction portions of the box when not so divided. Usually they would rush pell mell across the grill to the incentive portion of the box the instant they were placed at *A*, and after rushing to *C* paid no heed whatsoever to the incentive stimulus (food), but attempted to escape by jumping up against the glass panel covering the box or, perhaps crouched sullenly in a corner. Rats that had been starved for long periods of time (one to three days) responded in this manner. The animals seemed to be dominated by fear of the combined *A-B* situation rather than by hunger. They were apparently avoiding the entrance and obstruction sections rather than seeking the incentive section. This escape response would be reënacted promptly, over and over again, as soon as the animal was returned to the entrance end of the box, in preparation for a succeeding test.

When *A* and *B* were separated into compartments by partitions, and the animal restrained in compartment *A* by the door, d_1, for a brief interval of time between successive tests, no such behavior occurred. This suggests that the animal probably did not discriminate, in the former arrangement, between the entrance and obstruction portions of the box. *A-B* as a whole rather than just *B* was avoided, since the two portions were not sufficiently isolated to be readily discriminated as separate sources of stimuli which might serve as the basis of a differential response. The copper wires on the *B* portion of the floor constituted the only important difference between that and the *A* portion, and the line of junction between the two was not very marked since the grill had been made to fit very neatly into the floor of the box. A second possible explanation for this escape behavior can be made in terms of a conditioning process of the substitute stimulus type as indicated in the following diagram:

S (shock at *B*)------------------*R* (escape into *C*)

S (sight of *B* while
still in *A*) - ↗ *R*

This explanation assumes that the *A* portion of the box had been actually discriminated from the *B* portion at first, but later came to serve essentially as a single situation, by virtue of the substitution indicated above. At any rate the separating of the box into definite compartments, as in the present apparatus, together with the introduction of a brief interval of time between successive tests remedied the difficulty.

The tunnel across the grill is 4 inches wide, 4 inches high and 10 inches long. The two sides are of wood, and the top of plate glass to permit the same degree of illumination in this as in the other two compartments of the box. The use of the tunnel has the effect of making the response to the obstruction more definite, since there is only this single point of approach and contact. Accidental contacts with the grill are now much less likely to occur than under the old arrangement. Furthermore, the movements of different animals in passing over the grill are more uniform — they must either walk or run. This makes possible relative constancy in the amount of electrical stimulation involved in the act of crossing, this latter being the chief unit of measurement on the behavior side.

The incentive section includes a double compartment, *C-D*, as indicated in the diagram. The division into two parts would not be necessary with such incentives as different kinds of food, or other non-living stimuli. However, when the incentive stimulus is a living animal (such as a female in testing the male sex drive) it is usually necessary to confine the stimulus animal in order to prevent it from crossing over to the test animal. The method of restraining the stimulus animal should be as natural as possible so as not to arouse in it behavior likely to reduce its efficacy as a normal positive stimulus. The restraint should cease at the moment the test animal has completed the act of crossing the grill, and the release of the stimulus animal should be automatic, if possible.

Moss restrained the female rat in the incentive chamber by a simple door that must be raised by the experimenter, presumably as the animal was crossing the grill. Such a method involves the error of inexact timing — of opening the door either too soon or too late. Furthermore, the moving door might be a factor of central importance in the stimulation situation during the test. A door of ordinary opaque material would decrease the visual, and probably

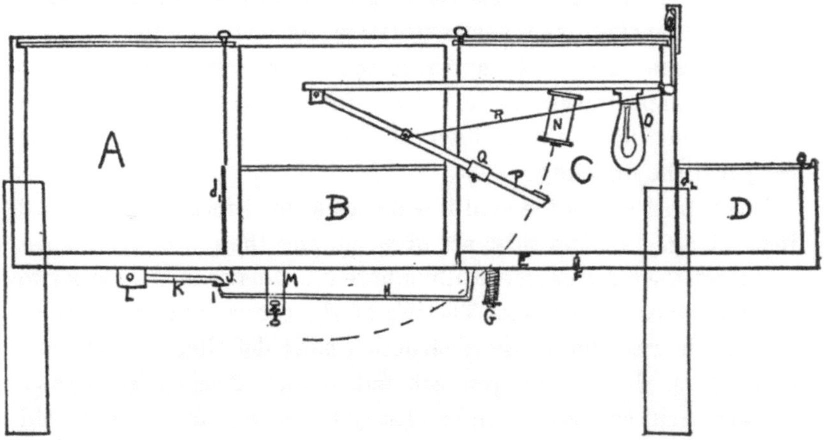

FIG. 2. DIAGRAM OF SIDE ELEVATION OF THE OBSTRUCTION BOX

A, entrance compartment; *B*, obstruction compartment; *C*, *D*, divided incentive compartment; *E*, release plate; d_1, manually operated door of entrance compartment; d_2, automatic door (operated by release plate) between two divisions of incentive compartment; *F*, a hinge and *G*, an adjustable coil spring, both supporting the release plate; *H*, a rod rigidly secured to the release plate; *I* and *J*, silver contacts; *K*, a rod pivoted at *L*, the adjustment of which determines the level of the release plate when the latter is not depressed; *M*, a block carrying the bolt adjusting the depth of depression of the release plate; *N*, electro-magnet supporting free end of the pendulum, *P*, prior to its release; *O*, in circuit with the magnet and acting as resistance and as a signal; *Q*, adjustable pendulum weight; *R*, cord conveying movement of pendulum to the door, d_2.

the olfactory stimulation in connection with the animal in the incentive chamber. The movement of the door would likely serve also as a distraction to the test animal.

We have endeavored to avoid all these difficulties by the use of a separate restraining chamber, *D*, the entrance to which is guarded by the translucent door, d_2 (Figs. 1 and 2) which is opened automatically by the test animal in the act of crossing the grill. The door is 4 inches square, constructed of celluloid $\frac{1}{8}$ inch in thickness and is therefore semi-transparent. Three small windows, each 1 inch square, cut in the upper half of the door, permit the free passage of olfactory stimuli, whether of food or of sex object, from the restraint chamber.

The mechanism of the automatic door release is illustrated in figure 2. An opening 4 inches wide and 6 inches long is cut in the floor of compartment *C* directly in front of the exit of the tunnel. In this opening the release plate, *E*, is set flush with the floor. The plate is of wood and painted a dull black so as to be distinguished with difficulty from the adjacent floor. It is supported by a hinge, *F*, and the adjustable coil spring, *G*. Fastened rigidly to the lower surface of the release plate is the L-shaped rod, *H*, at the tip of which is the silver contact point, *I*. The tension of the spring, *G*, is just sufficient to keep this point in contact with the small silver plate, *J*. This latter is mounted on the rod, *K*, which is adjustable in the vertical plane, being pivoted at the wooden block, *L*, The V-shaped wooden block, *M*, spans the rod, *H*, and in the base of this block is a rubber pad which can be raised or lowered by an adjustable nut. The rod, *H*, rests on this pad when the release plate is fully depressed.

In circuit with the points *I* and *J*, is an electro-magnet, *N*, of 1200 ohms resistance, and a 10-watt electric lamp, *O*, both of which are mounted on the underside of a 15-inch shelf fastened on the outside of the box. Suspended from cone bearings at the other end of the shelf is a pendulum, *P*, made of $\frac{1}{4}$ inch square brass tubing which carries an adjustable weight consisting of a brass collar held in place by a set screw. A small piece of iron is soldered to the pendulum at such a point that when the latter is raised the iron will make contact with the pole of the magnet, *N*, and be held in contact if the current is passing through the magnet. A piece of felt fastened over the pole of the magnet serves to deaden the click produced by this contact. By means of a cord and 3 small pulleys, the falling pendulum lifts the door, d_2. Any tendency to rebound is entirely overcome by the friction within the system.

The operation of this automatic door release is exceedingly simple and practically noiseless. At the first touch of the release plate by the animal in crossing the grill, the contact between *I* and *J* is broken, to be re-made only after the animal has entirely cleared the plate. The maximum depth of depression of the release plate is controlled by the device at *M* and need not be more than 2 or 3 mm., the rod, *K*, being always adjusted so that the plate is flush with the floor of the compartment. The tension of the spring, *G*, can be easily and quickly adjusted for animals of any weight. We have found that a tension of 40 grams is sufficiently delicate when 80-gram rats are being tested.

Even when the white rat dashes through the tunnel at high speed, the action of the automatic door release is so prompt that the animal never comes into collision with the rising door. Indeed the door does not seem to be noticed by the animal at all. The stimulus animal is visible in the restraint chamber through the windows of the celluloid door even when the latter is closed. The restraint chamber is 4 by 4 by 6 inches in size and is covered with plate glass. It can be removed by the simple turning of a thumbscrew for purposes of cleansing or deodorizing. All incentive stimuli, whether living animals or food, etc., are placed in the restraint chamber, so that the automatic door device is utilized in testing the strength of any drive.

The obstruction apparatus is placed for use in a dark room in which the only source of light is an illumination hood. This latter is made of wood and is 36 inches long and 4 inches wide and deep. It is heavily coated with flat white inside so as to make a good reflecting surface, and finished on the outer surface in dull black. The hood extends the full length of the control box at a point approximately 12 inches above the box in the central position. It is illuminated by 8 lights of 3 C.P., set at equal intervals (4 inches) along the top of the hood within. The light is diffused through a sheet of translucent paper lying above the "Celotex" screen that covers the lower open side of the hood facing the box. A piece of ground glass of the proper grade would likely do as well. The hood device gives a uniform illumination of the interior of all compartments of the box, the outside environment being, at the same time, relatively unilluminated. This arrangement permits the experimenter to observe the behavior of the animal freely during the test period, while remaining practically invisible to the animal.

2. The mechanism for the control of the electrical stimulation

The obstruction used in the present apparatus is an electric grill forming the floor of a tunnel 4 inches wide and 10 inches long, through which the animal must pass in order to obtain the incentive stimulus. The scientific value of the apparatus depends very largely upon the exactitude of control exercised over the current supplied to the grill in compartment *B*. It is necessary, in the first place, that the significant attributes of the physical stimulus be specified in quantitative terms, if duplication by other investigators is to be made possible. Among these are the form, intensity and frequency characteristic of the stimulus. The dependence of *stimulation value* of an electrical stimulus upon the rate at which the energy is expended is well known. Waller (20) in 1899 showed that stimulation value is dependent not only upon both voltage and amount of current, but that there is an optimal ratio of energy used to the rate at which it is used. In other words, there is an optimal minimal stimulus or "optimum gradient of mobilizing energy producing greatest excitation by least energy." Moss seems to have made the mistake of supposing that the voltage value is a sufficient specification of an electrical stimulus, for he fails to state the current values used in his work. In our own apparatus we have taken great care to avoid any possible error in providing a constant physical stimulus value in the grill.

A schematic diagram of the electrical system, arranged to provide for two control boxes to be placed side by side, is shown in figure 3. The system would not be complex except for the fact that it was considered desirable to provide for a series of 8 degrees of electrical stimulation, either of which could be switched into the grill by simply turning a button, as will be explained later. If only a single degree of stimulation were desired the apparatus would be very simple in construction.

The voltage used is generated in the secondary of a General Electric potential transformer shown at *T*, and voltage changes are secured by varying the input into the primary by means of the potentiometer, *P*. The electrical pressure can be measured at any time by means of a high tension voltmeter, *V*, by closing a snap switch, *K*.

Current regulation is secured by means of variable resistance units, each controlled by multipoint switches, *C*, and C_2. Each unit contains eight elements (*a*, *b*, *c*, etc.), each having a resistance of 1,250,000 ohms. A common resistance, *r*, of 100,000 ohms, is in

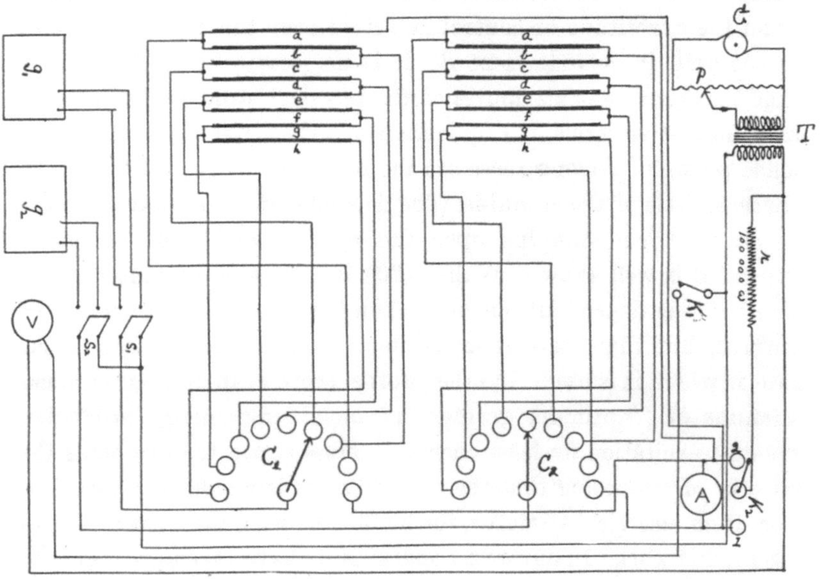

FIG. 3. DIAGRAM OF MECHANISM FOR PRODUCTION OF ELECTRICAL STIMULI

G, *A.C.* generator; *P*, potentiometer for controlling the voltage impressed upon the primary of the potential transformer at *T*; K_1, switch used for taking voltage readings off voltmeter, *V*; *A*, microammeter for measuring stimulus current (the current through either grill, g_1 or g_2, is measured by throwing the knife switch K_2, to either pole *1* or pole *2*); C_1 and C_2, multipoint switches for throwing different amounts of resistance (*a, b, c, d, e, f, g, h*) into the stimulus circuit; S_1 and S_2, switches for closing the stimulus circuit through the grills.

circuit with both resistance units, so that when all the elements (*a*, *b*, *c*, etc.) are cut out there will remain approximately 100,000 ohms of resistance in the circuits. The multipoint switches enable the operator to vary the resistance cumulatively in equal steps from 100,000 to 10,000,000 ohms. The elements are liquid resistance in glass tubes 2 feet long. Some of the better types of standard radio resistance were first tried, but these were found to be too unstable to be of service for the present purpose.

The current is measured by a micro-ammeter located at *A*. The method of wiring is such that by means of the double-throw knife switch, *K*, the current in either grill (g_1 or g_2) can be measured by one fixed measuring instrument, thus insuring the same amount to both control boxes, when they are used at the same time. The measuring instrument used adds only about 30 ohms of resistance to the circuit, and this amount is so small in comparison to the total resistance of the circuit that it may be altogether disregarded.

In using the apparatus, the voltage is first determined by closing the switch, *K*, and any necessary adjustment to bring the voltage to the point desired is then made at *P*. The current values for each step can be measured after first short circuiting the grills. The current is thrown into either grill by closing the switches S_1 or S_2.

This arrangement enables us to specify the form, intensity, and frequency characteristics of the physical stimulus in quantitative terms. The writers mention form here because commercial alternating current is approximately sinusoidal. Stimulation value ought to be practically constant, therefore, for a given voltage since amperage is kept practically constant throughout.

The problem arises in this connection regarding the factor of individual difference in the resistance of the skin of different animals, and of the same animal from moment to moment. In order to be certain that we are obtaining an adequate measure of motivation in terms of the obstruction overcome by the animal, not only must the electrical stimulus in *B* be kept constant, but the effect of the electrical stimulation on the animal must be controlled as definitely as possible.

Moss attempted to insure a constant, uniform stimulation by filling compartment *A* with water, or partly filling it. This would make certain a more ready contact with the grill, but does not adequately control the factor of individual difference in skin resistance. Furthermore, the presence of water in the entrance compartment complicates the drive factor under investigation, especially since there was no water in the incentive compartment. The water may have aroused an avoiding reaction which operated along with the hunger drive. Moreover, the effect of the water may have been different for different animals, and for variations of the hunger drive in the same animal.

We have attempted to control the factor of difference in skin resistance by the use of high terminal voltages and by introducing large resistances into the circuit to reduce the current values to the extent required to give the appropriate stimulation value in the grill. The logic of this procedure is fairly obvious. If we let r be the variable skin resistance of the rat, and R the remaining resistance in the circuit, we have the equation:

$$I = \frac{E}{R + r}$$

where I is the current and E the impressed voltage. If R is very large in comparison to r, it follows that a relatively large change in r will produce relatively small changes in I.

Let us suppose, for example, that we are operating the system at a pressure of 500 volts through a resistance of 4,000,000 ohms, and that the skin resistance of a sample of rat varies from 100 to 5000 ohms. Then the current flowing for a skin resistance of 100 ohms would be 124.99687 microamperes, and that for 5000 ohms would be 124.84394 microamperes. The maximal change in current value is only 0.122 per cent. In our work last year we used 1200 volts. In such a case the differential in skin resistance becomes a negligible quantity. It is not necessary to resort to the use of water to insure uniform contact.

Moss used a current of 28 volts or less in his work, and in such case the difference in skin resistance would likely be an important factor in determining the results. Let us consider, for example, the values for a current of 28 volts, with a secondary coil resistance of 100 ohms, within the limits stated above. In this case a skin resistance of 100 ohms gives a current of 0.14 amperes, and a skin resistance of 5000 ohms gives a current of 0.00549 amperes. This represents a change in current value of 96.07 per cent. If the E.M.F. is 20 instead of 28 volts the corresponding value will be 99.6 per cent. This argues very clearly for the use of high voltage in this type of work where a uniform stimulation value is one of the primary requisites.

The quantity of energy that strikes the excitable tissue each time is determined by the voltage, amperage, and frequency of the alternating current. As noted above, Waller has shown that stimulation value is dependent upon both the amount of energy that strikes the excitable tissue, and the rate of impact. Knowledge of the voltage, amperage, and frequency furnishes us a means of determining both the amount of energy and the rate with which it acts. The direction of flow of the current changes at every cycle, so that the excitable tissue is struck once at every cycle. Most probably the effect of the initial blow differs somewhat from that of succeeding blows, inasmuch as the latter are conditioned to some extent by the refractory period of the nerve.

The procedure to be followed in using the Obstruction apparatus is exceedingly simple. The electrical supply to the grill is first set at the point desired by an adjustment of the potentiometer. The incentive is placed in chamber *D* and all doors and panels of the control box closed. The test animal is then placed in compartment *A*, and after a brief interval allowed to the animal for adjustment to the situation, d_1 is raised and the stopwatch started. When the animal crosses the grill he automatically releases d_2, and thus obtains access to the incentive stimulus. The animal should be conditioned to the box without the electric stimulation on each of several days preceding the test, the incentive stimulus being present in this preliminary work.

Several units of behavior may be secured in using the apparatus. The number of times contact with the grill is made, without actually crossing within a test period of specified length, and the number of successive crosses within such a stated interval are among the most important data obtainable. Jumping and similar activity apparently connected with the escape motive can also be quantitatively determined by the experimenter. It will be seen, furthermore, that by keeping the stimulation value of *B* constant, the general problem of individual differences in a given drive may be attacked, and different drive situations may also be directly compared. This is possible because of the fact that the motivation of the animal is in all cases measured in terms of its response to an identical obstruction stimulus.

Perhaps a word should be said concerning the control of the intra-organic conditions that represent the physiological basis of animal drives. The use of standardized environmental conditions, and especially that of nutrition, offers the only possible method of insuring a normal animal in any experimental work, and the chief requirement in studies of motivation is simply for a more careful and rigorous control of these objective conditions. The present apparatus offers no special difficulties in this direction, and the application of general physiological principles and results furnishes the only sound basis of control of this complex factor.

Systematic variation of the physiological state underlying a given drive cannot be had, of course, in any very ideal sense. The best method of isolating a specific drive, in so far as this is possible, and of obtaining varying degrees of it for measurement, is to deprive the normal animal of the appropriate incentive stimulus for varying periods of time, keeping the general physiological state of the organism as constant as possible. For example, an objective index of the hunger drive can be had by varying the interval of food deprivation. The same method will probably apply in studying the male sex drive. The female sex drive, on the other hand, must be investigated with reference to the oestrus cycle, and the determination of the latter requires the use of appropriate histological technique. In all cases objective criteria should be employed in determining the physiological status and of obtaining varying degrees of the same.

II. The Choice Method Apparatus

The ground plan of the apparatus designed for the study of motivation by the choice method is shown in figure 4. It consists essentially of an entrance section, *B*, and two incentive chambers,

L and *R*, together with the necessary connecting pathways. The control box measures 19½ inches in length by 10 inches in both width and depth (inside measurements). The body is built of wood and finished in dull black, and the partitions are of the same construction, except as otherwise stated.

The entrance compartment, *B*, is 3½ inches wide and extends forward 7 inches to the door, D_B. The latter is of thin sheet metal sliding in felt grooves and operated manually. It serves to restrain the animal in the entrance compartment until the release to the incentive compartments is desired. The pathways (A_L and A_R) leading to the incentive chambers are 3 inches wide. The incentive compartments are indentical in size (5 inches wide by 7 inches long) and in construction. The walls surrounding each compartment are made of wood except for a 3-inch strip of wire netting (¼ inch mesh) which forms the lower portion of the wall all around. The purpose of the netting is to allow the test animal a more adequate stimulation from the incentives in the two chambers. The smooth wooden sides above tend to keep the incentive animal (when such is used) from climbing out of easy range of the test animal in the entrance compartment. The incentive chambers have been made

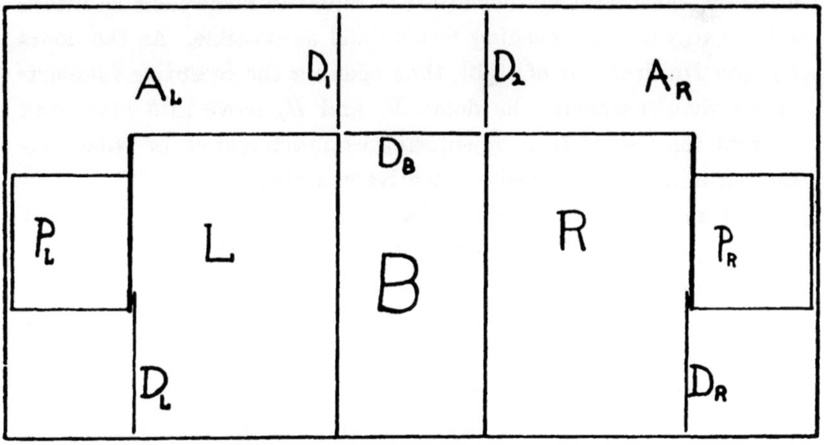

FIG. 4. DIAGRAM OF FLOOR PLAN OF THE CHOICE BOX

B, entrance compartment; *L* and *R*, incentive compartments; P_L and P_R, release plates for the pendulum, the operation of which opens the doors D_L and D_R and closes D_1 and D_2; A_L and A_R, runways from entrance compartment to incentive compartments; D_B, door for retaining animal in compartment *B*.

small for the purpose of bringing the incentive animal as near to the test animal as possible and keeping the relation between the two as constant as possible.

The incentive chambers are closed by the two doors D_L and D_R, which serve to keep the incentive animal (when such is used) within the chamber occupied. Both of these doors will be released automatically by the animal in running across the appropriate release plate (P_L or P_R) which operate noiselessly. At the points where the release plate is situated, a wire netting similar to that forming the lower portion of the side wall of the incentive chamber, is made an integral part of the plate. This arrangement prevents the incentive animal from operating the release plate by crawling about on the netting.

The mechanism of the release plate operates as follows: When the animal steps upon the plate, a circuit is broken through an electromagnet which releases a pendulum. To this pendulum 4 sectors are attached in such wise that they can move through slots in the floor of the box (D_L, D_R, D_1 and D_2). When the pendulum drops, the attached sectors are carried along with it so that the door (D_L or D_R) drops out of sight through the floor. The doors are made of translucent mathematical celluloid to make them as inconspicuous to the approaching test animal as possible. As the doors D_L and D_R drop out of sight, thus opening the incentive compartments simultaneously, the doors D_1 and D_2 move into place and prevent the animal from re-entering compartment *B*, or from running around to the opposite incentive chamber.

This automatic door release has the merit of being practically noiseless, and we believe it to be superior to any of the usual sliding door devices. An electric light placed in series with the magnet goes out when the animal steps on the release plate. The release plates rest firmly upon levers attached to a common beam which operates in cone bearings mounted on the under side of the box. This type of mounting is preferred because the bearings are far enough removed to prevent particles from the box from falling into them and hindering the free movement of the mechanism.

The control box is equipped with plate glass panels on top to prevent the escape of the animal from the various compartments.

It is used in the dark room and the lighting controlled by an illumination hood similar to that employed in connection with the Obstruction apparatus. The experimenter is kept outside the visual range of the animal while the latter can be conveniently observed by the experimenter during the test.

The technique of operating the choice method is simple. The test animal is placed in *B* and, after a specified period of stimulation, is released into the runways leading to the two incentive chambers by opening the door D_B. If the animal turns to the right and stops on the plate P_R in approaching compartment *R*, the door D_1 leading to the opposite compartment is closed. At the same time D_R opens permitting the animal to enter the incentive chamber and secure the reward. If the animal turns to the left when released from the entrance compartment, a similar mechanism is set in operation.

The control of the intra-organic condition of the organism is less satisfactory under this method than under the Obstruction Method because of the fact that simultaneous stimulation by two incentive stimuli always occurs. This means that systematic variation of both drive conditions must be made previous to the test in which the two incentive stimuli are utilized. Both food and sex starvation in measurable amounts must be brought about during the period previous to the test, if food and sex object are to be used as incentive stimuli. Naturally the control cannot be as satisfactory when two sets of intra-organic conditions are varied instead of only a single one. For this, and other reasons, the choice method seems much less promising than the Obstruction Method, as a means of investigating animal motivation.

2. AN EXPERIMENTAL ANALYSIS OF THE OBSTRUCTION METHOD OF MEASURING ANIMAL DRIVES [1]

C. J. Warden and H. W. Nissen [1]

The increasing use of more or less complicated apparatus in the study of animal behavior brings with it the very serious problem as to the extent to which the data obtained are a function of the particular experimental conditions imposed rather than of the variable selected for investigation by the method. Perhaps not enough attention has been given to this obviously important source of error in our laboratory studies. The attitude is all too common of assuming that a given apparatus adequately serves the purpose for which it has been designed if it seems to do so in the opinion of the experimenter, or other human censor, without recourse to thoroughgoing experimental tests to determine its actual fitness upon the animal to be studied, or upon the specific function to be measured.

The general purpose of the present study was to determine how adequately the Obstruction Method, as finally standardized for our project, gives a valid measure of drive-incentive behavior in the white rat. The hunger drive in the male was chosen for study largely as a matter of convenience. One is able to avoid altogether thereby the complicating factor of activity rhythms associated with the oestrus cycle in the female. That the hunger drive is thoroughly typical is clearly shown by results already published or awaiting publication. The present study should be thought of as methodological and as applying to the measurement of drives in general rather than to the hunger drive in particular.

The same general conditions as those obtaining in our regular drive work were maintained as far as possible. Younger and less expensive animals were used, however, and therefore the absolute

[1] Reprinted with modifications from *The Journal of Comparative Psychology*, 1928, 8: 325-42.

indices secured cannot be directly compared. In the drive project we have used throughout the experimental colony strain of the Wistar Institute at 185 days of age, whereas the rats employed in our present study were approximately 120 days of age and were secured from a local dealer. Our regular time in testing drives is from 8 p.m. to 3 a.m., the white rat being more active then, as a rule, than in the early forenoon (6 a.m. to 9 a.m.) when our present tests were made. McCollum's standard diet, to which was added bi-weekly rations of greens was used throughout except that when our procedure required a long series of starvation periods alternating with short feeding periods this diet was supplemented by milk-soaked bread. The animals were kept in the laboratory for at least ten days before being tested. The regular diet (McCollum's) was always used as the incentive. Each group was starved for forty-eight hours immediately preceding the test in cages similar to the regular living cage with an ample supply of water. Unless specifially stated otherwise the standard shock as used in our regular drive work was employed. This shock is produced by an alternating current of 60 cycles, with a terminal pressure of 475 volts, an external resistance of 10,100,000 ohms in the circuit and a current of 0.047 milliamperes. The shock produced is considerably lower than the lowest degree used by Holden and was selected after much painstaking effort to find the shock most suitable to be employed as a constant throughout the drive project. A higher degree of shock (Shock II) was made use of in a few instances in this investigation (475 volts, 8,750,000 ohms, 0.054 milliampere).

We have attempted to isolate by the usual experimental methodology the two most important features of the apparatus in order to determine the influence of each in the measurement of drive behavior: the obstruction, or shock and the incentive arrangement. The method must be judged in the last analysis by the adequacy with which these parts of the apparatus serve the functions for which they were designed, when used in the proper manner. A third problem attacked is concerned with the effect of practice when groups are re-tested at regular intervals, and the validity of the method in the study of individual differences in drive. These topics will be discussed in the order mentioned.

THE OBSTRUCTION (SHOCK)

The value of the obstruction placed between the test animal and the incentive depends upon its effectiveness in bringing out a reasonably wide range of differential behavior. It represents a constant against which the individual, various drives, and different conditions of the same drive are placed for direct comparison. Our reasons for selecting a high voltage electric shocking device in preference to other possible types of obstruction have been fully set forth in a former paper (4). The standard shock was determined upon after a large amount of preliminary testing of animals of the age and weight that we had planned to use in the project. While our testing out of many different degrees of shock was fairly systematic, we have omitted the detailed data from this report. The important fact is that the standard shock finally selected has proven highly satisfactory in the measurement of all drives so far studied. In connection with the method as a whole, and for the mature white rat, it fulfils exceedingly well the fundamental requirement of a test; it brings out an adequate range of individual and group differences in the functions which it supposedly measures, and yields valid indices.

TABLE 1

Showing the effect of the standard shock, with and without incentive

GROUP	NUMBER OF ANIMALS	OBSTRUCTION CONDITION	INCENTIVE CONDITION	APPROACHES		CONTACTS		CROSSINGS	
				Average	Range	Average	Range	Average	Range
A	8	No shock	No Food	13.0	1-26	3.2	0-5	3.4	0- 5
B	7	No shock	Food	13.1	10-16	1.1	0-4	24.0	12-38
C	8	Standard shock	Food	22.9	15-30	3.6	1-7	11.4	0-22

Only one animal in each of groups A and C did not cross at all during the test period.

Our primary problem, in the present instance, was to determine the extent to which the standard shock as administered operates as a deterrent to the normal tendency of the animal to approach the incentive when placed in the entrance compartment. Accordingly two groups were tested under standard conditions except that in the case of one group the shock was omitted after the preliminary conditioning. The data are given in table 1, in which is also included that of a group which was tested with both shock and incentive omitted.

All of our studies indicate that the number of actual crossings furnish the best index of the drive function. It will be noted that when neither shock nor incentive is included in the set-up the tendency to cross over and explore compartment *C* is very weak. When an incentive (food) is present an average of 24.0 crossings occur when no shock is involved. Only 11.4 crossings occurred in a comparable group with the standard shock in operation, a reduction of somewhat more than one half. The P.E. of the difference is 4.299 and the obtained difference (12.6) is 2.93 times the P.E. of the difference, so that the difference is quite reliable although the small size of the groups should not be overlooked. These results illustrate the normal operation of the obstruction; the shock inhibits without stopping altogether the tendency of normally motivated animals to approach the incentive. It is doubtful whether the standard shock affords much real "punishment" to the animal in crossing since the stimulation value is quite low, being about the same as we have used in visual discrimination work with mature white rats in the Yerkes-Watson apparatus.

Instead of using a shock of constant value as we have consistently done in our drive studies it would be possible with the apparatus as constructed to begin with a low degree of shock and increase the same either during a given test, or at successive tests, to the point where the animal would refuse to cross. By means of a multipoint switch (4) it is possible to vary the resistance in the circuit cumulatively in eight equal steps each representing 1,250,000 ohms. Two of the higher steps had been used by Holden and were found to be too severe at 1200 volts. However, it appeared to be worth trying to begin with the first step (standard shock at 475 volts) and increase by one step at each successive test. This method of using the apparatus proved to be wholly unsatisfactory. A group of four animals showed an average of 15.2 crossings on the initial test when the standard shock was employed, but not one of them crossed when re-tested with the switch set at shock II. The plan had to be abandoned, especially since it was found that the animals still refused to cross even after eight or more successive tests at this setting were made. They seemed to be very strongly conditioned against the grid. Of course a finer gradation of increasing steps could be arranged but this method of testing does not seem to offer much. Even if a high degree of shock did not inhibit the activity of the animal altogether, as seems to be quite generally true, it would be likely to disturb the drive in some way so that the results would tell us little concerning the normal drive. The use of a mild shock kept constant throughout the test period and consequently throughout the study of a given drive, or series of drives is much to be preferred, especially since it has been found to be highly satisfactory in actual application. In brief, we feel the data here presented justifies our use of a constant versus an increasing shock as an obstruction in the apparatus, and indicates that our standard shock, while not necessarily the best possible one, does fulfil in a highly satisfactory manner the purpose for which it was designed in connection with the method. No attempt has been made as yet to work out a series of standard shocks for the white rat at various age or weight levels, although this could be done readily enough and would add greatly to the general usefulness of the obstruction method. Our project as planned is limited to the measurement of drives in the mature animal and the age-weight factor has been kept constant within very narrow limits.

The Incentive (Food, Water, Sex-Object, Etc.)

In an apparatus designed to test the drive-incentive aspect of behavior the incentives should be so placed with respect to the animal under test as to be highly stimulating if not actually dominating in the test situation. This idea was effectually carried out in developing the obstruction method. The smaller compartment *D* which contains the incentive until the test animal crosses over the grid and automatically opens the door, d_2, is situated directly in line with the tunnel. Visual stimulation by the incentive is further aided by the fact that the door itself is made of translucent mathematical celluloid, and the whole apparatus is flooded with diffused light during the test by the illumination hood, contrasting sharply with dark room conditions outside. The door is also amply perforated to permit the free passage of odor from the incentive which naturally passes through the tunnel into the entrance compartment since all compartments are closed, or nearly so during the test. Every opportunity is thus afforded for normal stimulation of the distance receptors directly.

The nature of the incentive is further emphasized by the plan of having 4 preliminary crossings without shock and 1 with shock

TABLE 2

Showing the effect of introducing incentive on group previously tested twice without incentive

GROUP	NUMBER OF ANIMALS	OBSTRUCTION CONDITION	INCENTIVE CONDITION	APPROACHES		CONTACTS		CROSSINGS	
				Average	Range	Average	Range	Average	Range
D—Initial test	8	No shock	No food	13.0	1-26	3.2	0-5	3.4	0-5
D—Re-test 1	8	No shock	No food	7.5	1-18	0.6	0-1	3.0	0-13
D1—Re-test 2	4	No shock	No food	7.7	2-12	0.7	0-2	1.2	0-2
D2—Re-test 2	4	No shock	Food	13.7	8-21	1.0	0-3	13.7	7-22

After the second test, group D was dvided into 2 sub-groups of 4 animals each, one of which was tested with, the other without incentive. In 2 later re-tests of D2, crossings rose to 21.2 and 26.2 respectively.

TABLE 3

Comparison of incentive and non-incentive group scores (Warner — sex drive)

GROUP	INCENTIVE CONDITION	APPROACHES					CONTACTS					CROSSINGS				
		Median	Range	Average	Standard deviation	Coefficient of variabilty	Median	Range	Average	Standard deviation	Coefficient of variabilty	Median	Range	Average	Standard deviation	Coefficient of variabilty
20 males	Female (cornified)	1.5	0-10	2.3	2.35	102.2	1.1	0-6	1.5	1.5	100.0	15.0	4-20	13.45	4.03	29.9
20 males	No incentive	1.0	0-3	1.2	1.03	85.8	1.0	0-4	1.05	1.18	112.3	3.0	0-7	2.95	1.8	61.0
21 females (cornified)	Male	5.0	0-14	5.43	3.63	66.9	3.0	0-9	3.14	2.73	86.8	16.0	2-24	14.14	5.14	36.0
22 females (cornified)	No incentive	3.0	0-10	3.23	2.7	83.6	2.0	0-6	2.27	1.91	84.1	5.0	1-9	5.05	2.55	50.5

precede immediately the test period proper in every case. Doubtless the animal has been well conditioned to the specific incentive, as well as to the electrified obstruction by this time. The more or less frequent crossings during the test must result in a cumulative enhancement of the specific incentive since contact with the incentive stimulus is allowed after each crossing. It is difficult to see how a better arrangement could be devised to bring into relief the specific incentive factor since in the present case these successive re-conditionings are aided by the direct stimulation of the appropriate distance receptors throughout the test.

That the incentive does operate as the dominant factor in causing the animal to cross into compartment *C*, both when the grid is electrified and when it is not, can be demonstrated easily enough.

Certain data of table 1 show the value of the incentive, in this case food, when no shock is administered during the test. The normal exploratory activity of group A resulted in an average of

TABLE 4

Showing reliability of the difference between incentive and non-incentive scores (average number crossings, Warner — sex drive)

GROUPS	DIFFERENCE	STANDARD DEVIATION OF THE DIFFERENCE	DIFFERENCE / S. D. OF DIFFERENCE	CHANCES IN 100 OF A TRUE DIFFERENCE GREATER THAN 0
Males: incentive—non-incentive	10.5	1.14	9.2	100.0
Females: incentive—non-incentive	9.1	1.35	6.74	100.0

TABLE 5

Comparison of temporal distribution of crossings during test, percentile scores, of incentive and non-incentive groups (Warner — sex drive)

GROUP	INCENTIVE CONDITION	0-5 TH MINUTE	6TH-10TH MINUTE	11TH-15TH MINUTE	16TH-20TH MINUTE
20 males	Female (cornified)	28.2	23.4	22.3	26.0
20 males	No incentive	47.6	28.9	15.3	8.5
21 females (cornified)	Male	19.2	14.7	26.4	39.7
22 females (cornified)	No incentive	43.2	32.4	15.3	9.9

only 3.4 crossings into compartment *C* whereas group B crossed to the food incentive 24.0 times. Further evidence on this point is furnished by the results presented in table 2. The number of crossings steadily decreased at each re-test when food was not present in the incentive compartment, whereas crossings jumped to 13.7 when food was suddenly introduced at the second re-test of group *D2*, and rose steadily to 21.2 and 26.2 at the third and fourth re-tests of this group. This shows that the incentive compartment calls out little exploratory activity in the absence of some sort of specific incentive.

The dominating influence of the incentive even in the presence of the standard shock has been clearly indicated by the work of Warner (22) on the sex drive, and tables 3, 4, and 5 of our report have been arranged from his data. We will discuss here only the scores covering crossings inasmuch as such scores have been found to be by far the most significant indices of drive obtained by the obstruction method. The incentive factor represents the sole difference in the general and test conditions of the groups compared, and hence the differences in score may be taken as a measure of the incentive factor in the standard test set-up. The facts in the case are so clear cut that little discussion is necessary. Both male and female animals were tested when in the physiological condition at which the normal sex drive has been shown by Warner to be at its maximum, and the incentive animals as used likewise represented the best possible condition for giving the maximum sex stimulation.

Under these conditions, the males crossed on the average 13.45 times when the proper incentive was present in the compartment and only 2.95 times when no incentive was present. The relative variability of the former group is also very much lower. The difference between incentive and non-incentive groups is almost as great in terms of average number of crossings for the females, although in this case the tendency to cross even when no incentive is present is stronger than in the males. This latter fact, as Warner has suggested, may be due to some special oestrous determinant in the female which also shows itself in the higher absolute score in crossings when the incentive is present. The relative variability is also measurably lower in the case of the females for the incentive group. The validity of the difference between incentive and non-incentive scores for both male and female test groups is extremely satisfactory as a glance at table 4 will show.

In table 5 the temporal distribution of crossings during the test period, when the latter is divided into 4 equal divisions of five minutes each, are given. It will be noted that the percentage of crossings shows a marked tendency to decrease as the testing proceeds when no incentive is present in the case of both male and female groups. When an incentive is present, the percentage of crossings steadily rises as the testing proceeds for the females and remains practically constant for the males. This would seem to indicate that the more or less frequent contact with the incentive following crossings during the test period tends not only to maintain but also to increase the stimulation value of the specific incentive being used at the time. Another possible interpretation of the data of table 5 would be an increasing tendency to become negatively adapted to the grid as the test proceeds. Probably both factors are at work here and are jointly accountable for the temporal distribution of crossings and the opposite tendency shown under incentive and non-incentive conditions. The behavior of the test animal must result from some sort of balancing of the positive influence of the incentive and the negative effect of the obstruction. The two factors are so interpenetrated in their actual operation that it is difficult if not impossible to evaluate them independently in a satisfactory manner. The important fact is that, as both are used in our standard technique, valid indices of the drive-incentive aspect of behavior can be secured as in no other apparatus so far devised.

Effect of Re-Testing

It seemed desirable, in connection with this general study of the obstruction method, to determine the effect of repeating the test on the same group a number of times. Two questions arise at this point. The first is that regarding the practice effect when a given test period is considered a practice period and a number of these are given in succession as in a learning experiment. We should expect the animal to become adapted to the experimental conditions, particularly the grid, each successive test or practice period. But at any rate the speed at which this adaptation takes place is a matter of some concern. The second question concerns the validity of the apparatus as a measure of individual differences in drive behavior. Do the individual indices obtained in an initial test hold their relative positions in later re-tests under the same general conditions? If so, then this is so much evidence in favor of the validity of the method as a means of securing dependable individual as well as group scores of drive.

The re-test procedure was in general the same as that of the initial test although certain important modifications were necessary. The starvation period was kept constant for each re-test at forty-eight hours by running these on every third day, food being supplied on the day following each test period. The test animal was placed in the starvation cage, however, for at least one hour immediately after testing before being transferred to the regular living cage, in order that no direct connection should be formed

TABLE 6

Showing effect of repeated re-testing

	TEST GROUP								CONTROL GROUP*							
			Approaches		Contacts		Crossings				Approaches		Contacts		Crossings	
	Number of animals	Average weight	Average	Range	Average	Range	Average	Range	Number of animals	Average weight	Average	Range	Average	Range	Average	Range
		gm.								gm.						
Test	8	148	22.9	15–30	3.6	1–7	11.4	0–22	8	143						
Re-test 1	8	141	14.4	5–25	1.5	0–4	19.6	10–25	8	139						
Re-test 2	8	134	8.2	3–20	2.0	0–8	25.0	6–31	8	132						
Re-test 3	8	132	2.7	0– 8	0.5	0–2	33.0	18–40	8	131						
Re-test 4	8	129	3.6	0–14	0.6	0–3	36.0	24–42	8	126						
Re-test 5	8	126	1.5	0– 6	0.4	0–2	40.1	38–43	8	124						
Re-test 6	8	124	0.4	0– 1	0.1	0–1	41.4	39–44	8	122						
Re-test 7	8	124	1.4	0– 6	0.6	0–1	42.2	40–44	8	118						
Re-test 8	8	121	0.4	0– 2	0.1	0–1	42.2	40–44	8	115	25.9	17–35	1.0	0–3	23.1	17–27
Re-test 9	8	121	0.0	0	0.0	0	42.9	42–44	8	113	9.5	2–23	0.4	0–1	35.9	24–41
Shock II A 10	4	125	15.5	8–22	5.2	2–7	0.5	0– 1								
Shock II B 10	4	113	11.0	7–13	2.7	2–3	22.7	22–23								

between the test and the 23-hour feeding period. The schedule in re-testing was thus as follows: starved two days — initial test — starved one hour, feeding allowed twenty-three hours; starved 2 days — second test — starved one hour — feeding allowed twenty-three hours, etc. Inasmuch as the normal food was a powder (McCollum's diet) which does not lend itself to gorging, opportunity to feed for a day (twenty-three hours) between successive starvation periods seemed necessary. And even this precaution did not prove sufficient to keep the animals up to their normal weight as will be noted later.

A second difference between test and re-test involved the omission of the regular preliminary crossings except for the fifth, or conditioning crossing. It did not seem necessary to repeat the series of 4 crossings without shock after the initial test since the purpose of these crossings is to establish a connection between the incentive and the act of crossing and this connection has been well formed by the end of the initial test. Furthermore the animals have been negatively conditioned to the grid in the first test and will not readily pass over it even when it is not electrified in later tests. Each re-test proper began then, after a single re-conditioning crossing with shock and lasted the usual twenty minutes. Shock II was used in the tenth re-test, the standard shock being employed in the first 9 re-tests. Food was always present in the incentive compartment.

If the results of table 6 be taken at their face value then it is quite clear that the animal becomes rapidly adapted to the grid so that it no longer inhibits him from crossing. However, the results given in the first section of the table are complicated by the fact that, quite unintentionally on our part, the animals were being seriously underfed. This would mean, of course, that the physiological basis of the drive under investigation was being varied from test to test and that the effects of chronic starvation were probably influencing the results. The animals were losing weight on our schedule which allowed one day of normal feeding between every two-day starvation period (see table 6) whereas they should have been growing and taking on weight. After the sixth test milk-soaked whole wheat bread was added to the diet of the animals on the interpolated feeding day but even this did not stop altogether the losing of weight, although a few remained practically at this level thereafter.

The results in the first section of table 6 involve not merely the factor of practice growing out of successive re-tests but also that of quantitative stunting. In order to determine the influence of the factors separately a control group was given the same feeding schedule for twenty-seven days as the test group had had and then tested on what corresponded to the eighth re-test period of this latter group. The loss in weight was approximately the same in both test and control groups. It is quite apparent that the effect of the stunting was to increase the drive, since the control group crossed 23.1 times instead of 11.4 times as the test groups show on their first test. It appears then that the factor of stunting was in no small degree responsible for the increase in number of crossings at successive tests in the case of the test group. The effect of stunting and of practice, as involved in re-testing, is, obviously, to run up the score.

When due allowance has been made for the influence of stunting on the test group, the data of table 6 show clearly that adaptation to the obstruction does take place rather rapidly under our conditions of re-testing. The average number of crossings jumps from 11.4 on the initial test to 19.6 on the first re-test made only three days later. Also the effect of practice is evident from the fact that the crossings rise from 23.1 to 35.9 on a second test of the control group. The most convincing evidence, however, is the fact that the first test of the control group shows a score of only 23.1 crossings as compared with 42.2 crossings for the corresponding period (re-test 8) of the test group. The latter value represents both practice effect and stunting, while the former is a measure of the stunting factor alone.

It is interesting to note in passing that the average number of crossings came finally to approach very closely the theoretical limit for a twenty-minute test period in the case of the test group. That is, under the influence of stunting and re-testing conditions the animals crossed continuously as soon as replaced in compartment *A*. The theoretical limit is 44 crossings in a twenty-minute test period and is based upon the speed of the animal in crossing and the time required to shift the animal back to the entrance compartment again. This maximum number of crossings was actually reached in many individual cases and became the upper limit of the range in re-tests, 6, 7, 8, and 9. The fact that the average number of crossings was above 40 on re-tests 5 to 9 shows that the obstruction did not inhibit the tendency to cross over at this stage of practice under the conditions of stunting then obtaining. It does not mean that complete adaptation to the obstruction would follow a half dozen re-tests under normal feeding conditions. Adaptation to the grill probably does take place rather rapidly in successive testings and on this account we recommend that the data of the initial test only to be taken as measures of drive. The learning factor entering into re-test scores make them less and less valuable as exact measures of drive if the method be thought of as a test rather than a learning technique. In our opinion the learning method of measuring drives, as used by Simmons, Szymanski and others gives a much less adequate index of drive function than the obstruction method. The directness and simplicity of the response in the latter reduces immeasureably the large element of chance factors that is always prominent in maze learning and similar learning situations. There is every reason for using a test rather than a learning set-up in quantitative studies of drive, although interesting corroborative data may be secured by determining the relative value of various incentives in habit formation.

On the ninth re-test the animals of the test group were crossing the grid continuously and we decided to give them shock II on the next re-test. The 8 animals were divided into 2 groups, A and B. Shock II was given to group A at the beginning of re-test 10 and to group B at the middle of the corresponding re-test period. In both cases the higher shock practically inhibited all crossings, the

TABLE 7

Rank of individual animals on successive tests

ANIMAL	A	B	C	D	E	F	G	H
Initial test	1	2.5	2.5	4	5	6	7	8
Re-test 1	1	2.0	4	3	5	6	7	8
Re-test 2	1	3.5	3.5	2	6	5	7	8
Re-test 3	1	4.5	4.5	2.5	6	4.5	7	8
Re-test 4	1	4	4	2	4	6	7	8

value recorded for group B representing the number of crossings that occurred before shock II was thrown in, i.e., during the first half of the test period. Both groups were given a re-test, using shock II, three days later but not one of the 8 animals crossed the grid. Four of the 8 rats were held over for thirty days and given another re-test and again not one of them crossed into compartment *C*. These results show that shock II is altogether too severe for practical use in connection with the grid, or obstruction. It also argues against the technique of increasing the shock in a graded series of steps in testing the same animal and getting the strength of drive in terms of the degree of shock that inhibits crossing altogether, unless much finer gradations of shock were used than the series afforded by our present apparatus. A shock strong enough to condition the animal completely against the grid is also likely to bring about a shift in physiological condition and in the drive under investigation, and thus defeat the purpose of the test altogether.

We come now to the question of the validity of the method as a measure of the drive function in the individual. No extensive investigation of this aspect of the method has been made but such data as we are able to present are all favorable. In table 7 we give the ranking of a group of 8 animals on the initial test and on four successive re-tests. The rat which crossed the largest number of times was assigned the rank of I, etc. It will be seen that animals A, G, and H kept the same relative position throughout, while there was comparatively little shifting around in rank among the other animals of the group. So far as the evidence goes, the method seems to have a relatively high self correlation, although such a conclusion should finally rest upon a larger body of data.

Our general conclusion must be that the obstruction method as standardized and used in connection with our project on animal drives fulfills most adequately the purpose for which it was designed. In its present form it can only be used in the measurement of drive-incentive behavior in the white rat. However, we plan to work out such modifications of the animal control section as may be necessary in order to extend its use to a wide variety of animal forms of different sizes and habits. Inasmuch as the essential response is that of normal locomotion, and the set-up relatively simple so far as the animal is concerned, its use as a method in comparative psychology would seem to be almost unlimited so far as adaptation to divergent morphological types goes. We plan in the near future to build and standardize the apparatus of the proper size to be used on dogs, cats, raccoons and monkeys.

Bibliography

(1) Dashiell, J. F. 1928. Fundamentals of objective psychology. New York, Houghton, Mifflin, pp. 257, 258.

(2) Graber, V. 1884. Grundlinien zur Erforschung des Helligkeits — und Farbensinnes der Tiere. Prag, Tempsky, 322 p. (Excellent summary of preference method work of Bert, Graber, Lubbock, etc.)

(3) Holden, Frances. 1926. A study of the effect of starvation upon behavior by means of the obstruction method. *Comp. Psychol. Monographs*, 3, No. 17, 45 p.

(4) Jenkins, T. N., L. H. Warner, and C. J. Warden. 1926. Standard apparatus for the study of animal motivation. *J. Comp. Psychol.*, 6: 361-82.

(5) Lubbock, J. (Avebury). 1882. Ants, bees, and wasps. New York, Appleton, 448 p.

(6) Lubbock, J. (Avebury). 1882. Ants, bees, and wasps. New York, Appleton, p. 247.

(7) Morgan, J. J. B. 1922. The measurement of instincts. Proc. 31st. Ann. Meeting of the Amer. Psychol. Assoc., Cambridge, Mass., p. 94.

(8) Moss, F. A. 1924. Study of animal drives. *J. Exp. Psychol.*, 7: 165-85.

(9) Murphy, G. 1929. An historical introduction to modern psychology. London, Kegan Paul, p. 378.

(10) Richter, C. P. 1922. A behavioristic study of the activity of the rat. *Comp. Psychol. Monographs*. 1, No. 2, 55 p.

(11) Richter, C. P. 1927. Animal behavior and internal drives. *Quart. Rev. Biol.*, 2:307-43.

(12) Slonaker, J. R. 1907. The normal activity of the white rat at different ages. *J. Comp. Neurol.*, 17: 342-59.

(13) Slonaker, J. R. 1908. Description of an apparatus for recording the activity of small mammals. *The Anatomical Record*, 2: 116-22.

(14) Slonaker, J. R. 1912. The normal activity of the albino rat from birth to natural death, its rate of growth and the duration of life. *J. Anim. Behav.*, 2: 20-42.

(15) Stewart, C. C. 1898. Variation in daily activity produced by alcohol and by change in barometric pressure and diet with a description of recording methods. *Amer. J. Physiol.*, 1: 40-56.

(16) Szymanski, J. S. 1912. Modifications of the innate behavior of cockroaches. *J. Anim. Behav.*, 2: 81-90.

(17) Szymanski, J. S. 1920. Aktivität und Ruhe bei Tieren und Menschen. *Z. allg. Physiol.*, 18: 105-61.

(18) Tsai, C. 1925. The relative strength of sex and hunger motives in the albino rat. *J. Comp. Psychol.*, 5: 407-15.

(19) Turner, C. H. 1912. An experimental investigation of an apparent reversal of the responses to light of the roach (Periplaneta orientalis L.). *Biol. Bull.*, 23: 371-86.

(20) Waller, O. D. 1899. The characteristic of nerve. *Proc. Royal Soc.*, 65: 207.

(21) Wang, G. H. 1923. The relation between "spontaneous" activity and oestrous cycle in the white rat. *Comp. Psychol. Monographs*, 2, No. 6, 27 p.

(22) Warner, L. H. 1927. A study of sex behavior in the white rat by means of the obstruction method. *Comp. Psychol. Monographs*, 4, No. 22, 67 p.

(23) Wodsedalek, J. E. 1912. Formation of associations in the may-fly nymphs (Heptagenia interpunctata Say). *J. Anim. Behav.*, 2: 1-19.

PART II

THE HUNGER DRIVE

Part II

THE HUNGER DRIVE

C. J. Warden

The two papers included in Part II deal with very different aspects of the hunger drive. The first covers a repetition of the work of Holden, using the Obstruction Method in its final form in the measurement of the normal hunger drive in the male and female white rat. By normal hunger drive we mean the tendency to approach and secure food after a given period of starvation when the general physiological condition of the animal has been kept undisturbed, except in the matter of starvation. By controlling other possible drives in the manner indicated below, the hunger drive was made dominant during the time of the test — though, perhaps, it was not isolated from all other tendencies in the strict sense.

The thirst drive was eliminated by the simple procedure of allowing the animal to have access to its regular water supply up to the moment when it was taken from the cage to be tested. The sex drive offered more difficulties because the method of control for the different sexes would necessarily be different. The sexual factor in the males was eliminated as far as possible by using animals that had had no opportunity to copulate for 35 days, at which time the male, while normally active, shows little or no tendency to be stimulated by the female in oestrum (Part IV, 1). For the females, selection of only those in dioestrum for testing solved the problem of control of the sex factor. The additional precaution was taken to test animals of the same sex only on a given day, and to cleanse the apparatus thoroughly after the series of tests for each day had been given. The maternal drive was not involved, inasmuch as pregnant females and females with litters were not included. General exploratory behavior could not be strictly controlled. It was, however, kept as constant as possible by testing all animals at the same period of the day, and at the time (between 9 P.M. and 4 A.M.) when the white rat is known to be most active. The small size of the incentive compartment should have the effect of making the incentive object stand out and thus would tend to discourage exploration in the main. This general procedure of standardization of both living and testing conditions was effective, we feel confident, in isolating to a high degree the hunger drive—at least as far as such is possible without introducing controls so artificial that they themselves would be a disturbing factor important enough to invalidate the indices of drive behavior obtained.

Part II, 2, opens up a large field of motivation as tested by this method, i.e., the effect of immediacy of reward and strength of drive. In all other work included in the main body of this report the animal is given access to the incentive as soon as he passes over the obstruction and enters the incentive compartment. In the present section, however, the animal is held up after making the usual attempt to secure the incentive (food) for various intervals of time. The logic of such procedure might be either that the delay tends to weaken the drive or that the delay tends to intensify the stimulation value of the incentive situation in a cumulative manner. This study attempts to determine the matter so far as short intervals of delay are concerned. The thirst, sex, maternal, and exploratory drives were controlled in the same manner as indicated above in connection with the normal hunger drive. In selecting a certain degree of hunger on which to study the effects of delayed incentive, the 48-hour starvation period was decided upon because this had been found (Part II, 1) to represent the stage at which the hunger drive was at its maximum under normal conditions. The experiment is thus limited to the study of the effect of delayed incentive when the hunger drive is at its maximum. Further work covering the effect of delayed incentive on other stages of the hunger drive, should be carried out before wide generalization is warranted on this aspect of motivation.

The term "hunger drive" as used in connection with this project has reference to the tendency aroused in the animal by starvation to approach food in the incentive compartment. In our work the strength of this tendency is indicated by the indices of behavior obtained. Since no independent check was had of actual stomach conditions, no implications are intended as to the presence or absence of "hunger contractions" during the state of arousal. The term "incentive" refers to the food itself, while "incentive-value" or "incentive-index" denotes the stimulation effect of the incentive in terms of the behavior score. None of the terms used carry any implications whatsoever regarding the subjective state of the organism.

1. A STUDY OF HUNGER BEHAVIOR IN THE WHITE RAT BY MEANS OF THE OBSTRUCTION METHOD[1]

L. H. Warner

In a recent monograph (13) were published the results of an experimental study of sex behavior in male and female white rats. The method used in that study has been termed the obstruction method (4). The most important data consist of the records of the number of times in a test period of constant duration that an animal will cross an electric grid to reach a sex object (an animal of the opposite sex). The effect upon this tendency to approach a sex object or, in brief, the effect upon the sex drive of variation in physiological condition was studied. In the female it was found that the sex drive was dominant during only a rather restricted phase of the entire oestrous cycle, that during this period (the cornified stage) the crossing was decidedly and reliably more frequent than during the dioestrous stages. The average number of crossings per unit time made by the animals tested when in the cornified stage, at the peak of sex activity, can be considered an index of the sex drive in the normal female.

In the case of the male the factor influencing most profoundly the sex drive is the length of time which has elapsed since the last previous mating period. Of the seven sex deprivation intervals studied the twenty-four hour interval resulted in the maximum drive. The results for this group may be considered an index of the sex drive in the normal male.

A second fundamental type of behavior to be studied is that directed toward food. Such behavior has previously been studied by Moss (6) and by Holden (2) using the obstruction method. The results of neither of these investigations can be compared with those of the sex study mentioned above because of decided dissimilarities in method. These will be enumerated in a later section.

[1] Reprinted from *The Journal of Comparative Psychology*, 1928, 8: 273-99

The present article is a report of the experimental study of food-seeking behavior made under such conditions as to render the results directly comparable with those presented in the monograph on sex behavior. It also includes a comparison of these results with those on the sex drive.

METHODS AND PROCEDURE

The animals used were of the same stock and raised under the same conditions in every particular as were those used in the study of sex behavior. The apparatus and general procedure were the same. There will be taken up here consideration of only such matters as refer especially to feeding conditions. For further details see Warner (13).

The animals had from the time of weaning been supplied with a diet which was adequate qualitatively and quantitatively. The details of the diet during their first hundred and fifty days during which they were in the care of the Wistar Institute cannot be given, but the long and successful experience of this organization in the care and breeding of normal animals is ample guarantee of its adequacy. During the thirty-five days each animal was in this laboratory prior to testing the following diet was supplied it continuously:

	Per cent
Whole wheat flour	70.0
Whole milk powder	16.6
Casein	11.0
Sodium chloride	1.4
Calcium carbonate	1.0

This was mixed, dry, in an electric mixer for at least ten minutes. Greens were supplied in abundance once a week. There was always an ample supply of water from inverted bottles which rendered fouling impossible. More than a sufficient supply of food was kept in the cages at all times (except, of course, during the starvation periods mentioned below). These feeding conditions differed in no way from those which held in the study of the sex drive but are described here since they are closely related to the hunger drive here under investigation.

Just as in the study of sex behavior the physiological condition of the animal as related to food consumption was kept as constant as possible (and in that condition accompanying a minimum of food-seeking activity), so in this study we have determined and kept constant the animal's condition as this related to sex behavior and have used those conditions accompanying a minimum amount of sex behavior. In no other investigation of either of these types of behavior has such control been attempted or, perhaps, even the need for it recognized. This criticism holds for even the more recent work (Simmons (9), Moss (6), Tsai (11), Holden (2)).

The most dominant factors affecting sex activity in the female are, apparently, (1) the present stimulus situation and (2) the oestrous condition. To take up the first of these factors: there was no element in the experimental situation which seemed at all related to, and thus likely to arouse sex behavior. Although it seems well nigh impossible to control the olfactory situation when animals of the two sexes are used on the same apparatus we at least spared no effort in attempting to do so. Animals of the opposite sex were not used in the apparatus on the same night (all experimentation took place between 9 P.M. and 4 A.M., as in the case of the sex study).

Immediately at the conclusion of a period of experimentation the apparatus was thoroughly washed out with soap solution and allowed to air. At no time was an animal given the opportunity of associating the apparatus with another animal. In other words, but one animal at a time was in the apparatus, the object in the incentive compartment being not another animal but food. The second factor was controlled by determining the oestrous condition of the animal at the time of its testing. This determination was based upon the histological characteristics of the vaginal secretion. Sampling of the secretion was never made until after the animal had been tested; thus there was eliminated the possibility that this process might interfere with the normal behavior of the animal during testing. Sex activity in the female is rather rigidly restricted to a relatively brief portion of the oestrous cycle. During the remainder of the cycle the animal rarely shows any interest in the male as a sex object and will, in fact, actively avoid mating in most cases. By using females in the inactive period or dioestrum the sex factor is reduced to the minimum. The data reported herein concern only such females as were in the following stages as defined in the study on sex behavior: Recuperative, Early inactive, Inactive and Late inactive. Several females were in oestrum at the time of testing (as determined afterward by smear examination) in spite of the fact that the females were selected for testing on the basis of behavior observation which seemed to indicate that they were in dioestrum. Data on such animals were discarded. There was not a sufficient number of them to make it worth while to consider them as a separate group.

The dominant factors affecting sex behavior in the male are (1) the present stimulus situation and (2) the period of time which has elapsed since the last previous mating period. The first of these was controlled as in the case of the female. The second was kept constant in the case of all males used. The sex deprivation interval immediately preceding testing was thirty-five days. This long period was determined upon after careful consideration of the results of the study on the sex drive. When studying food-seeking behavior it was not merely our purpose to keep conditions affecting the sex drive constant but so to arrange them that the sex drive should be at its minimum at the time of the testing. This cannot be done as simply with the male as in the case of the female since the male is not subject to a cycle involving a period of sex inactivity. It might be suggested that we use males immediately after they had been given a period of mating, after sex "satiation" had resulted. While it is true that the sex drive appears to be at its minimum at this time our observations upon rats tested (in the obstruction apparatus with a female in oestrum in the incentive compartment) when in this condition indicate that their behavior is so profoundly affected that a fair measure of the food-seeking drive could not be obtained. Such males were found to be extremely inactive. The usual type of behavior displayed was sleep (or what appeared to be such).

Again it might be suggested that we use males whose sex drive had been partially satiated. But casual laboratory observations appear to indicate that the drive remains at a high point under this condition although it is displayed rather more spasmodically.

Nor would the use of males after a short deprivation interval prove satisfactory. As has been shown in the sex study, recovery of the sex drive is very rapid. It is apparently well along the road to recovery by the time (varying with different animals) when the male becomes aroused from the lethargy following a prolonged period of mating. It thus seems that if we should use a rather short sex deprivation interval we would not be dealing with a homogeneous group of animals but with a group containing both lethargic and active animals.

A sex deprivation period of one day has been shown to result in sex behavior at or near the high point. But a glance at figure 4 in the sex monograph shows that beyond this point there is a gradual drop in the drive. The longest sex deprivation interval used in that study was one of twenty-eight days. A decided falling off in the strength of the drive was found in this group of twenty males. Just as significant for our present purpose is the consideration of the behavior displayed by these twenty-eight day sex-deprived animals when placed in a situation involving sex stimulation. Of the twenty males so tested only two crossed to the female during the first minute. This is the lowest number recorded for any group, being even lower than that for the control group (no sex object or other incentive). This indicates that these two crossings were very probably determined by other stimulation than that related to sex. During the first five minutes but sixteen crossings were recorded for the twenty-eight day group as contrasted with seventy-six for the one day group. After the first five minutes the activity became progressively greater indicating what might be described as a reawakening of interest in the sex object.

As has been pointed out in the sex study, the sex drive in the male appears to be more determined by the external stimulation and less by the internal stimulation than in the case of the female. Thus sufficiently long deprivation periods in the male produce a drop in the drive such as would be much less true of the female because of the rhythmic occurrence of an internal stimulus situation in the latter. It is true that the male after a long period of sex deprivation can recover in a few minutes its normal interest in a sex object when it is placed in the presence of a female in oestrum. This fact does not, however, prevent such males from being the ideal animals for our present purpose since there is nothing in the situation involved in the measurement of the hunger drive which would directly, or by association arouse the sex drive.

Thus by using long time (35-day) sex deprived males in the present study of the hunger drive we feel that we are keeping constant and at a minimum the sex drive. These animals may in fact be considered somewhat analogous to females in dioestrum.

The incentive stimulus in the present study was food, about twenty grams of the mixture described above. This food was contained in a black metal receptacle firmly fixed in the smaller section of the incentive compartment, i.e., where the incentive or stimulus animal was placed in the study of the sex drive. (See either of the following references for cuts and description of the apparatus used: (4) or (13)). The automatic door, which had for its purpose the restraint of the incentive animal in the study of sex, operated in just the way it did in that study although it was of course not necessary to restrain the food. In brief, except for the fact that a food object was substituted for a sex object the experimental conditions were no different from those which held in the study of the sex drive.

The behavior which determined when the animal should be returned to the entrance compartment after crossing consisted in the taking of a nibble of food. This criterion was adhered to with the following exceptions. If during the preliminary training period the animal did not take one nibble of food within one minute it was returned anyway. This period was reduced to thirty seconds during the test period. This exception had a particular bearing on the groups of animals which were run immediately after being taken from cages containing food, i.e., after no starvation period. These animals, as would be expected, failed to respond to the food any more than to other details of the incentive compartment. As a matter of fact, crossing was far less frequent in animals of this group even though they were sometimes returned prior to sampling the food.

It was originally planned that a control group corresponding to that group showing the greatest hunger drive should be tested. This group was to have differed from the test group only in that there would no food or other incentive object in the incentive compartment while the animals were tested. Our results showed, however, such reliable differences between the behavior of animals recently fed and those starved for a period that it was felt that additional indication that the differential behavior was a function of the animal's condition as related to food was unnecessary. It might be noted parenthetically that the control groups run in the sex study cannot serve as controls for this investigation in spite of the fact that these groups consisted of animals tested with no object in the incentive compartment. This is true because the female control consisted of females in oestrum (whereas the females used in the present study were in dioestrum) and the male control consisted of males taken immediately from cages supplied with food (and would thus not be comparable with the four day starvation group, which in the present study showed the hunger drive at its height in the male).

TABLE 1

Showing grouping of animals and the number of each sex in group

LENGTH OF STARVATION PERIOD	NUMBER OF ANIMALS	
	Male	Female
days		
0	10	10
2	10	10
3	10	10
4	10	10
6	10	10
8	10	10

The periods of starvation first studied in this investigation were 0, 2, 4, 6 and 8 days. The 3-day group was later added as a result of data already obtained since these seemed to indicate the possibility that the peak of food-seeking activity for the two sexes combined might lie between 2 and 4 days. Standard groups of twenty animals each were used as was done in the study of the sex drive and as is being done in other studies on drive to be published from this laboratory. The grouping is tabulated in table 1.

Results

Table 2 gives the distribution of the approaches, contacts and crossings for the various groups. The medians, ranges, averages, standard deviations, and coefficients of variability are found in table 3. Table 4 gives the reliability of the differences between the average crossings for various groups. The approaches, contacts and crossings as distributed minute by minute throughout the twenty minute test period are given in table 5. In table 6 these data are combined in five-minute units and are also reduced to percentages. Figure 1 shows graphically the effect of variation in the starvation period upon the average number of approaches, contacts and crossings, the data for the two sexes being combined. Figure 2 is like figure 1 except that the two sexes are treated separately. Figure 3 graphically demonstrates the effect of variation in the starvation period upon the temporal distribution of crossing during the test period. The writer is deeply indebted to Dr. Frances Holden for the statistical treatment of the data and the construction of the tables and figures.

The results will be discussed in four main sections dealing with (1) the effect of variation in the length of the starvation period upon the hunger drive, (2) sex differences in the hunger drive, (3) a comparison of the results here reported with previously reported data on food-seeking behavior, and (4) a comparison of the strength of the hunger drive as herein measured and the sex drive as measured in the work reported previously by the writer (10).

1. The effect of variation in the length of the starvation period upon the hunger drive

The groups will be considered in order from the shortest to the longest starvation period.

The O group consists of animals taken immediately from the living cage, where food was always available, to the obstruction apparatus where they were tested. About ten minutes elapsed between their removal from the cage and the commencement of the

TABLE 2

Distribution table covering approaches, contacts and crossings for the various periods of starvation of male and female rats

INTERVAL	0						2 DAYS						3 DAYS						4 DAYS						6 DAYS						8 DAYS					
	Approaches		Contacts		Crossings		Approaches		Contacts		Crossings		Approaches		Contacts		Crossings		Approaches		Contacts		Crossings		Approaches		Contacts		Crossings		Approaches		Contacts		Crossings	
	Female	Male	Female	Male	Female	Male	Female	Male	Female	Male	Female	Male	Female	Male	Female	Male	Female	Male	Female	Male	Female	Male	Female	Male	Female	Male	Female	Male	Female	Male	Female	Male	Female	Male	Female	Male
0	3	5	1	2	3	5	3	2	1	1						1				1						1	3	3			1	1	2	1	2	1
1	3	3	2	2	1	1	1			1			1	2						2		1			2	2		1				2	1	3	1	
2	2		2	3	3			3		1						1			1			4				2		1			1	1	1			
3			1		1		2	1			1	1	1	1		1			1	1		2				1	1	1		1		1				
4			1	2	1	1	1	1	2	1			1						2	2	1	2	1		2	1		1	2					2		
5		1	1				1	1	2	3	1		1	2		2			2	2	4	1			1	1		1			1	1			1	2
6	1		1	1		1	2	1	1				4	4	4	1			2	1	3				1		2	1			1			1	1	
7	1	1			1	1				1			1		2	2		1		1	1						1					2			2	1
8									1	1		1	1		1				1		1				1		1				3		1		1	
9						1		1								1	1		1						1	2		1			1					
10									1						2			1						2			1		1	1	1	2	1	1	1	
11												1		1		1	1	1							1					2			1			3
12									1	1							1						1		1				1	1	1		1			
13																											1		1				1			
14															1								1						1	1						1
15									1		1	1					1	1					1	1					1					1		1
16																	1												1	1			1	1	1	
17											1							1					1													
18											1	2					1						1	2												
19												1					1	1												1						1

TABLE 2 — *Continued*

Distribution table covering approaches, contacts and crossings for the various periods of starvation of male and female rats

Interval	0						2 days						3 days						4 days						6 days						8 days					
	Approaches		Contacts		Crossings		Approaches		Contacts		Crossings		Approaches		Contacts		Crossings		Approaches		Contacts		Crossings		Approaches		Contacts		Crossings		Approaches		Contacts		Crossings	
	Female	Male	Female	Male	Female	Male	Female	Male	Female	Male	Female	Male	Female	Male	Female	Male	Female	Male	Female	Male	Female	Male	Female	Male	Female	Male	Female	Male	Female	Male	Female	Male	Female	Male	Female	Male
20																		2					1													
21											1	1											1	1						1						
22																	1							1												
23												1											1	1												
24																																				
25											1	1												1					1	1						
26																	1						1													
27																													1							
28											1																									
29											2							1						1												
30																		1																		
31																																				
32																																				
33																																				
34																																				
35																																				
36																	1																			

TABLE 3

Showing results for each starvation period

STARVATION PERIOD	SEX	APPROACHES					CONTACTS					CROSSINGS				
		Median	Range	Average	Standard deviation	Coefficient of variability	Median	Range	Average	Standard deviation	Coefficient of variability	Median	Range	Average	Standard deviation	Coefficient of variability
days																
0	Female	1	0–7	2.0	1.34	67	2	0–6	3.5	1.55	44	2	0–7	2.1	2.07	98
	Male	0	0–7	1.5	1.34	89	2	0–6	2.2	1.20	55	1.5	0–9	2.7	3.37	125
	Total	1	0–7	1.75	1.36	79	2	0–6	2.85	1.42	50	2	0–9	2.4	2.78	116
2	Female	3	0–6	3.4	1.56	46	5	0–15	6.9	1.87	27	18	3–29	19.0	8.91	47
	Male	2	0–9	3.3	1.47	44	5	0–12	4.9	1.59	32	18	3–25	16.1	6.56	41
	Total	3	0–9	3.35	2.81	84	5	0–15	5.9	3.92	66	18	3–29	17.6	7.78	44
3	Female	6	1–8	5.2	1.25	24	7	6–14	8.0	1.41	76	16	9–36	18.4	7.63	41
	Male	5	1–11	5.0	1.41	.28	5	0–10	5.7	1.58	28	17	7–30	18.0	7.52	42
	Total	6	1–11	5.1	2.46	48	6	0–14	6.85	2.93	43	17	7–36	18.2	7.58	42
4	Female	5	2–9	5.2	2.04	39	6	4–8	5.7	1.10	19	18	4–26	17.0	5.92	35
	Male	4	0–7	3.6	2.20	61	3	1–5	2.8	1.17	42	19	10–29	19.1	5.87	31
	Total	5	0–9	4.4	2.37	54	5	1–8	4.75	1.91	40	18	4–29	18.1	5.98	33
6	Female	5	1–12	6.1	1.77	29	6	0–13	5.3	1.80	34	13	4–27	14.0	7.18	51
	Male	2	0–9	3.6	1.59	44	2	0–9	3.0	1.55	52	12	3–25	14.2	5.98	42
	Total	4	1–12	4.85	3.58	74	3	0–13	4.15	3.61	87	13	3–27	14.1	6.61	47
8	Female	8	0–12	6.8	1.69	25	8	0–16	7.3	2.29	31	6	0–16	6.0	4.69	78
	Male	3	0–10	4.6	1.79	39	4	0–16	5.5	2.18	40	11	0–19	9.8	5.33	54
	Total	7	0–12	5.7	3.77	66	4	0–16	6.4	5.71	89	7	0–19	7.9	5.07	64

TABLE 4

Showing reliability of the differences between the average crossings of each group

GROUPS	STANDARD DEVIATION OF THE DIFFERENCE	DIFFERENCE BETWEEN THE AVERAGES	DIFFERENCE / S. D. OF DIFFERENCE	CHANCES IN 100 OF A TRUE DIFFERENCE
0 and 2 days	2.43	15.1	6.21	100
2 days and 3 days	2.36	.7	.3	62
3 days and 4 days	2.17	.1	.05	52
4 days and 6 days	2.00	4.0	2.0	98
6 days and 8 days	1.86	6.2	3.33	100
0 and 3 days	2.40	15.8	6.58	100
3 days and 6 days	2.22	4.1	1.85	97

twenty-minute test period. This time was occupied by the preliminary training in the apparatus. These animals crossed to the incentive compartment almost as readily as did starved animals during this preliminary training (during which the grid was not electrified). But once in the incentive compartment they were occupied mainly with exploring, not with eating. With the beginning of the test period (during which the grid was electrified) their behavior was greatly altered. They continued to cross occasionally, but only after much hesitation. During the first five minutes of the test period a total of only 31 crossings was recorded for the twenty animals while 28 contacts were recorded for the same period showing that the animals retreated from the grid after touching it almost as often as they continued on across. The second five-minute period shows only 9 crossings, the third, 5, and the fourth, 3. Thus there is a continuous diminution in the tendency of the animals to cross. The average number of crossings for this group was only 2.4, approaches, 1.75 and contacts, 2.85. This group represents the low point for all three forms of behavior. No animal in this group crossed more often than nine times.

The two-day group presents a striking contrast to the preceding. The average number of approaches increases to 3.35, the average number of contacts increases to 5.9 while the average number of crossings leaps from 2.4 to 17.6. The difference between the average number of crossings for the 0 and the two-day group is highly reliable being over six times the standard deviation of the difference. Animals of this group were, obviously, decidedly more active. There was still occasional hesitancy as is shown by the approaches and contacts, but for the most part the animals recrossed the obstruction thirty to sixty seconds after being replaced in the entrance compartment. The crossing was more uniformly distributed throughout the entire test period than in the case of the 0 group although this activity was still rather more frequent during the early part of the period.

The three-day group does not differ markedly from the two-day. There is a slight further increase in all three forms of behavior recorded but, especially in the case of crossings, the differences are not reliable. The chief difference between this and the preceding group is to be found in the temporal distribution of crossings during the test period. These are now almost uniformly distributed throughout the period.

The four-day group shows a slightly lower average number of approaches and contacts while the crossings remain practically the same. The crossings are still rather uniformly distributed.

The average number of approaches and contacts for the six-day group is almost the same as for the four-day group but there is a definite drop in the number of crossings, the average being 18.1 for the four-day group and 14.1 for the six-day group. This difference being twice the standard deviation of the difference is not entirely reliable, the chances of a true difference being 98 in 100. When it is noted that the succeeding group shows a further drop there seems reason to suppose that this difference is a real one.

The eight-day group shows a further and decided reduction in the number of crossings. Animals of this group were very weak as shown by their gait and general behavior. There is some selection in the composition of this group since three females and two males died in the course of filling out the group which as usual consisted

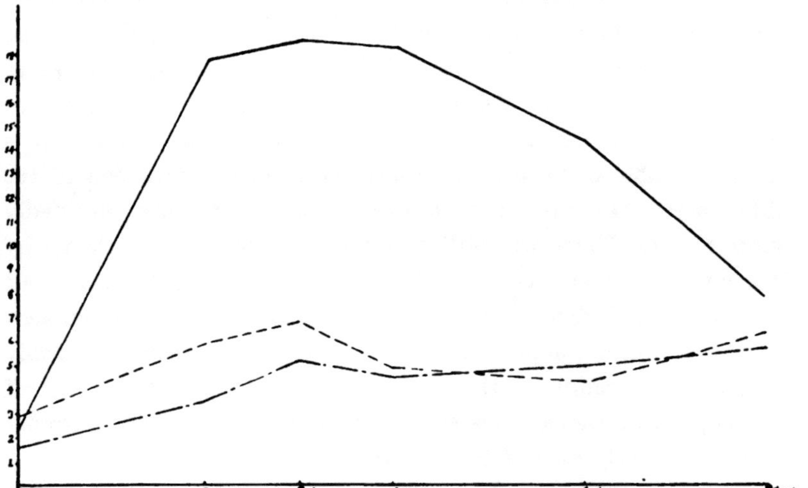

FIG. 1. Showing average number of crossings (solid line), contacts (dash line) and approaches (dot-dash line) for each interval of starvation, the data for the two sexes being combined. Each point represents the average of twenty animals. ("Control" should read "O," i.e., no starvation period.)

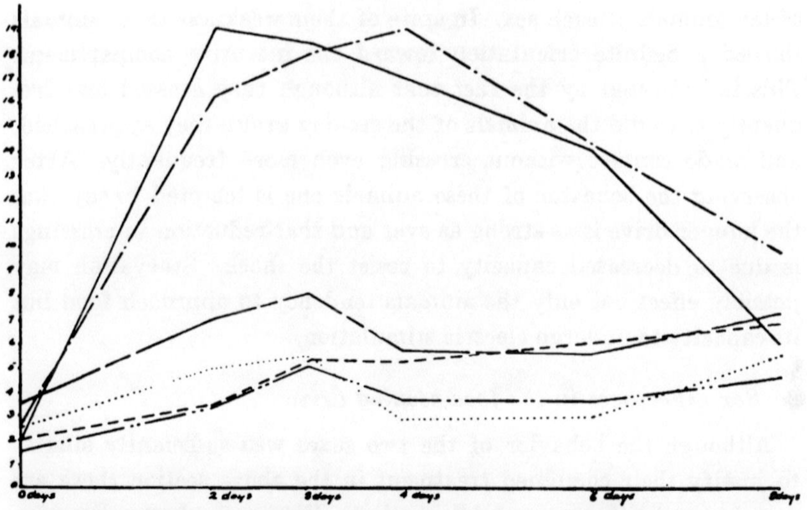

FIG. 2. Showing diagrammatically the same data illustrated in Fig. 1, but here the two sexes are treated separately. The abscissa represents length of starvation period. Average number of crossings for the female (solid line), for the male (dot-dash line); average number of contacts for the female (long dash-short dash line), for the male (dotted line); average number of approaches for the female (short dash line), for the male (dash-two dot line).

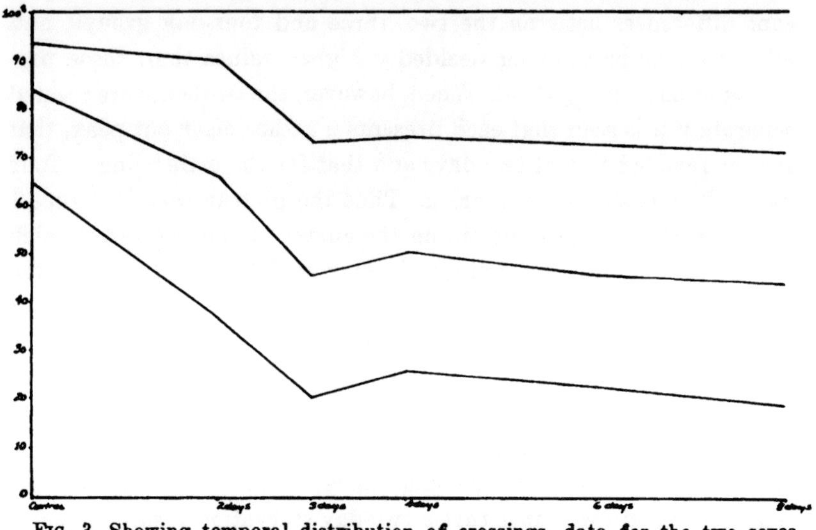

FIG. 3. Showing temporal distribution of crossings, data for the two sexes being combined. The upper line represents the percentage of crossing which occurred during the first five minutes of the test period. The middle line represents the percentage which occurred during the first ten minutes. The lower line represents the percentage which occurred during the first fifteen minutes. ("Control" should read "O," i.e., no starvation period.)

of ten animals of each sex. In spite of their weakness these animals showed a definite orientation toward the incentive compartment. This is indicated by the fact that although they crossed less frequently than did the animals of the six-day group they approached, and made contact without crossing even more frequently. After observing the behavior of these animals one is tempted to say that the hunger drive is as strong as ever and that reduction in crossings is due to decreased capacity to resist the shock. Starvation may possibly effect not only the animals tendency to approach food but its capacity to undergo electric stimulation.

2. Sex differences in the food-seeking drive

Although the behavior of the two sexes was sufficiently similar to justify their combined treatment in the above section there are certain sex differences worth noting. Figure 1 shows diagrammatically the effect of variation in the starvation period upon the approaches, contacts and crossings, the two sexes being considered together. In terms of crossings, the most significant data, the peak of activity extends from two to four days, there being but insignificant differences between the two, three and four-day groups, and all three groups showing decidedly higher values than those preceding or succeeding them. When, however, the two sexes are treated separately it is seen that each presents a rather clear cut peak, that for the female being at two days and that for the male being at four days. This is shown in figure 2. Thus the plateau seen in figure 1 is merely the result of combining the curves of the two sexes which show peaks at different points. It may well be supposed that the drop for the longer starvation periods is due to the lowering of vitality through the lack of nourishment. If so, then it might be said that this weakening influence effects the female both sooner and more severely than it does the male. Starvation periods of three and four days lead to reduction of crossing in the female whereas in the male the crossings increase up to four days. This does not explain why it is that a two-day starvation results in somewhat greater activity in the female than in the male.

A sex difference is also seen in the data concerning contacts and approaches.

Approaches. The 0, two and four-day groups show almost no sex difference. From this point on, however, the female shows decidedly more activity of this sort than does the male.

Contacts. The trend in the number of contacts with various degrees of starvation for the male and the female runs almost exactly parallel, with the female at all times showing considerable more activity of this sort than the male.

In terms of approaches and contacts it can thus be seen that the male at no time displays more activity than the female, and, in fact, that the reverse is usually the case, especially in the groups of longer starvation periods. It might then be said that the orientation of the female was just as surely directed toward the food as in the case of the male; furthermore, that the more decided drop in crossings in the case of the female might therefore be due not to less internal stimulation but to a greater sensitivity to the shock, due to a more rapid lowering of physical vigor.

Data on the temporal distribution of activity during the test period show no significant sex differences and so are not considered here.

3. A comparison of the results here reported with previously reported data on food-seeking behavior

The present work cannot be compared in any great detail with any of the work done previously since none have made use of technique sufficiently like that used here. Holden's study is the most nearly comparable.

Moss (6) used so few animals that his quantitative data are of little significance. The procedure which he adopted is not well suited to the problem of the rapidity of the increase in the hunger drive and the period of starvation resulting in a maximum. His longer starvation groups were composed merely of those animals which had not crossed the obstruction when tested after shorter periods of starvation. Nevertheless he appears to conclude that the drive constantly increases with an increase in the period of starvation. Presumably we can conclude, on this principle, that it reaches its maximum at the moment the animal dies of starvation. The present study however, clearly indicates an optimum followed by a decline.

Szymanski (10), studying the effect of the length of starvation upon the rate of learning a problem involving a food reward found increased efficiency with increased starvation periods. The longest period he used was, however, twenty-four hours so it seems doubtful whether his studies included the optimal period.

Dodson (1), studying the effect of variation in the length of starvation upon the formation of a visual habit found an optimal point at 41 hours. Animals starved for this period learned more rapidly than those starved for periods of 24, 31 or 48 hours. The differences, however, are not great. Furthermore he made use of animals of but 78 days of age. It might well be expected that such young animals would feel the deleterious effects of starvation sooner than would the 185 day old animals used in the present study.

The effect of starvation periods upon the amount of activity registered by a rat in a freely revolving drum has been noted by Richter (8). Of the eight animals used in the starvation experiment four were deprived of water as well as food and are therefore not comparable to ours. Of the other four animals, two showed a peak of activity after two days of starvation, one after one day while the last animal showed constant decrease in activity after the removal of the food. Unfortunately neither the ages nor the weights of the animals are given.

Nicholls (7) using the same method on the guinea-pig found the peak of activity on the second day followed by a decrease.

The work of Holden (2) is more nearly comparable to the present investigation than any other. Since her results are quite different it seems worth while to enumerate the differences in methodology between her work and this.

The apparatus as used in the present work included the following features not found in Holden's: a tunnel-shaped obstruction compartment provided with manually operated doors at the end, a non-shortcircuiting stimulation grid, a restraining compartment and automatic door built for the purpose of restraining the stimulus animal in the study of the sex drive (but used throughout the present work), an illumination hood providing a constant source of light which threw no shadows.

The animals used in the present work were of the experimental colony strain of the Wistar Institute of Anatomy. They had been raised until the age of 150 days with the sexes unsegregated. They were fed while in this laboratory on a modification of McCollum's standard mixture, which mixture was used as the incentive in the experimentation. They were 185 days of age when tested. Holden's animals were purchased from Douredoure, Philadelphia. They were raised with the sexes segregated. They were fed on bread and milk, which food was used as the incentive in the experimentation. They were 50-60 days of age when tested.

Several differences in procedure should be noted. Holden gave the animals preliminary training on three successive days prior to the test, rather than immediately preceding it, and to groups of ten animals at once rather than to each singly. The test period was ten minutes rather than twenty minutes in length. Holden used three degrees of shock, the lowest of which was much greater than that used in the present study, 1200 volts rather than 475 (the external resistance in the circuit being the same in both cases). Only those results of Holden's concerned with the lowest degree of shock will be considered here since the other degrees of shock did not yield reliable data from the present standpoint.

Using starvation intervals of 0, 12, 24, 36, 48, 60 and 72 hours Holden found a constant increase in the number of crossings up to the 36 hour period which represents the peak of the curve. Further increase in the interval gave constantly decreasing crossing. The absolute values reported are entirely different from those found in the present investigation. For example, the animals of the 36 hour group crossed on the average 40.3 times in the ten-minute test period. Using test periods *twice* this long we found that the most frequently any animal of any group crossed was 36 times and that the highest group average was 19.1. The decided difference in absolute values reported must be due to the radical differences in apparatus and procedure. That Holden found the peak of activity at 36 hours and almost complete reduction of activity for those starvation periods which in the present investigation resulted in the greatest activity is probably accounted for by the fact that the animals she used were immature while those used in our work were full grown.

4. A comparison of the hunger drive (on the basis of the data here reported) and the sex drive (on the basis of data previously reported)

No effort has been spared in these two studies to obtain data which could justly be compared. Animals of the same stock and age were used. The general conditions, feeding, social and so on were the same. The same apparatus was used and no differences in procedure or technique were introduced. All the data for both studies were taken by the same investigator and he made every effort to handle the animals and the apparatus in a uniform way throughout. The method of handling animals in experimental work has been given little consideration in the past. Perhaps in studies of maze learning it matters little whether the rat is lifted by the tail or otherwise. It is the writer's opinion that in the study of animal drives too much care cannot be taken on this point. Slight emotional disturbances may not seriously prolong the rat's learning of a maze but it will serve decidedly to alter the behavior of the animal toward an incentive object. A criticism to be raised against the obstruction method in particular is that results obtained by it are too readily influenced by the technique of the individual investigator in handling the animals. If comparable data are to be obtained by two investigators it seems almost essential that they work together for a time in order that they may adopt the same manner and rhythm of handling the animals. The details of this technique are difficult to transmit exactly in writing. An attempt has been made to enumerate the important points in the report of the study of sex behavior.

The value of the obstruction, i.e., the physical characteristics of the current in the grid, was the same in the study of the two drives. In brief, the only variable consisted in the type of incentive used, food in the one case and sex object in the other.

In the study of the hunger drive the physiological condition most closely related to this phenomenon (the effect of starvation) was varied in order that the optimal condition might be determined. The same holds for the study of the sex drive although the physiological conditions most intimately involved in this case were not the same for the male and the female. With the male it depended upon the length of the sex deprivation interval, with the female it depended upon the oestrus condition.

In the study of the hunger drive the sex factor was kept constant and at the minimum in the two sexes, in the case of the male by

TABLE 5

Showing temporal distribution of approaches, contacts and crossings during the twenty-minute test period, in one-minute intervals

GROUP	ACTIVITY	1ST MINUTE	2ND MINUTE	3RD MINUTE	4TH MINUTE	5TH MINUTE	6TH MINUTE	7TH MINUTE	8TH MINUTE	9TH MINUTE	10TH MINUTE	11TH MINUTE	12TH MINUTE	13TH MINUTE	14TH MINUTE	15TH MINUTE	16TH MINUTE	17TH MINUTE	18TH MINUTE	19TH MINUTE	20TH MINUTE
days																					
0	Approaches	2	4	5	1	3	0	3	2	1	4	0	1	2	4	0	0	1	0	2	0
	Contacts	7	11	3	5	2	3	6	3	2	4	1	0	0	3	0	0	2	3	1	1
	Crossings	10	6	8	3	4	5	0	2	1	1	1	0	2	0	2	1	0	0	2	0
2	Approaches	7	2	4	1	3	2	0	3	1	2	4	2	8	2	2	4	4	4	7	3
	Contacts	4	3	2	3	3	6	3	4	7	3	7	9	15	7	4	5	5	8	12	7
	Crossings	29	34	19	14	34	30	22	24	18	8	11	23	26	10	16	13	7	9	3	1
3	Approaches	6	11	13	4	8	8	4	7	1	5	4	3	6	4	0	2	6	3	2	5
	Contacts	12	8	11	8	15	11	8	13	4	1	7	3	4	5	9	6	2	4	3	3
	Crossings	24	10	14	15	12	17	14	31	13	16	19	24	24	20	13	17	21	26	26	8
4	Approaches	1	8	6	4	4	6	8	5	7	8	3	0	4	1	2	5	4	1	7	4
	Contacts	15	11	7	4	4	5	4	8	1	4	2	0	4	1	1	4	5	0	3	2
	Crossings	20	18	24	21	11	15	21	20	18	15	13	17	18	16	23	22	26	19	14	10
6	Approaches	4	7	9	3	5	4	2	7	6	1	1	3	7	4	3	5	3	9	5	9
	Contacts	6	5	2	2	3	2	3	0	4	5	3	4	1	1	4	4	10	7	4	8
	Crossings	14	11	9	15	15	13	8	16	18	10	17	22	13	16	9	7	11	14	21	23
8	Approaches	2	8	6	11	3	6	8	2	7	5	3	4	1	3	5	8	6	14	12	0
	Contacts	8	3	4	7	5	8	4	6	7	4	5	1	8	7	6	8	7	12	11	8
	Crossings	6	5	9	4	7	6	9	14	7	4	7	11	8	4	12	13	10	9	5	8

using a long deprivation interval and in the case of the female by using only animals in dioestrum. Similarly in the study of the sex drive the food-seeking factor was kept constant and at the minimum by testing the animals immediately after removal from a cage in which there had always been sufficient food.

a. Comparison of the two drives in the male animal. Fairly to compare the two drives in the male we must compare those two groups which represent each drive at the optimum, i.e., those groups which displayed the maximum activity directed toward the incentive in each case. Naturally, it would not do to compare a food-seeking group of a given starvation period with a sex group of a sex deprivation period of the same length, since our results show that the two drives do not recover their strength after a period of satisfaction at the same rate. The sex drive recovers much the more

TABLE 6

Showing temporal distribution in five-minute intervals of approaches, contacts and crossings during the twenty-minute test period

STARVATION PERIOD	ACTIVITY	0-5TH MINUTE		6-10TH MINUTE		11-15TH MINUTE		16-20TH MINUTE		TOTAL
		Number	Percent	Number	Percent	Number	Percent	Number	Percent	
days										
0	Approaches	15	42.8	10	28.6	7	20.0	3	8.6	35
	Contacts	28	49.0	18	31.6	4	7.0	7	12.6	57
	Crossings	31	64.5	9	18.8	5	10.4	3	6.2	48
2	Approaches	17	26.2	8	12.6	18	27.6	22	33.8	65
	Contacts	16	13.5	23	19.5	42	35.5	37	31.4	118
	Crossings	130	37.0	102	29.1	86	24.5	33	9.4	351
3	Approaches	42	41.3	25	24.5	17	16.7	18	17.6	102
	Contacts	54	39.4	37	27.0	28	20.4	18	13.1	137
	Crossings	75	20.6	91	25.0	100	27.5	98	26.9	364
4	Approaches	23	26.2	34	38.5	10	11.6	21	23.8	88
	Contacts	41	48.2	22	25.9	8	9.4	14	16.5	85
	Crossings	94	26.0	89	24.7	87	24.2	91	25.2	361
6	Approaches	28	28.8	20	20.6	18	18.6	31	32.0	97
	Contacts	18	23.1	14	17.9	13	16.7	33	42.3	78
	Crossings	64	22.7	65	23.0	77	27.4	76	26.9	282
8	Approaches	30	26.3	28	24.6	16	14.1	40	35.1	114
	Contacts	27	20.6	29	22.2	29	22.2	46	35.1	131
	Crossings	31	19.6	40	25.3	42	26.6	45	28.5	158

rapidly, reaching its maximum again in about 24 hours, whereas the hunger drive does not reach its peak until about the fourth day. It is the peaks of the two curves which should be compared or at any rate those two groups which, of those tested by us, represent the drives at their maximum. Our interest is in the limits of behavior as influenced by each of these factors and how these limits compare. The four-day starvation group in the case of the hunger drive and the one day sex deprivation group in the case of the sex drive represent these limits in the male. These data are brought together in tables 7, 8 and 9. It will be seen that the average number of crossings to the sex object was 13.45, to the food, 19.1.

TABLE 7

Distribution table of those groups which represent the sex drive and the hunger drive in the male and in the female at the optimum. Sex drive in the male is represented by the one-way sex deprivation group; sex drive in the female, by the group of animals tested when in oestrum (the stage of cornified cells); hunger drive in the male, by the four-day starvation group; food-seeking drive in the female, by the two-day starvation group. Distributions for the two groups representing the sex drive have been combined as have also the distributions for the two groups representing the hunger drive. These combined distributions have been placed side by side in the center for ready comparison.

INTERVAL	SEX DRIVE IN MALE	SEX DRIVE IN FEMALE	SEX DRIVE, TWO SEXES COMBINED	HUNGER DRIVE, TWO SEXES COMBINED	HUNGER DRIVE IN MALE	HUNGER DRIVE IN FEMALE
0						
1						
2		1	1			
3				1		1
4	1	1	2			
5		1	1	1		1
6	1		1			
7						
8						
9	2	1	3			
10				2	2	
11	2	1	3			
12	1		1			
13	2		2			
14		3	3			
15	4	2	6	2	1	1
16	3	3	6			
17	1	2	3	1		1
18	2	1	3	3	2	1
19		2	2			
20	1		1			
21		2	2	2	1	1
22				1	1	
23				1	1	
24		1	1			
25				2	1	1
26						
27						
28				1		1
29				3	1	2

This difference is quite a reliable one, being 2.8 times the standard deviation of the difference. We feel justified, then, in saying that, using the obstruction method, the hunger drive is stronger than the sex drive in the male animal.

b. Comparison of the two drives in the female animal. Here again we must compare those groups representing the two drives at their high points. These are the one-day starvation group in the case of the hunger drive and the cornified group in the case of the sex drive (see tables 7, 8 and 9). As in the case of the male there is a difference in favor of the hunger drive. This difference is not

TABLE 8

Showing the number of animals in each group, the average number of crossings and the standard deviations for those groups taken to represent the sex and the hunger drives in the two sexes

GROUPS	NUMBER OF ANIMALS IN THE GROUP	AVERAGE NUMBER OF CROSSINGS	STANDARD DEVIATION
Sex drive in male	20	13.45	4.3
Sex drive in female	21	14.14	5.5
Sex drive in two sexes combined	41	13.8	4.9
Hunger drive in male	10	19.1	5.87
Hunger drive in female	10	19.0	8.91
Hunger drive in two sexes combined	20	19.05	7.6

TABLE 9

Showing reliability of differences in the average number of crossings for the sex and the hunger drives in the two sexes taken separately and combined

GROUPS	STANDARD DEVIATION OF THE DIFFERENCE	THE DIFFERENCE	DIFF./S.D. OF DIFFERENCE	CHANCES IN 100 OF A TRUE DIFFERENCE
Sex drive in male and hunger drive in male	2.0	5.65	2.8	99.74
Sex drive in female and hunger drive in female	2.9	4.91	1.7	96
Sex drive in the two sexes and hunger drive in the two sexes	1.86	5.25	2.8	99.74

entirely reliable being 1.7 times the standard deviation of the difference (chances of a true difference: 96 in 100).

c. Comparison of the two drives, data for the two sexes being combined. As would be expected, when the data for the male and the female groups are combined the difference between the strength of the two drives as here measured is a reliable one. The hunger drive is clearly stronger than the sex drive (tables 7, 8 and 9).

This conclusion is in accord with that of Tsai (11) who used the choice method and with that of Simmons (9) who used the learning method. Since all data, in so far as they are interpreted in terms of general tendencies such as drives, are functions of the method used there is good reason for attacking problems by means of any and all methods at our disposal. Nevertheless we feel that the results here reported are important not merely because they have been obtained by a different method but also because they have been obtained by a more reliable method than any others which have been employed.

Using the learning method as Simmons did the problem to be solved is a distraction which serves to reduce the prominence of the incentive object in the entire stimulus situation. This reduces the possibility of various incentives producing clear cut differences in the animal's behavior. A minimum amount of learning is involved in the obstruction method as here used. A very simple association is formed in the preliminary trials, that between the incentive object and the incentive compartment. In the course of developing the technique used in these two investigations data were obtained which indicated that the brief preliminary training period used is quite as effective as is a much longer one. As tested by the obstruction method the animal is presented with a stimulus situation in which the incentive is dominant. It is not distracted by the complexities of threading a maze or solving other problems, before being stimulated by the object which will call out the drive under investigation.

The choice method does not suffer so much from involving considerable learning. Two simple associations must be formed, however, instead of the one demanded by the obstruction method, and the actual movement of the animal is less simple. Furthermore it seems probable that chance or uncontrolled factors must always play an important part in the choice method. The animal frequently responds simply to whichever incentive happens to stimulate it first in spite of the experimenter's efforts to have the two incentives stimulate it simultaneously and equally. The choice method suffers from a further and more serious difficulty which was apparently not recognized by either Moss or Tsai. To obtain a choice which is to represent a fair comparison between the two drives each acting at its optimum it would be necessary to control, simultaneously, the related physiological conditions. In the case of the female this would involve prediction of oestrous condition, no simple task, and one in which manipulation of the animal prior to testing could scarcely be avoided. In the case of the male it is true that, perhaps on the basis of the results reported herein, one could start the period of starvation at one time and the period of sex deprivation, in the same animal, at another in such a way that theoretically the two drives would reach their optimal points simultaneously. It is improbable, however, that the physiological conditions effecting the two drives are entirely independent of each other. Starvation would effect sex behavior; mating would effect food-seeking behavior. Each drive would be tested under the artificial conditions imposed by the control of the other.

Summary

1. The tendency of the male white rat to approach food as measured in terms of the number of times it will cross an electrical obstruction within a given period of time is at its low point when the animal is tested immediately after being removed from a cage containing food. This tendency increases with an increase in the length of the starvation period, up to a period of four days, and from this point on, decreases.

2. In the case of the female white rat this tendency reaches its high point much earlier. The tendency after one day of starvation is apparently almost as strong as it is in the male after a four-day period of starvation. After one day of starvation the tendency in the female declines gradually and then more rapidly, never after the second day being as strong as in the male.

3. Comparison of the groups which, of those tested, represent the hunger and the sex drives at their maximum, indicates that the tendency for a white rat to approach a food object is stronger than its tendency to approach a sex object. This is true of rats of both sexes although the difference is not so great in the case of the female.

2. THE EFFECT OF DELAYED INCENTIVE ON THE HUNGER DRIVE IN THE WHITE RAT (EXPERIMENT 1: THE OBSTRUCTION METHOD)[1]

E. L. Hamilton

I. Introduction

In experimental studies of animal motivation, the organism is usually required to perform some task in order to obtain the incentive. Ordinarily, the latter is accessible to the animal immediately upon completion of the performance. This is perhaps the best procedure for most experimental work on animals, since a more favorable emotional adjustment to the artificial laboratory situation tends to be made under this condition. An interesting problem from the standpoint of general motivation, however, is that of separating the incentive stimulus from the required performance by various intervals of time, thereby introducing the factor of delayed incentive. The question arises as to the effect of various intervals of delay between the performance of the task and the incentive on the strength of the drive under investigation. Is the drive weakened, strengthened, or in no way affected when the reward does not immediately follow the performance?

In the previously reported studies of drives (hunger, thirst, normal sex, segregated sex) under this project, the matter of delayed incentive has not been considered. In all of these investigations the animal was allowed contact with the incentive object as soon as it crossed the grill and entered the incentive compartment. The present study is an attempt to measure the effect of delayed incentive on the strength of one of these drives — the hunger drive. In Appendix 3, experiment 2, making use of the maze and the

[1] Reprinted with modifications from *Genetic Psychology Monographs,* 1929, 5, No. 2, 137-66. For balance of monograph, covering experiment 2 on maze learning, see Appendix 3.

learning method of measuring the influence of delay on behavior, will be found. Although the obstruction method has, we believe, many advantages over the learning method as a means of determining the effect of delay upon strength of drive, it will be seen that the results obtained by the two methods show marked agreement in general.

II. Experiment

The obstruction method

Method and procedure. The apparatus employed in experiment 1 was the Columbia Obstruction Apparatus which has been described in detail in Part I, 1, of this volume. It consists of four compartments: A, the entrance compartment; B, the obstruction compartment; C, the reaction compartment; and D, the incentive compartment. In the present study, the animal was delayed in compartment C for various intervals of time as described later. The electrically operated grid device, developed by Dr. T. N. Jenkins (see Part IV, 2) was used in giving the animal the initial shock on the fifth trial. The standard shock was, of course, employed in the present experiment. For Fig. 1 of this monograph, showing apparatus, see Part I, 1, of this volume.

A total of 100 albino rats of the "Experimental Colony strain" of the Wistar Institute of Anatomy were tested. Until shipped to us at about 150 days of age, they were kept in litter groups unsegregated as to sex, an approximately equal number of males and females making up each group. Upon arrival at our laboratory, they were segregated to the extent of placing males and females in separate cages in which condition they were kept until they reached the age of about 185 days. All animals fell within the age range of 175 to 196 days at the time of experimentation. The age range at the time of segregation was 143 to 157 days. The segregation period extended from 32 to 39 days. This period provided ample time for the animals to become accustomed to the laboratory conditions and also insured the availability of non-pregnant females for use in the experiment, since litters are born about three to four weeks after the animals have become pregnant.

The animals were fed a well-regulated diet before they were

received by us. In our laboratory they were fed McCollum's Standard Diet (see 11, p. 27) with the addition of a weekly ration of greens. More than a sufficient amount of food was kept in the cages to insure uniform conditions with respect to hunger. We could be practically certain that, when the starvation period began, the animals were all at about the same stage of hunger. A small dish of McCollum's diet was also used as the incentive in the experiment. Fresh water was supplied daily in inverted bottles with nozzles to prevent fouling. The animals were thus always supplied with water, even during the 48-hour starvation period preceding the test. At the beginning of a starvation period, fresh sawdust was provided to insure complete absence of food. The general laboratory conditions were precisely the same as those obtaining in all other studies coming under the project.

Four periods of delay, 15 seconds, 30 seconds, 1 minute, and 3 minutes, were investigated, groups of 20 animals being used in each cage. The 48-hour starvation group representing the maximum strength of the hunger drive (14) was employed as a control. The exactitude of the procedure followed, in connection with the Obstruction Apparatus in this series of researches clearly warrants the use of the scores from one experiment to another as here employed (12). The grouping of the animals is shown in Table 4. It will be seen from the table that, although 11 males were used in the delay groups, only 9 females were employed in these groups. The reason for this is a practical one. The females could be used only when in dioestrum, a stage in the oestrom cycle, determination of which will be discussed below in connection with procedure. This stage could not be detected before testing since the smears were not made until after the animal was tested. A large number of females which had been run could not be used. Only 36 females out of 101 tested proved to be clearly in dioestrum. Naturally, the records of those in some stage of oestrum had to be discarded since the scores on such animals involved both hunger and sex drive. This difficulty does not arise with the males and hence male groups could be easily made at any desired size.

The starvation period was constant for all groups, consisting of a 48-hour interval during which the animals received no food, directly preceding the test period. This interval was employed, since it represented the maximum strength of the hunger drive, as above stated, and individual and group differences should best be brought out by the use of maximal drive conditions.

TABLE 4

Grouping of Animals

LENGTH OF DELAY PERIOD	NUMBER OF ANAMALS		
	Male	Female	Total
0 (Control)	10	10	20
15 Seconds	11	9	20
30 Seconds	11	9	20
1 Minute	11	9	20
3 Minutes	11	9	20

Procedure. The sex drive was controlled in the males by use of the segregation period of 35 days, which insured uniform sex deprivation. Since Warner (13) found that the sex drive in the male is at its minimum with the longest interval of sex deprivation he studied (28 days), we assumed that with a still longer interval of 35 days it would probably be still weaker or at least as weak as at 28 days. This would insure little interference or complication by the sex drive. In the case of the females, the matter of dioestrum was determined by the use of the smear technique developed by Long and Evans (5) and employed in the drive studies in the Columbia laboratory. Only those animals in the inactive period (dioestrum) were used. As mentioned above, this necessitated discarding the data on a large number of females. Determination of the oestrous condition was made from both the quick smear and the permanent slide by the experimenter. Dr. Warner, who has had considerable experience with the technique, checked the determinations. In all cases the determination was made on the basis of the histological picture and without consideration of the behavior data. To eliminate olfactory stimulation connected with the opposite sex, animals of only one sex were run on any one night and the box was washed with soap and water and allowed to air for each night's experimentation.

Time of day was controlled by testing all animals between the hours of 9 P.M. and 4 A.M. Since the laboratory was practically deserted at this time, noise distractions were also reduced to a minimum. The temperature of the laboratory was kept fairly constant by thermostatic regulation.

The method of handling the animals was made uniform by previous training in which the method of picking up the animal and transferring it at a constant rate to the entrance compartment was reduced to a habit. The time for moving the animal from the incentive compartment back to the entrance compartment was about 30 seconds. Naturally, the fact of emotional arousal in motivation situations makes it necessary to take every precaution to avoid distractions of every sort.

The preliminary training given the animals did not differ from the usual procedure with the Obstruction Apparatus. The animal was placed in the entrance compartment and after 10 seconds was allowed to cross the grid, which was not electrified, to the food in the incentive compartment. After the animal had obtained a nibble of food, it was removed and again placed in the entrance compartment. After four crossings to the food in this manner, the animal was placed in the entrance compartment, but this time, upon clearing the door between the entrance and obstruction compartments, the animal received a shock from the grid. The shock was given by means of the indirect method of electrifying mentioned above in connection with description of apparatus, the closing of the door d_1) automatically causing current to flow in the grid. After the animal had obtained a nibble of food on the fifth trial, it was removed from the incentive compartment and the test period began. An exception to this procedure was made in the case of such animals as did not take a nibble of food upon reaching the incentive compartment but engaged in exploratory activity by climbing up the sides of the box, etc. If the animal had not taken a nibble of food at the end of 60 seconds, it was removed, nevertheless, and replaced in the entrance compartment. The exception was also observed in the study of the normal hunger drive (14) from which our control group was taken.

The procedure for the test period for the control group consisted in placing the animal in the entrance compartment and after 10 seconds releasing it into the obstruction compartment as in the case of the preliminary training. However, the animal now received a shock as soon as its feet made contact with the grid, since the direct method of electrification described in the preceding section had been set in operation for this and all later tests. The following types of behavior, which are characteristic for the Obstruction Method, were recorded on suitable blanks minute by minute: approaches, contacts, crossings.

An "approach" refers to orientation of the animal before the grid with its head projecting into the obstruction compartment followed by withdrawal without touching the grid. A "contact" refers to the touching of the grid by the animal followed by immediate withdrawal again. A "crossing" refers to the crossing of the grid by the animal and its entrance into the delay compartment. Quantitative data on these three types of behavior during a period of 20 minutes were recorded. Whenever a crossing occurred, the animal was removed from the incentive compartment after the usual nibble of food and returned to the entrance compartment. It was not detained for 10 seconds before being released into the obstruction compartment as before, but was allowed immediate access to the grid.

The procedure for the test period in the case of the delay groups was similar to that for the control group with the exception that the animal was not allowed immediate access to the food. Upon crossing the grid, the animal was held in the delay compartment for an interval of time, at the end of which it was permitted access to the food. In order to retain the animal during the interval of delay, d_3 was disconnected from the automatic delease (*E*) and operated manually. The operation of the door by the experimenter, after the period of delay, soon became mechanical, so that as soon as the stop-watch indicated the end of the delay period, the door was opened immediately. The animal was allowed a nibble of food and was returned to the entrance compartment after each crossing and delay period.

Quantitative behavior in the form of approaches, contacts, and crossings for a period of 20 minutes was recorded, as in the case of the control group. This time was exclusive of the delay periods. A cumulative stop-watch was used to measure the time of the 20-minute period. A second stop-watch was used to time the delay

TABLE 5

Frequency distributions covering approaches, contacts, and crossings for the various periods of delay

GROUP	0						15 SECONDS						30 SECONDS						1 MINUTE						3 MINUTES					
Number of responses	Approaches		Contacts		Crossings		Approaches		Contacts		Crossings		Approaches		Contacts		Crossings		Approaches		Contacts		Crossings		Approaches		Contacts		Crossings	
	Female	Male	Female	Male	Female	Male	Female	Male	Female	Male	Female	Male	Female	Male	Female	Male	Female	Male	Female	Male	Female	Male	Female	Male	Female	Male	Female	Male	Female	Male
0	3	2	1	1					1												1					1		4		
1	1			1						2						1	3				1	1					2	3	3	5
2	0	3		1					1	1					1		2	1	1		3	1	2	1				2	1	
3	2	1			1	1					1	2				4			1		2						3	1	1	1
4	1	1	2	1		0			2								1						1						1	
5	1	1	2	3	1				1	3	1	2			2	2		2			1	3		1			1	1		
6	2	1	1	0	0	0				2					2	1	1			1		1	1	1						
7	0	0		1	0	0		1			2				3	1					1	1	1	1	1		1			2
8			1	1		1	1	1		1				1	1	1		1	1			1	1							2
9		1				0		1	1					1						1			1			2				
10			1				1		1		1			1			1	1						1	2	2	1		2	
11						1		1		2	1	1		1			1	1					1	1		1				
12			1	1					1		1		1			1						1		1		1	1		1	1
13								1				2						1		2				1						
14								1				3								1		1								
15			1		1	1								2				1		2		1								
16							2				1	1								1				1						
17					1		1	2										1	2				1	1						
18					1	2					1			1						1					2					
19						1	1																		1					
20								1	1									1	1	2				1						
21					1	1							1	1				1												
22													1												1					
23						1	1																							
24												1	2																	
25					1	1																			1	2				
26																														
27																														
28					1																									
29					2								1	1																
30								1					2	1																
31																			1											
32																														
33																														
34								1											1											
35																														
36							1																							
37																										1				
38							1							1																
39													1																	
40																														
41																										1				
42																														
43																														
44																														
45																			1											
46																														
47																									1					
48																														

periods. The cumulative watch was started at the beginning of the test period. When an animal entered the delay compartment, the watch was stopped and the second watch started. At the end of the delay period the second watch was stopped and the cumulative watch started again. When the latter registered 20 minutes, the test period was considered completed. Thus, regardless of the number of delays, the time during which the test animal was exposed to the incentive and might cross to it was the same for all groups, including the control group. In addition to the quantitative data, notes were made on the behavior of the animals during the delay period. Each animal's weight was taken and recorded before the period of starvation and again after testing.

Results

The results are presented in Tables 5, 6, 7, 8, and 9 and in Figures 2 and 3. Table 5 gives the frequency distributions covering approaches, contacts, and crossings for the various periods of delay. It reads as follows: In the control or 0-delay group, there were 3 females and 2 males who made no approaches, 1 female and 1 male who made no contacts, and no females or males who made no crossings, etc.

Table 6 represents the measures of central tendency and variability for each delay period, including the median, range, average, standard deviation, and co-efficient of variability of approaches, contacts, and crossings for each sex and for the combined (male and female) groups.

Table 7 gives the measure of reliability of the difference between the average crossings of various groups.

Table 8 shows the distribution of approaches, contacts, and crossings during the 20-minute test period, for the various groups, minute by minute. It reads as follows: The control group during the first minute made 7 approaches, 4 contacts, and 29 crossings, etc.

Table 9 gives the distribution of approaches, contacts, and crossings during the 20-minute test period in 5-minute intervals, in terms of absolute scores and percentages.

The graph presented in Figure 2 shows the average number of

TABLE 6

Results for each delay period

Period of delay	Sex	APPROACHES					CONTACTS					CROSSINGS				
		Median	Range	Average	S. D.	Coefficient of variability	Median	Range	Average	S. D.	Coefficient of variability	Median	Range	Average	S. D.	Coefficient of variability
0	Female	3	0–6	3.4	1.56	46	5	0–15	6.9	1.87	27	18	3–29	19.0	8.91	47
	Male	2	0–9	3.3	1.47	44	5	0–12	4.9	1.59	32	18	3–25	16.1	6.56	41
	Total	3	0–9	3.35	2.81	84	5	0–15	5.9	3.92	66	18	3–29	17.6	7.78	44
15 seconds	Female	17	10–38	20.3	9.85	49	5	0–20	7.3	5.79	79	10	3–18	9.9	4.68	47
	Male	14	7–34	16.4	8.36	51	5	1–11	5.5	3.31	60	13	3–16	10.1	4.78	47
	Total	16.5	7–38	18.2	9.28	51	5	0–20	6.4	4.68	73	11	3–18	10.0	4.67	47
30 seconds	Female	24	12–39	25.7	7.10	28	6	2–8	5.9	1.68	28	2	1–11	4.2	3.71	88
	Male	15	8–38	18.5	9.45	51	5	1–12	5.1	2.94	58	11	2–21	11.5	6.01	52
	Total	21.5	8–39	21.8	9.19	42	5.5	1–12	5.5	2.48	45	7	1–21	8.3	6.27	76
1 minute	Female	17	2–45	19.7	13.81	23	2	0–7	2.7	1.99	74	7	2–17	7.3	4.47	61
	Male	15	6–20	14.5	4.08	28	6	1–15	7.3	4.39	60	11	2–20	10.8	5.27	49
	Total	15.5	2–45	16.8	10.08	60	5	0–15	5.3	4.17	79	8.5	2–20	9.3	5.22	56
3 minutes	Female	18	7–48	19.7	11.50	58	3	1–12	5.0	3.68	74	3	1–12	4.9	4.23	86
	Male	11	0–41	17.2	12.36	72	1	0–5	1.4	1.51	108	3	1–12	4.5	3.77	83
	Total	15	0–48	18.3	12.04	66	2	0–12	3.0	3.18	106	3	1–12	4.7	3.99	85

crossings, contacts, and approaches at each period of delay for the combined (male and female) groups.

Figure 3 presents graphically the distribution of crossings during the 20-minute test period in 5-minute intervals.

If we examine the scores on average crossings (Table 6) for the combined groups and the graph (Figure 2), we note that there is a decided drop with a 15-second delay. An animal subjected to such

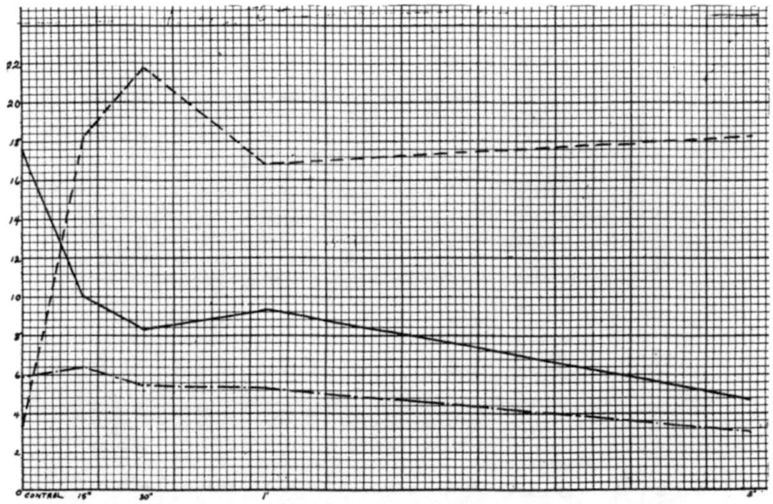

FIG. 2. Average number of crossings (solid line), contacts (dot-dash line), and approaches (dash line) at each period of delay for the combined (male and female) groups.

a small amount of delay crosses on an average of 10.0 times, while an animal which is allowed to proceed immediately to the incentive crosses about 17 times, indicating a decrease in the drive with delay.

The scores on average approaches show the following tendency: An animal with 15 seconds' delay approaches about 18 times, while one which is not delayed approaches about 3 times. This tendency toward an inverse relation of approaches to crossings has been found in the other studies on drives with the Obstruction Apparatus. The reason the number of approaches is larger for the delayed group which crosses fewer times is that the control group was probably affected by the incentive in such a manner that activity in the direction of the incentive compartment nearly always resulted in the full response of crossing, while the conditions of delay resulted in the activity's taking the less complete from of approaching. Stated in terms of drive, the hunger drive was decreased by the conditions of delay, or the incentive value was not great enough to elicit the complete response of crossing the grid so often.

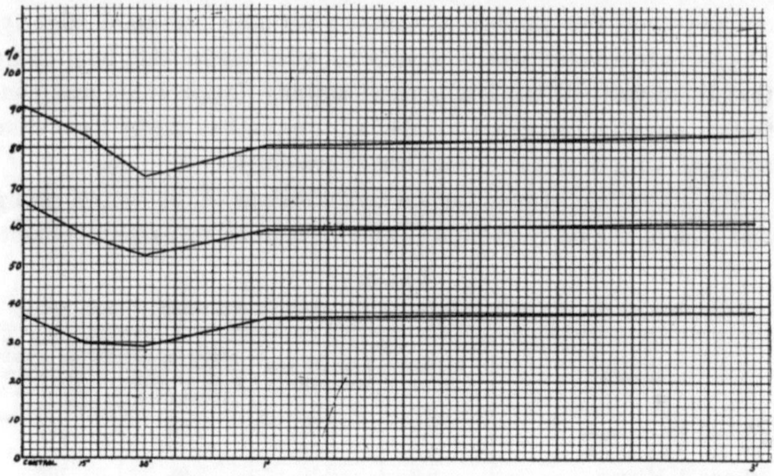

FIG. 3. Temporal distribution of crossings for the combined (male and female) groups. The lower line represents the percentage of crossing which occurred during the first 5 minutes of the test period. The middle line represents the percentage which occurred during the first 10 minutes. The upper line represents the percentage which occurred during the first 15 minutes.

The scores on average contacts were not materially different under conditions of delay than under the normal incentive situation, indicating that th delay did not influence significantly partial crossings.

The scores on average crossings for the 30-second and 1-minute groups are only slightly different from those of the 15-second group. The differences may perhaps be regarded as fluctuations due to chance factors, so that we may conclude that there is little difference between the effects of 15-second, 30-second, and 1-minute delays. The scores on average approaches and contacts tend to support this conclusion.

However, there is a decided drop in average crossings with the 3-minute group, an animal delayed for this length of time crossing on an average of only about 4 times as compared with 10 for the smaller periods of delay and 17 for the 0-delay group. This indicates a marked decrease in drive with delays as long as 3 minutes. There is some indication that the animals in the 3-minute group were less active than the other groups, since the number of approaches did not increase appreciably with the further decrease in crossings and the number of contacts actually decreased slightly. The incentive value was evidently so weak under this long delay that there was elicited less general activity (approaches and contacts), since an increased number of approaches might have been expected with decrease in number of crossings.

The difference between the average crossings of the control group and each of the delay groups (Table 7) is in all cases very reliable.

TABLE 7

Reliability of the differences between the average crossings of each group

GROUPS	DIFFERENCE	S. D. OF THE DIFFERENCE	DIFFERENCE / S. D. OF THE DIFFERENCE	CHANCES IN 100 OF A TRUE DIFFERENCE GREATER THAN 0
0 and 15 Seconds	7.6	2.03	3.74	100
0 and 30 Seconds	9.3	2.23	4.17	100
0 and 1 Minute	8.3	2.10	3.95	100
0 and 3 Minutes	12.9	1.95	6.62	100
15 Seconds and 30 Seconds	1.7	1.74	.98	84
15 Seconds and 1 Minute	.7	1.57	.45	67
30 Seconds and 1 Minute	1.0	1.82	.55	71
15 Seconds and 3 Minutes	5.3	1.37	3.87	100
30 Seconds and 3 Minutes	3.6	1.66	2.17	98.6
1 Minute and 3 Minutes	4.6	1.44	3.19	100

We find a tendency for the reliability to increase as the period of delay increases in length. There is also a tendency for the reliability of the difference between the groups to increase as the difference between periods of delay increases. Thus, when we compare delay groups in which the difference in length of period of delay is relatively small, as in comparisons between the 15-second, 30-second, and 1-minute groups, little reliability is found. This supports our conclusion that there is little difference in effect between delays of 15 seconds, 30 seconds, and 1 minute. In delay groups in which the difference in length of the periods of delay is relatively large, as in the comparisons between the 3-minute group and the

TABLE 8

Temporal distribution of approaches, contacts, and crossings during the 20-minute test period in 1-minute intervals

DELAY PERIOD	ACTIVITY	1ST MINUTE	2ND MINUTE	3RD MINUTE	4TH MINUTE	5TH MINUTE	6TH MINUTE	7TH MINUTE	8TH MINUTE	9TH MINUTE	10TH MINUTE	11TH MINUTE	12TH MINUTE	13TH MINUTE	14TH MINUTE	15TH MINUTE	16TH MINUTE	17TH MINUTE	18TH MINUTE	19TH MINUTE	20TH MINUTE
	Approaches	7	2	4	1	3	2	0	3	1	2	4	2	8	2	2	4	4	4	7	3
0	Contacts	4	3	3	3	3	6	3	4	7	3	7	9	15	7	4	5	5	8	12	7
	Crossings	29	34	19	14	34	30	22	24	18	8	11	23	26	10	16	13	7	9	3	1
	Approaches	20	27	19	30	26	17	17	20	19	24	16	18	24	17	8	9	8	15	15	14
15 Sec.	Contacts	7	9	7	10	12	12	3	5	6	7	5	9	4	4	1	2	10	4	5	5
	Crossings	14	12	11	15	8	14	9	8	11	14	13	8	8	10	12	3	7	8	5	10
	Approaches	33	17	22	25	29	22	37	24	17	30	30	19	18	21	17	16	19	10	11	18
30 Sec.	Contacts	10	8	9	12	9	8	6	12	2	2	9	2	2	5	3	3	2	0	2	13
	Crossings	15	9	12	4	8	7	6	11	7	7	6	9	5	8	6	9	8	10	9	9
	Approaches	18	27	14	17	14	18	12	31	18	18	14	12	17	13	11	16	15	18	13	20
1 Min.	Contacts	3	8	5	7	5	1	6	7	4	2	4	4	6	4	5	7	4	5	9	9
	Crossings	18	12	13	11	13	8	8	8	8	11	5	9	10	6	9	8	7	5	7	9
	Approaches	25	23	26	8	16	13	11	18	24	27	22	17	17	21	16	16	13	17	13	23
3 Min.	Contacts	10	4	13	7	0	4	1	2	6	3	2	2	0	1	2	0	2	0	1	0
	Crossings	13	5	8	4	6	3	5	4	6	3	5	3	3	6	5	4	5	1	3	2

TABLE 9

Temporal distribution of approaches, contacts, and crossings during the 20-minute test period in 5-minute intervals

DELAY PERIOD	ACTIVITY	0-5TH MINUTE		6TH-10TH MINUTE		11TH-15TH MINUTE		16TH-20TH MINUTE		TOTAL
		Number	Percent	Number	Percent	Number	Percent	Number	Percent	
	Approaches	17	26.2	8	12.6	18	27.6	22	33.8	65
0	Contacts	16	13.5	23	19.5	42	35.5	37	31.4	118
	Crossings	130	37.0	102	29.1	86	24.5	33	9.4	351
	Approaches	122	33.6	97	26.7	83	22.9	61	16.8	363
15 Seconds	Contacts	45	35.4	33	26.0	23	18.1	26	20.5	127
	Crossings	60	30.0	56	28.0	51	25.5	33	16.5	200
	Approaches	126	29.0	130	29.9	105	24.1	74	17.0	435
30 Seconds	Contacts	48	44.0	30	27.5	21	19.3	10	9.2	109
	Crossings	48	29.1	38	23.0	34	20.6	45	27.3	165
	Approaches	90	26.8	97	28.9	67	20.0	82	24.4	336
1 Minute	Contacts	28	26.7	20	19.0	23	21.9	34	32.4	105
	Crossings	67	36.2	43	23.2	39	21.1	36	19.5	185
	Approaches	98	26.8	93	25.4	93	25.4	82	22.4	366
3 Minutes	Contacts	34	56.6	16	26.7	7	11.7	3	5.0	60
	Crossings	36	38.3	21	22.3	22	23.4	15	6.0	94

other delay groups, we find that the differences are highly reliable. Hence the drop in the curve with the 3-minute group seems to represent a genuine tendency.

There is little evidence of sex differences from the scores on average crossings (Table 6) with the exception of one group. In the case of the 30-second group, the average of the females is considerably lower than that for the males. Reference to Table 5 shows that there were three females in the 30-second group which crossed only once. In view of the high average score, this is an unusually large percentage of animals with only one crossing. The notes on behavior indicate nothing unusual in the behavior of these animals. Their general behavior was perfectly normal and they were quite active in exploring the entrance compartment. This apparent sex difference for the 30-second group would seem to be due to some accidental factor rather than to a genuine difference between the sexes. At any rate, the other three groups show no such differences so that no consistent sex differences can be claimed as resulting from the delayed incentive factor.

Tables 8 and 9, which show the temporal distribution of approaches, contacts, and crossings during the 20-minute test period, and Figure 3, showing crossings, do not reveal any significant differences between the control group, which may be called the "normal" incentive group, and the delayed incentive groups. Table 9 and Figure 3 indicate that the approaches, contacts, and crossings occurred rather uniformly throughout the 20-minute period for all of the groups. That is, delay apparently had no effect in bringing about a change in the relative number of crossings, contacts, and approaches during the early or later stages of the test.

Summary of Results

(1) The hunger drive, as measured by the Obstruction Method, was markedly decreased by a period of delay as short as 15 seconds between the crossing of the grid and the obtaining of the incentive (43 per cent decrease in number of crossings).

(2) Longer delay periods of 30 seconds and 1 minute did not further decrease the drive index.

(3) A period of delay as long as 3 minutes effected a further marked decrease in number of crossings (73 per cent from control group).

(4) The data on approaches and contacts corroborated the conclusions based on crossings. The effect of delay tended to increase the number of approaches as would be expected since the number of crossings was decreased. The number of contacts remained approximately the same for all the groups.

Bibliography

(1) Dodson, J. D. 1917. Relative values of reward and punishment in habit formation. *Psychobiology,* 1: 231-76.

(2) Holden, F. 1926. A study of the effect of starvation upon behavior by means of the obstruction method. *Comp. Psychol. Monographs,* 3, No. 17, 45 p.

(3) Jenkins, M. 1928. The effect of segregation on the sex behavior of the white rat as measured by the obstruction method. *Genet. Psychol. Monog.,* 3: 455-571.

(4) Jenkins, T. N., L. H. Warner, and C. J. Warden. 1926. Standard apparatus for the study of animal motivation. *J. Comp. Psychol.,* 6: 361-82.

(5) Long, J. A., and H. M. Evans. 1922. The oestrous cycle in the rat and its related phenomena. *Mem. Univ. Calif.,* 6: 148 p.

(6) Moss, F. A. 1924. A study of animal drives. *J. Exp. Psychol.,* 7:165-85.

(7) Nicholls, E. E. 1922. A study of the spontaneous activity in the guinea-pig. *J. Comp. Psychol.,* 2: 303-30.

(8) Richter, C. P. 1922. A behavioristic study of the rat. *Comp. Psychol. Monog.,* 1, No. 2, 55 p.

(9) Simmons, R. 1924. The relative effectiveness of certain incentives on animal learning. *Comp. Psychol. Monog.,* 2, No. 7, 79 p.

(10) Szymanski, J. S. 1919. Die Abhängigkeit der Lerngeschwindigkeit von der Antriebsstärke. *Pflüg. Arch. f. d. gesamte Physiol.,* 173: 141-48.

(11) Tsai, C. 1925. The relation of the sex and hunger motives in the albino rat. *J. Comp. Psychol.,* 5: 407-15.

(12) Warden, C. J. and H. W. Nissen. 1928. An experimental analysis of the obstruction method of measuring animal drives. *J. Comp. Psychol.,* 8: 325-42.

(13) Warner, L. H. 1927. A study of sex behavior in the white rat by means of the obstruction method. *Comp. Psychol. Monographs,* 4, No. 22, 68 p.

(14) WARNER, L. H. 1928. A study of the hunger drive in the white rat by means of the obstruction method. *J. Comp. Psychol.*, 8: 273-99.

(15) ——— 1928. A study of the thirst drive in the white rat by means of the obstruction method. *J. Genet. Psychol.*, 35: 178-92.

PART III

THE THIRST DRIVE

Part III

THE THIRST DRIVE

C. J. Warden

The primary purpose of the project was to secure comparable norms of the more important drives in the white rat, and in the case of thirst no more than this could be done. In Part III, 1, will be found the indices representing a series of thirst conditions within such limits as to exhibit the maximum thirst drive in the white rat. The hunger drive was controlled by allowing the animal access to its usual diet (McCollum's powdered mixture) during the periods of water-deprivation. The sex, maternal, and exploratory drives were controlled in the same manner as was indicated (see Warden, The Hunger Drive). It was planned to determine the influence of delayed incentive on the thirst drive paralleling the study on the hunger drive (Part II, 2) but this had to be given up in order to carry through other work more in line with the general purpose of the project.

1. A STUDY OF THIRST BEHAVIOR IN THE WHITE RAT BY MEANS OF THE OBSTRUCTION METHOD [1]

L. H. WARNER

The effect of water-deprivation upon animals has not been subject to such extended investigation as has inanition following food-deprivation. Kudo (2, 3) has made an anatomical study of the effect of thirst upon the development of the white rat and especially upon the relative loss of weight by various organs and systems. His interest did not extend to any degree to the behavior of the animal. Richter (4) appears to be the only investigator to have observed the influence of water-deprivation upon the behavior of the white rat. Using the rotating wheel apparatus for measuring general activity he found that three animals "starved but permitted to have water all the time showed a definite increase in activity for the first two to three days after the beginning of starvation and then a steady marked decrease to a point of almost complete inactivity on the eighth day." In contrast to this it is to be noted that four animals deprived of both food and water showed a steady marked decrease in activity immediately. This group reached the point of complete inactivity already on the fifth day." The animals were presumably adult. Since no data are given on the food- and water-deprived animals after the fifth day it is supposed that they died soon after that point. Richter reports no data on animals supplied with food but deprived of water. In a more recent article (5) he makes a brief statement regarding the detection of a thirst cycle in the animal: "A recess large enough for the animal to enter and drink, but too small for it to lie down or turn around, was built on one side of the cage and an inverted watering tube was placed at the end away from the cage. The bottom of this recess was made of a piece of aluminum pivoted at one end and supported

[1] Reprinted from *The Journal of Genetic Psychology*, 1928, 35: 178-92.

on a tambour at the other, so that whenever the animal entered to drink, a mark was recorded on a smoked drum. The records . . . show that the rat drinks about ten times a day at intervals of two and a quarter hours. The periodicity of the thirst response is quite as remarkable as that of the hunger response, in view of the current conception that the rats eat and drink at very frequent and irregular intervals."

One other passage in the same article relates to thirst behavior as casually observed in the case of a single animal: "This animal had habitually deposited its feces in the water-cup. Usually the water was changed every day, but on one occasion, by some neglect, it was not changed for several days, so that the resulting odor became very unpleasant. At this point the animal started to cover the hole over the water-cup. It first removed part of the upper layer of the cardboard bottom of the large central cage, and dragged it into the water-box. It placed the cardboard over the cup and smoothed it down on all sides until the hole was perfectly covered. Then from the bottom of the central cage it lifted stones larger than its head three inches into the drinking cage and placed them over the cardboard cover. Besides the large stones numerous pebbles and sticks were used until the water-box was completely blocked. The animal had cut off its only water supply by this performance. Since we wished to see what it would do when it became very thirsty, the material was left undisturbed and no other water was given. After three days, the animal pushed all of the sticks and stones from the drinking cage into the large central cage, tore up the cardboard seal, and drank its fill of the polluted water."

The present paper reports the third of a series of studies on drives in the normal white rat. In two of the several investigations on animal drives undertaken in this laboratory recently, data were obtained on the sex and on the hunger drive in the normal male and female white rat (7, 8). What may, for convenience, be termed the thirst drive is the subject of the present report. Since the standardized method, apparatus, and animals have been thoroughly discussed in the above-mentioned articles, the writer will here restrict himself to a very brief description of such general conditions and will detail only such matters as relate specifically to the study of the thirst drive. The term "thirst drive" is used merely to describe the tendency of animals to approach water under certain conditions.

The Method

The method used in these studies has been termed the obstruction method (1). It rests on the principle that the energy released by internal stimulation can be directed against an obstruction of known value and measured by its ability to overcome this obstruction. As the method is applied in these studies the obstruction consists of the stimulation resulting from contact with an electric grid which is so placed that the animal must of necessity cross it in order to reach the object which will "satisfy" the drive under investigation. In the case of the thirst drive this object is, naturally, water. The most important data consist of the records of the number of times in a test period of constant duration that an animal will cross the grid and reach the object in question. Perfect specificity in our measure of a given drive cannot be obtained. It is, however, possible to arrange conditions so that, at the time of testing, the drive in question dominates the behavior of the organism. This is accomplished, first by the establishment of a physiological condition conducive to the operation of that particular drive and non-conducive to the operation of other drives. For example, in the study of the thirst drive, the animals are water-deprived for various periods of time prior to testing but are not deprived of food, nor tested in those conditions (oestrum in the female, relatively short period of sex-deprivation in the male) which previous study has shown to result in a strong sex drive. Besides this control of the internal condition of the animal, the external situation is so arranged as to call out the drive under investigation only, or, at any rate, dominantly. Thus in the present study the animal reached water upon crossing the electric grid but did not find an object appealing to any other strong drive, such as food, or a receptive animal of the opposite sex. In a preliminary training series the animal had the opportunity of associating water with the apparatus, but not any other incentive object. Needless to say, a given animal is used in the study of but a single drive.

Apparatus

The only modification made in the obstruction apparatus relates to the smaller section of the reward compartment (for cuts see 1 or 7). It was in this small compartment that food had been placed in the study of the hunger drive and a sex object in the study of the sex drive. A piece of wire mesh similar to that used in the animals' living cages was constructed across the end of this compartment shortening it about two inches. Hung on the outside of this was a small inverted bottle whose curved nozzle projected in through a hole in the mesh about one inch above the floor of the compartment. This exactly duplicated the drinking situation which the rats were accustomed to in their living cages. The copper nozzle was, however, so manipulated that the water would not flow from it as copiously when the end of it was licked by an animal as was the case ordinarily. The rat could little more than moisten its tongue on the nozzle's end.

In other respects the apparatus was similar to that used in former work.

Animals

White rats of the Experimental Colony Strain of the Wistar Institute of Anatomy were used. They were shipped to us from the institute at the age of 150 days and were used by us at the age of 185 days. While in this laboratory they were fed on a modified McCollum's mixture, plus greens. These greens were withheld only during the water-deprivation period described below. This was done because of their high water content. The dry food mixture was at no time withheld, being supplied in a more than sufficient amount even during the water-deprivation periods. The water content of this mixture was negligible. The animals were thus kept constantly provided with sufficient food up to the very moment of testing for the purpose of reducing the hunger drive to the minimum. Even though so provided, the food intake fell off decidedly in the longer deprivation periods. There seems no way of attaining water deprivation without involving a certain degree of food-deprivation. Forced feeding was impossible for practical reasons and would probably result merely in upsetting the animals. That the hunger drive was as a matter of fact operating far less effectively than the thirst drive in animals under these conditions was indicated beyond question by observations upon animals other than those actually used in this work that were similarly water-deprived but with food present. Such animals removed to a fresh cage containing both a food dish and a water bottle invariably fought for the privilege of licking the water nozzle and neglected the food during the first half hour — a period as long as the entire testing required.

The males were tested after a 35-day period of sex-deprivation. The females were tested only if in dioestrum, i. e., after an animal was tested, a vaginal smear was taken and examined under the microscope and the data on the animal were not used unless this examination showed it to be in dioestrum. For the reasons for adapting these conditions see (8).

Procedure

This varied from that used in the other studies with respect to the criterion for the return of the animal to the entrance compartment after a crossing, and here it was quite analogous to that used in the hunger study. Instead of being returned after one nibble of food it was returned after it had once licked the water nozzle. With respect to preliminary training, length of test period, use of the automatic door, amount of electric shock, etc., the procedure of the former studies was followed exactly. As before, an "approach" refers to orientation of the animal before the grid with its head projecting into the obstruction compartment followed by withdrawal without touching the grid; "contact" refers to the touching of the grid by the animal followed by withdrawal again; "crossing" refers to the crossing of the grid by the animal and its entrance into the incentive compartment.

Grouping was based upon length of the water-deprivation period and is indicated in Table 1. It will be noted that each group contained 20 animals, 10 of each sex. Deprivation periods were begun at about eleven in the evening by the removal of the water bottles. All animals were tested at night, between 9 P.M. and 3 A.M.

TABLE 1

Showing groups of animals and the number of each sex in each group

LENGTH OF WATER-DEPRIVATION PERIOD	NUMBER OF ANIMALS	
	Male	Female
0	10	10
1 day	10	10
2 days	10	10
4 days	10	10
6 days	10	10

TABLE 2

'istribution table covering approaches, contacts, and crossings for the various periods of water-deprivation or male and female rats

Interval	0						1 day						2 days						4 days						6 days					
	Approaches		Contacts		Crossings		Approaches		Contacts		Crossings		Approaches		Contacts		Crossings		Approaches		Contacts		Crossings		Approaches		Contacts		Crossings	
	Female	Male	Female	Male	Female	Male	Female	Male	Female	Male	Female	Male	Female	Male	Female	Male	Female	Male	Female	Male	Female	Male	Female	Male	Female	Male	Female	Male	Female	Male
0	2	1	5	4	1	1	1		2	1			2	2	3	1			4	1	5	4			4	4	3	5	3	1
1	2		2	2	3	1			1	2			1			1			1	2	2	2		1	2	4	3	3		1
2	2	4	3	2		3		1	1			1	1	1	1	2		1		1	1	1			1	2	1	1		
3		1				1	1	1		2	1			1	1	2		1	1	1	1		1	1			1		1	1
4	1	1			1	3	1		2				2	1	1	1	1	1		2	1	1	1		2		1	1	1	
5	1	3		1	1		2	1		1	1		1	1	2		1		1	3		2		1	1					1
6				1	2			2		1			1			1	1							1			1			1
7	1						1		1				1						1											
8							1	1	1	1				2					2										1	1
9	1							1						1	1	1	1												1	
10					1		1		1	1	1	1			1		1	1					2	1						1
11								2	1					1															1	1
12					1			1				1	1				1						1							1
13																								1						
14																													1	
15						1	1				1					1	1						1	1						
16							1											2												
17												1											1							
18																														
19										1	1						1													1
20											1	1						1					1	1					1	
21												1						1												
22												1											1							
23																														
24											1																			
25																								1						
26																														
27												1																		
28																								1						
30											1																			
31																	1													
32																							1							
33																														
34											1																			
35																		1												
36																														
37											1																			
38																														
39												1																		
40																		1												
41												1																		
42																	1													

Results

Table 2 gives the distribution of the approaches, contacts, and crossings for the various groups. The medians, ranges, averages, standard deviations, and coefficients of variability are found in Table 3. Table 4 gives the reliability of the differences between the average crossings for various groups. The approaches, contacts, and crossings as distributed minute by minute through the twenty-minute test period are given in Table 5. In Table 6 these data are combined in five-minute units and are also reduced to percentages. Figure 1 shows graphically the effect of variation in the water-deprivation period upon the average number of approaches, contacts, and crossings, the data for the two sexes being combined. Figure 2 graphically demonstrates the effect of variation in the water-deprivation period upon the temporal distribution of crossings during the test period.

It will be seen at once that the average number of crossings is affected markedly by the length of the water-deprivation interval. At first it was planned to use the intervals, 0, 2, 4, 6, 8 days, to correspond with the groups studied in the case of the hunger drive. The 8-day group had to be given up because of the high mortality rate. During the taking of the data on the other four groups casual observation of the animals during the deprivation period suggested that the point of greatest activity lay between the 0-, and the 2-day groups. For this reason the 1-day group was run and proved to represent the optimal interval of those studied, for the production of thirst drive. The average number of crossings for the 0 group is 4.15, somewhat but not reliably higher for the corresponding group in the hunger drive. The average leaps to 20.4 after one day of water-deprivation. This figure represents the highest average thus far obtained in the study of normal drives although it is not reliably higher than the high point for the hunger drive. These animals crossed and recrossed the grid to the water almost as rapidly as they could be returned, especially during the early part of the test period.

In fact the frequency of their crossing approaches the maximum possible under the conditions imposed by our procedure as determined recently in this laboratory (6). Approaches and contacts are also higher for this group than for any other.

The average number of crossings for the 2-day group shows a decided though not entirely reliable drop, a decline which is continued by the 4-day group. This result stands in decided contrast to that for the hunger drive which remains virtually on a plateau for 2, 3, and 4 days of starvation. In general appearance, too, there is a difference between animals water-deprived for 4 days and animals

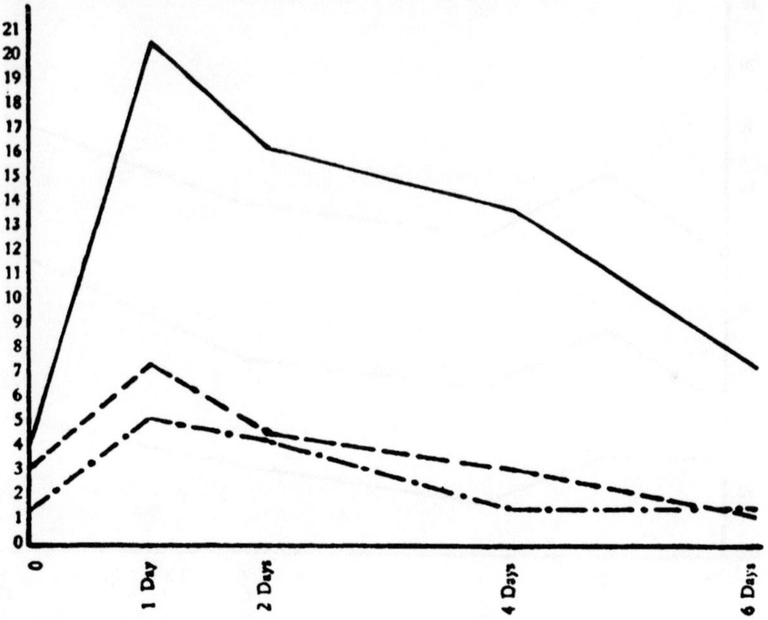

FIG. 1. Showing average number of crossings (solid line), contacts (dot-dash line), and approaches (dash line) for each interval of water-deprivation.

starved for the same period. The latter are the more active. The former customarily rest in the corners of their living cages with their heads curled up in under their chests. Whereas after but one or two days of water-deprivation they are often observed to stick their noses and tongues out of the opening where they had formerly found the water nozzle, after four days' deprivation such seeking behavior largely disappears. Small blood clots are often to be observed in their nostrils. After six days of water-deprivation the same symptoms are seen not merely in a few but in practically all animals. In three cases there was indication of semi-circular disturbance but whether this was due to the lack of water or not, cannot be said. Two deaths occurred in this group, two other animals being water-deprived and tested in order to bring the group up to the regular size. Had the dead animals been scored as not

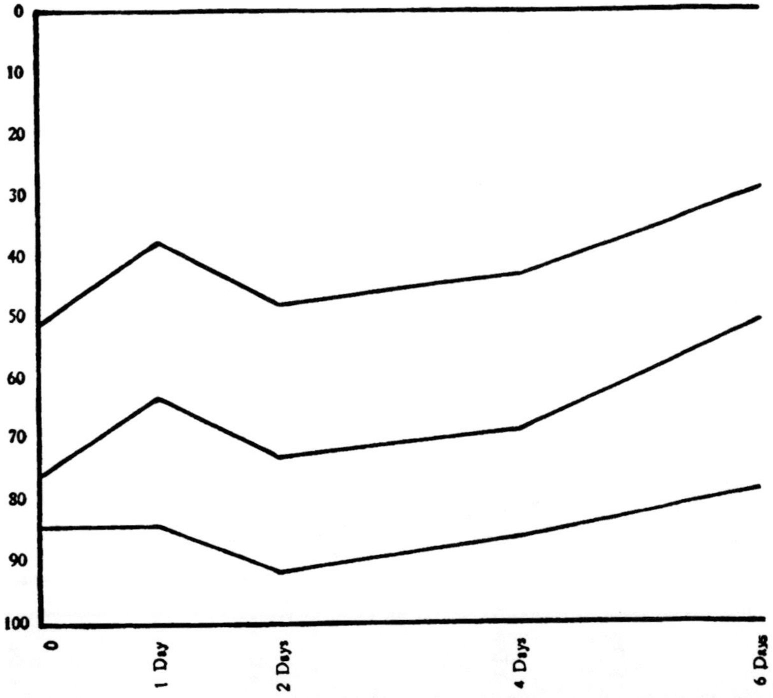

FIG. 2. Showing temporal distribution of crossings. The upper line represents the percentage of crossing which occurred during the first five minutes of the test period. The middle line represents the percentage which occurred during the first ten minutes. The lower line represents the percentage which occurred during the first fifteen minutes.

having crossed, the average would, of course, have been lowered. As it is, the average number of crossings for the 6-day group is reliably lower than that for the shorter period of water-deprivation. It is interesting to observe that there is a rather consistent decline in approaches and contacts with an increase in the deprivation period beyond one day. This is to be contrasted with the results

TABLE 3

Showing the results for each water-deprivation period

Period of water-deprivation	Sex	Approaches					Contacts					Crossings				
		Median	Range	Average	Standard deviation	Coefficient of variability	Median	Range	Average	Standard deviation	Coefficient of variability	Median	Range	Average	Standard deviation	Coefficient of variability
0	female	2	0–9	3.1	2.9	94	.5	0–2	0.8	0.9	112	4.5	0–12	4.6	3.8	83
	male	2.5	0–5	3.	1.6	53	1	0–6	1.7	2.1	123	2.5	0–15	3.7	4.	108
	total	2	0–9	3.05	2.3	75	1	0–6	1.25	1.3	104	3.8	0–15	4.15	3.9	94
1 day	female	6	0–16	7.3	4.7	64	4	0–11	4.7	3.9	83	19.5	3–37	19.7	11.1	56
	male	7	2–12	7.3	3.3	45	4	0–19	5.6	5.4	96	20.5	2–41	21.1	11.6	54
	total	6.5	0–16	7.3	4.1	56	4	0–19	5.15	4.7	91	20	2–41	20.4	11.4	56
2 days	female	4	0–12	4.	3.5	87	3	0–10	3.8	3.4	89	11	4–42	15.3	11.7	76
	male	4.5	0–11	5.	3.7	74	3	0–15	4.5	4.3	95	16	2–40	16.7	12.3	74
	total	4	0–12	4.5	3.6	80	3	0–15	4.15	3.9	94	13.5	2–42	16.	12.	75
4 days	female	2	0–8	3.2	3.3	103	.5	0–4	1.1	1.4	127	13.5	3–32	14.5	8.3	57
	male	3.5	0–5	3.	1.8	60	1	0–5	1.8	2.0	111	11.5	1–29	12.7	9.	71
	total	3	0–8	3.1	2.7	87	1	0–5	1.45	1.7	117	11	1–32	13.6	8.7	64
6 days	female	1	0–5	1.7	1.9	111	1	0–6	1.8	1.9	105	6	0–20	6.9	6.9	100
	male	1	0–2	0.8	0.7	87	.5	0–4	.9	1.2	133	7	0–19	7.5	5.5	73
	total	1	0–5	1.25	1.5	120	1	0–6	1.35	1.6	119	7.3	0–20	7.2	6.0	83

for the hunger drive. In the latter case after prolonged starvation, the number of crossings was reduced, but the number of approaches and contacts increased. This was interpreted as indicating a strong drive to get the food but probably a physical weakness rendered the animals unable to overcome the resistance offered by the shock.

The animals oriented definitely in the direction of the food but often hesitated to cross. In the case of the thirst drive a number of 6-day water-deprived animals did not even show this definite orientation after having crossed several times.

TABLE 4

Showing reliability of the difference between the average crossings of Adjacent Groups

GROUPS	STANDARD DEVIATION OF THE DIFFERENCE	DIFFERENCE BETWEEN THE AVERAGES	DIFFERENCE / S D. OF DIFFERENCE	CHANCES IN 100 OF A TRUE DIFFERENCE
0 and 1 day	2.69	16.25	6.0	100
1 and 2 days	3.8	4.4	1.16	87
2 and 4 days	3.4	2.4	2.70	76
4 and 6 days	2.36	6.4	2.7	99.7

It will be noted that the temporal distribution of crossings is uneven throughout. From 28% to 51% of the crossings always occurred during the first five minutes of the twenty-minute test period. Especially in those groups where the drive was operating at its strongest is there a predominance of crossing early in the period. This is quite different from what was observed for the groups in which the hunger drive was operating strongly. The writer is inclined to think that the internal stimulation at the basis of the thirst behavior was more readily rendered ineffective by the moistening of the tongue which accompanied each crossing than was the internal stimulation underlying the hunger behavior by the nibble of food which accompanied each crossing in that case. The sum total of the minute bits of food obtained would scarcely suffice to inhibit the hunger contractions of the stomach. If the stimulation related to the thirst behavior is at least partially derived from dryness of the mouth parts, several crossings would be sufficient to reduce the efficacy of such stimulation to a considerable degree. If this is the case, the conditions which we have set up operate in favor of the hunger drive. If we were to take as our measure of

TABLE 5

Showing temporal distribution of approaches, contacts, and crossings minute by minute

		1st minute	2nd "	3rd "	4th "	5th "	6th "	7th "	8th "	9th "	10th "	11th "	12th "	13th "	14th "	15th "	16th "	17th "	18th "	19th "	20th "
	Approaches	4	6	1	9	0	6	6	3	0	3	4	0	1	2	6	2	0	4	1	3
0	Contacts	4	0	1	1	2	0	0	3	1	2	0	0	1	3	0	0	1	0	0	6
	Crossings	9	12	8	7	6	4	5	2	6	4	0	4	2	1	0	0	4	2	6	1
	Approaches	21	26	11	9	7	8	6	3	9	5	5	2	4	0	11	6	3	4	3	3
1 day	Contacts	12	8	8	5	5	5	4	2	7	6	4	2	7	5	2	3	0	3	3	1
	Crossings	46	22	31	30	24	29	14	14	22	27	11	14	19	25	15	24	12	6	12	11
	Approaches	4	7	5	6	2	6	7	4	6	8	2	4	3	4	3	4	5	3	4	4
2 days	Contacts	3	11	6	3	5	4	3	8	2	5	2	0	4	0	8	3	0	6	4	6
	Crossings	41	30	36	17	29	22	11	19	19	10	14	9	16	11	10	4	6	8	3	5
	Approaches	4	9	2	1	2	5	6	2	0	4	6	4	2	4	4	0	2	1	3	1
4 days	Contacts	3	2	0	5	4	2	0	0	1	0	1	1	0	2	2	2	0	3	0	1
	Crossings	27	33	18	15	22	17	14	12	19	8	10	6	7	11	14	7	5	9	6	12
	Approaches	1	3	2	1	1	1	0	4	3	0	0	1	2	0	1	2	0	3	1	1
6 days	Contacts	3	4	2	1	0	0	2	1	0	2	1	1	0	1	1	2	3	0	1	1
	Crossings	14	12	5	3	5	8	7	5	3	8	10	2	6	9	15	6	5	5	5	11

TABLE 6

Showing temporal distribution in five minute intervals of approaches, contacts, and crossings during the twenty minute test period

Period of water-deprivation	Activity	0–5th minute		6–10th minute		11–15th minute		16–20th minute		Total
0	Approaches	20	33%	18	30%	13	21%	10	16%	61
	Contacts	8	32	6	24	4	16	7	28	25
	Crossings	42	51	21	25	7	8	13	16	83
1 day	Approaches	74	51	31	21	22	15	19	13	146
	Contacts	38	41	24	26	20	22	10	11	92
	Crossings	153	37	106	26	84	21	65	16	408
2 days	Approaches	24	26	31	34	16	18	20	22	91
	Contacts	28	34	22	26	14	17	19	23	83
	Crossings	153	48	81	25	60	19	26	8	320
4 days	Approaches	18	29	17	27	20	32	7	11	62
	Contacts	14	48	3	10	6	21	6	21	29
	Crossings	115	42	70	26	48	18	39	14	272
6 days	Approaches	8	32	8	32	4	16	5	20	25
	Contacts	10	37	6	22	4	15	7	26	27
	Crossings	39	28	31	22	42	28	32	22	144

the drive the amount of crossing during the first five minutes of the test period, the thirst drive would stand out as clearly stronger than the hunger drive. During the first five minutes the twenty animals of the 1-day water-deprived group made a total of 153 crossings and those of the 2-day group a like number. The largest number of crossings made in any five minutes by any hunger group was 130, the record for the first five minutes in the testing of the 2-day starvation group.

The results for the two sexes have not been separately discussed because, as the data show, the differences are negligible.

Summary

The tendency of the white rat to approach water as measured in terms of the number of times it will cross an electrical obstruction, as here employed, within a given period of time is at its slow point when the animal is tested immediately after being removed from a cage containing available water. Of the periods studied, that of one day of water-deprivation results in the greatest demonstration of this tendency. From this point on, the tendency diminishes constantly until the death of the animal. Sex differences appear to be negligible. This tendency appears to be at least as strong in the white rat as that to approach food as previously measured by the same method.

Bibliography

(1) Jenkins, T. N., L. H. Warner and C. J. Warden, 1926. Standard apparatus for the study of animal motivation. *J. Comp. Psychol.*, 6, 361-82.

(2) Kudo, T., 1921. Studies on the effects of thirst: I. Effects of thirst on the weights of the various organs and systems of adult albino rats. *Amer. J. Anat.*, 28, 339-430.

(3) Kudo, T., 1921. Studies on the effects of thirst: II. Effects of thirst upon the growth of the body and of the various organs in young albino rats. *J. Exper. Zool.*, 33, 435-461.

(4) Richter, C. P., 1922. A behavioristic study of the activity of the rat. *Comp. Psychol. Monog.*, 1, No. 2, 55 p.

(5) Richter, C. P., 1927. Animal behavior and internal drives. *Quart. Rev. Biol.*, 2, 307-43.

(6) WARDEN, C. J., and H. W. NISSEN. An experimental analysis of the obstruction method of measuring animal drives. Unpublished.

(7) WARNER, L. H., 1927. A study of sex behavior in the white rat by means of the obstruction method. *Comp. Psychol. Monog.*, 4, No.. 22, 68 p.

(8) WARNER, L. H., 1928. A study of hunger behavior in the white rat by means of the obstruction method. *J. Comp. Psychol.*, 8.

PART IV

THE SEX DRIVE

PART IV

THE SEX DRIVE

C. J. WARDEN

The sex drive offers a most fruitful field for intensive experimentation. The work of the project on this drive comprises three major studies — each a doctorate monograph. The first study, by Dr. Warner, is concerned with the determination of norms for the sex drive in male and female white rats which had been reared together from birth in litter groups and thus had had a normal development of sexual life. The second study, by Dr. Marion Jenkins, deals with the effects of segregating the sexes at various stages of the life cycle previous to the tests. The evidence is clear that early segregation (pre-puberty) tends to disturb the "normal" sex life and to introduce homosexual tendencies. The third study, by Dr. Nissen, also deals with variations from the normal sex conditions — gonadectomy, vasotomy and injections of prepared extracts of the male and female gonadal hormones.

In each of these studies the general methods of control were the same, so that direct intercomparisons can be made between the different special conditions introduced and the norms determined by the first study of the series. The hunger drive was controlled precisely as in the work on the thirst drive, by allowing the animal access to its usual diet (McCollum's powdered mixture) up to the moment of the test. The thirst drive was controlled in a similar manner by having the usual water supply in the cages continuously during the pre-test period. The maternal drive was eliminated altogether by using only non-pregnant females and those having no litters at time of test. The exploratory drive was kept as constant as possible in the manner described above (see Warden, The Hunger Drive). Variations in the male sex drive were obtained by a series of sex-deprivation intervals during which the animals were given no opportunity to copulate. This period began in the case of each group immediately following a period of copulatory satiation (2 hours) which reduced the sex drive to something like a zero level, so that the sex-deprivation intervals represent periods of recovery from a common physiological zero point. Variations in the female sex drive corresponded in general to the different stages of the oestrous cycle, these stages being determined in all cases by vaginal smears (quick and permanent) which were classified by experts in this field who knew nothing whatsoever of the behavior indices with which these stages were to be correlated.

The generous aid of Professor G. N. Papanicolaou of the Cornell Medical College in connection with the several studies on the sex drive deserves special mention. He very kindly consented to examine and classify the permanent smears in the first study of this group, and offered many valuable suggestions concerning techniques and procedures from time to time. He also gave the parts of the study dealing with the various special physiological conditions investigated a critical reading. Dr. E. F. Kinder was also kind enough to demonstrate operative techniques in connection with the work reported in Part IV, 3. Dr. Warner checked the classification of permanent smears made by Jenkins in her study of segregation, Part IV, 2.

The plan to carry out work on the effect of delayed incentive on the normal male and female sex drive, corresponding to the study of Hamilton (Part II, 2) on the hunger drive, had to be given up in order to complete other lines of work.

1. A STUDY OF SEX BEHAVIOR IN THE WHITE RAT BY MEANS OF THE OBSTRUCTION METHOD[1]

L. H. Warner

I. Introduction

The general purpose of this investigation was the study of the sex behavior of the normal male and female albino rat. More specifically, the attempt was made to determine the effect upon such behavior of varying the physiological conditions most intimately related to it.

The female mammal is subject to periodic changes in the reproductive tract which are, apparently, paralleled by definite changes in the behavior of the animal toward a sex object. In the female white rat ovulation is supposed to occur at rather regular intervals of from four to six days. Observation of behavior has shown that the animal "comes into heat," i.c., it will accept copulation, during only a few hours at a given time, and that these periods of heat, or oestrum, also recur rather regularly at intervals of from four to six days. At other times the female will avoid mating by fleeing or fighting.

Although the relationship between the changes in the reproductive tract and the animal's behavior has been noted it has not been worked out in great detail. It is not known, for example, whether the onset and termination of active sex behavior is abrupt or gradual. One reason for the lack of exact knowledge has been the lack of a method of observing any sex behavior other than that at the two extremes: Active mating and active avoidance of mating. A second reason has been the lack of an accurate method for the determination of the oestrous condition of the animal.

[1] Reprinted with modifications from *Comparative Psychology Monographs*, 1927, 4; No. 22, 68 p.

No rhythm analogous to the oestrous cycle (the periodic changes in the reproductive tract of the female) has been observed in the male rat. In the case of the male the effect upon behavior of variations in the physiological condition due to various periods of segregation involving sex deprivation has been studied. The data obtained show the speed of recovery of the copulative ability of the male after a period during which it has mated freely and may be said to be "satiated" sexually. So far as the writer knows no data bearing directly upon this aspect of male sex behavior have previously been reported.

II. Physiological Conditions Related to the Sex Drive

A. In the female white rat

The study of the sex drive in the female must take into consideration the well known fact that sex activity in the female mammal is not displayed, at least to the same extent, at all times but is confined rather rigidly to certain recurring periods. The regularity and period of this rhythm, the oestrous cycle, varies from species to species.

Heape (7) classified mammals from this standpoint as either monoestrous or polyoestrous. The former group is supposed to include those animals which experience a single oestrum separated by comparatively long periods of inactivity, that is, but a single oestrum in a single breeding season. Polyoestrous animals are those experiencing more than one oestrum in each breeding season. The latter group is subdivided according to the presence or absence of the anoestrum in the reproductive cycle. Those animals experiencing an anoestrum, or long period of sexual inactivity between breeding seasons, usually breed during but one period of the year. Those animals whose breeding seasons are not so separated breed throughout the year.

Heape's classification is not entirely satisfactory. The facts are not so clearcut and simple as his terminology seems to indicate. Not all mammals are found to fall into one group or another. The dog, for example, does not strictly belong to any of the above groups. However, the white rat, the animal used in the present investigation, may be said to belong to the group of polyoestrous mammals experiencing no anoestrum. This animal, under laboratory conditions, breeds throughout the year. Oestrum, in the normal, unmated female rat, recurs every four or six days from pubescence to menopause. The success of the present investigation depended upon a knowledge of the nature of the rhythm in the animal used and upon a reliable method for determining the condition of the animal with respect to the oestrous cycle.

Until recently, observations upon the periodic changes which take place in the generative organs of non-pregnant female mammals have been inexact and largely qualitative. In most of the early papers which have touched upon the oestrous cycle the interest in this phenomenon has been merely incidental. Among such papers are those dealing primarily with the problems of ovulation and of pregnancy. Studies of this kind were made for the most part upon the smaller mammals commonly found in laboratories. The fact that detection of the oestrous rhythm is much more difficult in the case of such animals than in the case of the larger domestic and wild animals is no doubt the main reason why exact knowledge regarding this phenomenon has been for such a long time delayed.

Several methods have played a part in the detection of an oestrous rhythm in the smaller laboratory mammals and in the determination of the length of the complete cycle and of the various phases of the cycle. The most important methods or types of observation will be taken up in turn, together with mention of representative investigators who have made use of each. The results obtained will be considered, especially in so far as they throw light upon the reliability of the various methods.

1. *Macroscopic observation of the living animal.* In many mammals the oestrous cycle is accompanied by changes in the amount and character of the vaginal secretion so pronounced as to be easily observable without instrumentation. There is often, also, such congestion that the vulva and other parts become noticeably swollen. Such signs are commonly observed among the larger animals but they are not so evident among the animals used in the laboratory. Nevertheless certain workers have found such signs convenient in the absence of more exact ones. Lataste (14), for example, says

that the approach of heat may be recognized in rodents by detection of swelling in the vaginal region:

Mais, à l'approche du rut . . . les bords de la vulve s'épaississent considérablement. Le changement d'aspect que subit alors cet organe est tellement caractéristique, qu'il me permettait de prévoir, a un jour pres le moment où une femelle, meme femelle meme jeune et précédemment unpubère, allait se trouver apte au rapprochement sexuell et à la fécondation.

Lataste was not primarily interested in the oestrous cycle. Nevertheless he apparently recognized its rhythmic nature although he gives no data on its exact duration in any single mammal. He gives it as his opinion that in rodents the cycle is usually about ten days in length.

Sobota (29) noted a thin septum closing the vaginal opening of the mouse between heat periods. This was ruptured when the vaginal region became distended upon the approach of heat. His estimate, however, that the cycle has a duration equal to that of gestation (twenty to twenty-one days) has since been found to be false as have most estimates based upon the general method of gross observation.

Rubaschkin (23) noticed closure of the vaginal opening of the guinea pig within three or four days after parturition when mating was not allowed during the oestrum which in most cases immediately follows parturition. This closure prevented mating until the next heat period, according to this writer. It might thus be taken as a sign of the dioestrous interval, but it was evidently not sufficiently clearcut to enable him to detect a typical length for the dioestrum. He disagrees, however, with Bischoff's report that the period averages about five weeks, since he noticed in certain cases an open and flushed condition of the vaginal orifice from ten to twelve days after parturition.

Another, and more recent worker who should be mentioned in this connection, since he has found useful a method of detecting the oestrous condition which involves no instrumentation, is Hartman (6). In the study of the reproductive cycle in the opossum he has found that by palpation of the mammary glands it is possible to recognize the pseudopregnant growth of these organs which occurs prior to each ovulation period. In conjunction with this method Hartman employed the method of microscopic examination of the vaginal secretion in the living animal (a method to be taken up in detail below). Hartman states that this latter method is the more accurate for the determination of the oestrum, but that "at other periods sometimes one, sometimes the other method has proved the safer guide." The palpation method, depending as it does on the unusual development of the mammary glands of the opossum, would scarcely be applicable to the smaller rodents where such development is lacking.

Probably the most skillful use of the general method of macroscopic observation has been at the hands of Stone (32, e). In the study of the initial copulatory response of the young female rat Stone was unwilling to use the vaginal smear method, although admitting it to be the most reliable, since he feared that the manipulation involved might affect the initial copulation. Out of twenty-one females he selected as being receptive, twenty mated, a tribute more to the skill of the experimenter than to the general reliability of the method.

The general conclusion regarding the method of macroscopic observation would seem to be that, while it possesses the advantage of necessitating a minimum amount of interference with the animal, it is not sufficiently reliable to be depended upon alone when dealing with the smaller laboratory mammals. It should be considered to be a supplementary method, helpful in the differentiation between the general oestrous and the general dioestrous periods only. As such it has been used in the present investigation in selecting from a large group of female rats those most likely to be in or approaching oestrum.

2. *Post-operative and post-mortem examination of the generative organs.* It is this general method which has been used in a large number of researches. In spite of this fact the results of these investigations would seem to indicate that the method is no more reliable than the preceding one. Using it, certain workers failed to detect the existence of a periodic cycle in the guinea pig, while others have claimed that there is such a cycle and yet do not agree among themselves as to its duration.

This method furnishes data at only one point in the oestrous cycle of an individual animal. It is thus necessary to determine accurately at least one other point, previous to the operation or to the sacrifice of the animal, in order to obtain data upon the duration of the cycle. It has often been observed that the guinea pig and the rat, among other mammals, experience oestrum and ovulation within a few hours after parturition. This first post-parturition oestrum has, therefore, been used as the starting point by most investigators. Different animals were operated upon or sacrificed at different intervals after this oestrum and the condition of the ovaries and related parts noted. Since this method offers no possibility of securing successive data upon a single animal and thus of tracing the cycle continuously in the individual it is useless in an attack upon many important problems. Successive cycles in the same animal might vary slightly or considerably and there still might be uniformity within the species as to the average duration of the cycle. On the other hand, although successive cycles within the individual might be remarkably uniform there might be notable individual differences regarding the duration of the cycle. These problems could scarcely be attacked by this method. Certain other problems could be, but only by using large numbers of animals since data upon but one cycle could be obtained from each. Problems of latter sort would include the effect upon the cycle of changes in the environment and in the organic condition of the animal.

One of the earliest investigators to use the method of post-mortem examination was Bischoff (2). He was primarily concerned with the problem of whether ovulation is spontaneous or dependent upon copulation. He was inclined to hold to the former view although he did not deny that copulation might have some effect upon the time of ovulation. Although in connection with this work he examined the reproductive tracts of a large number of guinea pigs killed at various intervals after the first post-parturition oestrum he was unable to detect an oestrous rhythm. He gave it as his opinion that the return of oestrum is dependent upon such factors as age, diet, time of year and individual differences.

Hensen (8), using the same animal, noted that if pregnancy did not follow the oestrum which ordinarily occurs shortly after parturition, ovulation took place in the four cases reported, 17, 18, 35 and 37 days later. We assume that an unfruitful copulation took place during the post-parturition oestrum. This would probably have influenced the time of subsequent ovulation periods. Like Bischoff, he concludes that the heat periods follow no sharp periodicity.

While the conclusions of these writers were based largely upon examination of the ovaries, the next group of workers based their observations upon post-mortem examination of the vaginal epithelium and secretion.

Morau (18) studied microscopically the vaginal condition of mice killed at various intervals of from one to twenty-one days after mating. In the cases of those females which had not become pregnant he noted histological changes which appeared to indicate a cycle of ten days duration. (Recent work has shown the cycle in the mouse to be about half that long.)

Retterer (22) made similar observations on the guinea pig but did not report estimates on the length of any cycle other than the first post-parturition cycle. He usually found cornification of the superficial layer of the vaginal epithelium on the fifteenth day following parturition.

Lataste (14) states clearly his opinion that the modifications in the secretion are independent of gestation:

> . . . tout au moins chez les Rongeurs de la famille des Muridés (rats et souris, gerbilles, merions, etc.) les modifications épithéliales du vagin sont directement liées à un tout autre phénomène que celui de la gestation. . . . C'est donc au rut, c'est-à-dire a l'ovulation . . . et nullement a la gestation, qu'est liée l'evolution de l'épithélium vaginal.

Lataste noted the cornified stage in the guinea pig from the fifteenth day to the twentieth day following parturition (mating being prevented). He carried his studies no farther than the twentieth day.

Königstein (13), although chiefly interested in the phenomena accompanying gestation, noted histological changes in the vagina of the non-pregnant rat during the first several days following parturition. These observations were largely incidental and were not followed up by more extensive work.

Smith (28) examined the ovaries of mice killed at various intervals after parturition, mating being prevented. He concluded that the average duration of the oestrous cycle was seventeen and a half days, but that the variability was great. Commenting upon this work Allen (1, a) states that "estimates of previous oestrous periods arrived at from histological comparisons of the corpora lueta are not reliable since . . . many mice when isolated from males do not ovulate spontaneously during oestrus."

This survey indicates that using the method of post-operative or post-mortem examination of the generative apparatus several investigators failed utterly to recognize an oestrous rhythm and that in no case was the duration of the rhythm determined with exactitude. These facts indicate the difficulty of using this method.

A type of observation which might have led to the discovery of the oestrous rhythm in the smaller rodents had it not been antedated by a method to be mentioned later is the observation of the spontaneous contraction of the uterine musculature. It is mentioned under this heading since it involves sacrifice of the animal. Blair (3) recorded the contraction of strips cut from the uteri of the rat and suspended in oxygenated Locke's solution at thirty-eight degrees Centigrade. He discovered the contractions to be rhythmic and noted that their period depended upon the oestrous condition of the animal. During oestrum the contractions occurred at the rate of twenty-four per hour, increasing to eighty per hour during the dioestrum with intermediate rates during the transition stages. Further work on the contraction of the uterine muscle, and also of the Fallopian tube has been reported by Keye (12) and by Seckinger (24).

3. *Observation of mating.* Since mating occurs in most animals only during oestrum the simplest method of studying the cycle would seem, at first sight, to be the analysis of data on the periods of mating. Unfortunately this method is rendered almost useless because of the frequent occurrence of pregnancies in any normal animals which are permitted to mate freely.

Bischoff, using the guinea pig, separated females from males during the first post-parturition oestrum. He then restored the males and noted the time of the first subsequent matings. These occurred at 40, 50, 51 and 51 days after parturition. Bischoff doubted the existence of periodicity. There seems no reason to suppose that this method would not have given at least an approximation of the length of the cycle. A much larger number of animals would have to be used and, since certain of the animals might first mate during the second post-parturition oestrum, some during the third, etc., it would be necessary to determine, if possible, the approximate largest common divisor for the intervals obtained. Supposedly this would represent the normal duration of an oestrous cycle.

A method resembling this to some extent was used by Danforth in the study of the mouse. This is reported by Allen (1, a) as follows:

> Danforth in unpublished observations in 1914-1915 attempted to check the length of the cycle by tabulating the number of days between two successive litters and computing the greatest common divisor of these intervals. In his records of sixty-six animals, twenty-three were excluded because of the possible complication due to the mother lactating while pregnant. The greatest common divisor of the modes of his curve was found to be four to six days.

Since the greatest objection to the method of the observation of mating has been the complication of pregnancies there have been attempts to avoid such complication. Two methods of avoiding pregnancy have been used: (1) by operation upon the female, and (2) by the use of males which have been rendered sterile. In addition to the usual post-mortem examination of the generative organs Loeb (15, a) has used observation of the mating reaction of female guinea pigs upon which double ligation of the tubes had been performed. Although this method may possess certain advantages over others it possesses the decided disadvantage of using other than strictly normal animals. Loeb concludes that the cycle in the guinea pig varies from thirteen and a half to nineteen days. This range of five and a half days is so much greater than that found by Stockard and Papanicolaou in an investigation to be reported below that it may reasonably be supposed that the extreme cases represent other than normal variations from the average.

The second method of preventing conception, and one which does not involve operation upon the female, was used by Long and Evans (16). In this case the mating of vasectomized males with normal females was observed. Long and Evans do not report in full this attempt to determine the normal oestrous cycle since it proved unsuccessful. Although the complication of pregnancy was avoided a condition which has been termed pseudopregnancy was found to follow mating with vasectomized males. The oestrous rhythm was disturbed, successive oestra being usually delayed. That this resulted from the mechanical stimulation involved in the mating act seemed indicated by the fact that a similar condition could be produced by stimulation of the cervical region by means of a spatula or glass rod. Wang's observations agree with those of Long and Evans on this point.

In conclusion it is seen that the disturbing influence of both fruitful and unfruitful copulation upon the oestrous cycle tends to invalidate the general method of observation of mating.

4. *Observation of behavior other than that of mating.* As long ago as 1883, Rein (21), working on the guinea pig, claimed that the heat periods could be recognized with great probability, though not with absolute accuracy, by observation of the animal's behavior even in the absence of the male. He describes running and jumping and the occurrence of acts resembling those of normal copulation. Nevertheless he states that he failed to notice a periodicity in the recurrence of heat.

More recently, Stone (32, a, e) has described in considerable detail the behavior of the female rat during oestrum but he does not claim that such observation was sufficient to permit positive determination of the oestrus condition of the animal.

In the limited experience of the writer such observation has been found sufficient in the case of many but not all rats to permit differentiation of the general oestrous period from the general resting period. But there seem to be decided individual differences with respect to the prominence of such overt manifestation of the physiological condition. Certain animals fail to display the behavior characteristics of heat at any time; an occasional animal has been found which displayed such characteristics to some extent at all times, not confining them to any one part of the cycle.

The type of behavior observation (other than that of mating) which reveals most clerly the oestrous rhythm is the measurement of the animal's voluntary or spontaneous activity. The activity wheel has been in use for many years, but only recently have studies been made which relate a rhythmic variation in activity to the oestrous cycle. Wang (34) found that normal mature female rats showed pronounced rhythms of from four to six days duration. This rhythm was absent in males, and in females which were pregnant, pseudopregnant (following artificial cervical stimulation), lactating, ovariectomized or sexually immature. Using the method of the examination of the vaginal secretion in the living animal (to be considered below) he found that the maximum activity occurred during the stage of cornified cells, i.e., the stage during which the animal will normally mate.

Slonaker (27, a) found an even more pronounced activity rhythm than did Wang and found that as a rat approached menopause the activity rhythm gradually disappeared. In general the results of the two investigators agree. A more detailed analysis of the activity at the various stages of the cycle has recently been made by Slonaker (27, b, c). This writer supports the activity method as the most satisfactory means of determining the oestrous rhythm in the rat.

> Our experience seems to indicate that the making of the vaginal smears, due to the necessary handling, etc., of the animal, tends to disturb the regularity of oestruation. We feel, therefore, that the normal sequence is more accurately shown by activity curves than by any other method advanced as it leaves the animal undisturbed by unavoidably varying conditions.

A very real objection to the activity method is that its use involves a separate piece of apparatus for each animal. Now that animals are used in larger numbers than in the past this would often lead to insurmountable practical difficulties. The first step toward a determination of the relative merits of the activity and the vaginal smear methods would be the careful comparison of results obtained from the two methods upon the same or similar animals. Since the evidence thus far obtained reveals very little difference between the results obtained by the two methods there seems no reason why the choice between them should not be governed by their relative convenience and economy.

5. Histological examination of the vaginal secretion in the living animal. As we have seen above the existence, and to some extent, the histological character of the vaginal flow has long been recognized. Sobotta (29) noted the secretion in the mouse during heat. Heape (7) makes this general statement regarding the phenomenon:

> Following the swelling and congestion of the external generative organs, there is, in most animals, a discharge from the generative canal. The discharge may consist merely of mucus from the uterine glands and from the glands of the cervex and from those in the neighborhood of the vaginal orifice, of the products derived from the breaking down of epithelial tissue and of fragments of small masses of pavement epithelium from the vagina; such a discharge is usually to be seen in the rat and mole. In addition, fragments or small masses of columnar uterine epithelium may be observed in various animals. Again, to the above, blood may be added for a large number of animals, some of which rarely, some frequently, and some always suffer from loss of blood. While, finally, more or less compact masses of uterine stroma tissue are included in the discharge of the primates and some of the lower mammals.

In only a few cases has the vaginal mucosa and epithelium been made a chief object of study. Examples of such cases are the papers of Morau and of Retterer, already cited. In no case was there made a systematic study of the secretion in the living animal until the work of Stockard and Papanicolaou (31, a, b). These investigators, stimulated by the need of a method for determining with exactitude the time of ovulation in the guinea pig, developed a method which has proven to be superior to the others we have mentioned. To quote from their earlier article:

> The observations were made by using a small nasal speculum which was introduced into the vagina and the arms opened apart by means of the thumb screw. The speculum permits an examination of the entire surface of the vaginal canal. In this way the vaginae of a number of females have been examined daily and smears made from the substance that happened to be present in the lumen.

Using this method a definite oestrous period of about twenty-four hours' duration was found to recur with striking regularity at intervals of fifteen or sixteen days.

According to the appearance and consistency of the secretion the following stages of the general oestrous period were distinguished:

I. Mucous secretion	6-12	hours or more
II. Cheese-like secretion	2-4	hours
III. Serous	4-6	hours
IV. Bloody discharge		?

Stage IV they considered to be sometimes lacking and perhaps not typical. The above stages compose the general oestrous period. During the early part of the dioestrous interval the secretion is very scant while during the second week following oestrum a slight mucuous discharge is noticed which increases as the new oestrum approaches.

Microscopic examination of the smears (stained with haemotoxylin and eosin) made during the various stages show the following types of cells to be present. During stages I to II squamous epithelial cells predominate, with "a certain number of elongate, cornified cells without nuclei" during the latter part of the first and the early part of the second stage. Stage III is marked by the appearance of leucocytes which seem to have a destructive influence upon the second stage cells, often entering such cells during the process of dissolving them. During stage IV there are found red blood corpuscles in addition to the cells of stage III. During the dioestrous interval the typical cells found in the scant secretion are leucocytes.

These findings were based upon the daily examination of twenty-six sexually mature virgin females, from one to seven cycles being observed in each. In all, sixty-seven complete cycles were observed. The average length of the cycle in the normal animal was found to be 15.65 days. The work was carried on throughout the year. The seasonal change was found to have very little effect on the length of the cycle, probably because the animals were kept in a room heated rather uniformly to seventy degrees Fahrenheit during the winter months. The individual differences in the length of the cycle were also very slight, the average length for a cycle ranging from fifteen to sixteen and a half days. In no case was a cycle of less than fifteen or more then seventeen days reported in a normal female. In their second paper Stockard and Papanicolaou state that in older females the cycle is occasionally as long as eighteen days, but that although they had records of observations upon several hundred cycles they had yet to find a cycle of less than fifteen or more than eighteen days. In this paper they also report that careful observations lead them to believe that copulation takes place only during stage I. The secretion at this time is a clear foamy mucous and contains no leucocytes. They state that "both the nature of the vaginal fluid at this time and the absence of leucocytes contribute to the success of copulation and fertilization."

Using the method introduced by Stockard and Papanicolaou Long and Evans (16) have made an extended investigation of the oestrous cycle and related phenomena in the rat. Records were made of the cycles of one hundred sexually mature virgin females. In all, the lengths of 1999 cycles were determined. While the average of these cycles was 5.4 days the distribution of the cases was such that Long and Evans feel justified in designating 4.6 days as the average of those cycles which may be considered normal. This is the average of those cycles (82 percent of the total), with a duration of from 4.5 to 6 days.

Changes in the cell content of the vaginal secretion as detected microscopically permits the subdivision of the cycle into five stages, according to Long and Evans. Table 1 lists these stages together with the various characteristics of each. It was compiled merely for my own convenience and is based partly upon the observations of Long and Evans and partly upon my own.

It will be seen that the changes correspond in a general way to those found in the guinea pig by Stockard and Papanicolaou. Long and Evans devote several pages of their monograph to a comparison of the cycle in the rat and in the guinea pig, noting that the chief differences lie in the length of the dioestrous interval (which is much longer in the guinea pig) and in the condition of the lumen during the period of receptivity (dry in the rat; characterized by abundant mucous in the guinea pig). An attempt to correlate in more detail Stockard and Papanicolaou's work on the guinea pig with Long and Evan's work on the rat has been made by Selle (25).

TABLE 1

Showing the outstanding characteristics of the subdivisions of the oestrous cycle

	STAGE (AS NUMBERED BY LONG AND EVANS)	APPROXIMATE DURATION	GROUPING USED IN THIS WORK	APPEARANCE OF MUCOSA	INSERTION OF SPATULA	DISLODGING OF SECRETION	CHARACTER OF SECRETION	QUANTITY OF SECRETION	CELL CONTENT			BEHAVIOR
									Leucocytes	Epithelial	Cornified	
		hrs.										
I.	Pro-oestrum	9–21	Early congestive	Dry; less transparent; some turgescence	With ease	Difficult, usually must be smeared	Translucent; jelly-like	Small drop	None	Small, round nucleated; very uniform, some times small sheets	None	Will not mate unless at end
II.	Oestrum (II–III)	27–51 (II–III)	Cornified	Very dry; opaque; maximum turgéscence	Often some resistance	Usually may be tapped off	Opaque whitish granular	Scant	None	None	Thin, transparent non-nucleated, scale-like, single	Actively solicits mating; excitement
III.			Late cornified	As above	Less resistance	Same as above	Same, but more dry and cheesy	Profuse	None	None	Same, but in sheets	Usually will not mate
IV.	Metoestrum	3–15	Post-ovulative	Less turgescence	With ease	As above	More fluid; creamy	Usually rather less profuse	Quantities of polymorphonuclear cells appear	Variable number may appear	A few still remain	Will not mate
V.	Dioestrum	42–75	Recuperative. Early inactive, Inactive, and Late inactive	Clear, moist glistening, translucent, pinkish	With ease	As above	Gruel-like	Variable	Predominate especially in early stages	Small, irregular shaped always single more numerous in later stages	None	Will not mate

Long and Evans' technique in obtaining and identifying the sample of the vaginal secretion differs somewhat from that used by Stockard and Papanicolaou. A narrow spatula is inserted into the vagina and withdrawn with a small drop of the secretion adhering. This is dislodged in a small drop of physiological saline solution and is immediately examined under eighty diameters. Under these conditions samplings from individuals in different stages present sufficiently different pictures to permit quick identification (at least in most cases).

Allen (1, a) has applied the method of vaginal smear examination to the study of the oestrous cycle of a third laboratory rodent, the mouse. This investigator finds that the histological changes in the vaginal secretion during the cycle are not markedly dissimilar from those previously observed in the guinea pig and rat. He finds for the length of the oestrous cycle a mode of four and a half days and an average of from four to six days. Data on 563 complete cycles were taken.

This review of the more important works upon the oestrous cycle serves to emphasize the significance of the method of examination of the vaginal secretion from the living animal. Prior to the introduction of this method there was disagreement among investigators concerning even a fact of such primary importance as the duration of the normal cycle in the various laboratory animals. Since the introduction of this method not only has the length of the normal cycle been determined rather exactly in several mammals but the cycle has been subdivided according to the histological picture. Furthermore, the means of determining the oestrous condition of the animal has led to the study of the relation of this phenomenon to various other internal and external conditions, with the promise that work of this kind will be widely extended in the future. Among such conditions may be mentioned normal pregnancy, pseudopregnancy, lactation, temperature, humidity, diet, hormones and internal secretion.

This method is of value to the animal psychologist as well as to the physiologist. It seems inevitable that behavior data of any sort taken on the mature female animal are affected by the oestrous cycle. The dominating influence of the oestrous condition upon activity has been demonstrated by Wang. It would not be surprising to find that learning, and perhaps discrimination, are also affected by this phenomenon.

It is obvious that without the use of this method the present study of the sex drive in the female would have been carried out with much more difficulty and with less promise of significant results.

B. In the male white rat

If there exists a rhythmic fluctuation of the sex activity in the male analogous to the rhythm in the female it should be recognized and controlled in a study of this kind.

Heape (7) made the following classification of males with respect to their mating periods:

1. Those which rut
2. Those which are sexually capable but which do not rut throughout the year either because:
   (1) The females of the species have no anoestrum (prolonged period during which mating does not take place), or because:
   (2) The anoestra of the females are asynchronous

This classification leads one to assume that mating habits in the male are determined by the type of oestrous cycle in the female rather than by some rhythmic change within the male. If this is the case, then males which normally rut but once a year would mate at any time during the year if it were possible to bring the female in heat at other than the one period. There is apparently no experimental work bearing directly on this problem. Animals commonly brought under experimental observation belong to the second class, i.e., they do not rut. The guinea pig and the rat belong to II A, the dog belongs to II B. Thus we cannot say that the existence of a sexual cycle in the male has been disproved but merely that none has been observed. By sexual cycle we mean here a periodic fluctuation in the intensity of the sex drive and the capacity of the male to copulate, this fluctuation being caused by changes within the male and independent of the female.

It has been noted that captivity tends to modify the rutting season. This may be due to the effect of the more uniform environment upon the cycle in the female rather than to its effect upon the male directly.

Sexual behavior in the male albino rat has been the subject of careful investigation by Stone (32, a). His particular interest has been in the following points:

1. The pattern of the sex response
2. The development of this pattern in the young male
3. The age at the first mating
4. The senses involved in the arousal of the act
5. The number of copulations during a given period

Stone reports nothing which seems to indicate a periodic fluctuation of the sex drive in the male.

Using the activity method neither Wang nor Slonaker observed periodic variations in spontaneous activity in the male such as are characteristic in the normal, mature female. The males show much less activity than do the females, and the slight irregularities which appear do not seem to be rhythmic. A recent article of Slonaker's (27, c) presents data hinting at, but not definitely establishing a fluctuation in the amount of activity which follows a two to four month cycle. It is very unlikely that such a rhythm is related to the reproductive process in the male since it was found in both sexes and since, in the male, it was found to continue on into the period of senility.

It seems, then, that we are justified in assuming that rhythmic, internally determined fluctuation in the strength of the sex drive of the white rat in captivity either does not exist or is so slight that it would not materially affect the results of the present investigation.

III. Method and Procedure

A. Age and care of animals

Since the purpose of this experiment was the study of the sex behavior in the normal male and female white rat, care was taken to use animals which had from birth been raised under normal sexual conditions. Such animals were raised especially for us by the Wistar Institute of Anatomy, Philadelphia. The animals (of the strain known by them as the "Experimental Colony Strain") were raised in groups of 5 to 10, the sexes being unsegregated, there being an approximately equal number of males and females in each group. As a rule 2 litters were thrown together at the time of weaning and raised together until they reached the age of 150 days at which time they were shipped to us. Upon their receipt the sexes were segregated. Segregation, or sex deprivation intervals involved merely the elimination of all tactual contact between an animal and one of the opposite sex. Olfaction was not controlled, it being impossible to obtain separate rooms for the two sexes. The animals were not used until about 5 weeks after we had received them and segregated the sexes, or until they had reached the age of about 185 days. No animal under 175 or over 196 days of age was used (except the males of the 28-day group which averaged 28 days older than this). The segregation period prior to the experimentation varied in length from 33 to 39 days. This 5-week period permitted the animals to become accommodated to the laboratory conditions. It also gave them time for all pregnant females to come to term and for the new litters to reach an age at which they could be weaned. Assuming a female to have become pregnant just prior to the segregation of the sexes, her litter would be born at the end of 3 weeks and in 2 weeks more could be weaned. (It is better however, not to wean until a litter is 3 weeks old.) These litters were not used in the present experiment but were added to the laboratory colony for future work.

The Wistar rats were used as being the best and most uniform animals available. It was felt that the variables studied in this experiment might have such slight effects that every care should be taken to rule out sources of error which might cloud the results. The individual variation in the animals was slight. The range in weights was small and in general appearance the animals were highly uniform. There seems no reason to suppose that these animals were anything but normal from the standpoint of sex experience. They were sexually mature and had copulated freely as was evidenced by the large number of litters produced. Being raised in groups containing never less than 2 animals of each sex all animals, it may be assumed, had had normal sex experience from maturity.

The writer feels that the most serious criticism that is to be raised against this work concerns the 35-day period of segregation of the sexes just prior to the experiment. In defense of this procedure the following points can be made:

1. The animals had matured approximately 3 months prior to segregation and had all, doubtless, mated many times. The normal sex drive was thus given full opportunity to develop.

2. Only rarely was behavior noted among the segregated groups which might have been interpreted as homosexual.

3. For practical reasons we were forced to segregate the females in order that they be non-pregnant when used. In order to keep the conditions constant for the two sexes it was necessary to segregate the males for the same interval. Several means of avoiding this procedure were considered:

A. A colony of vasectomized males could have been established. Such animals living with the females during this period would mate without the ensuing complication of pregnancy. On the basis of the work of Long and Evans, however, it is assumed that such mating, even though not resulting in pregnancies would have produced irregularities in the oestrous cycle.

B. The animals could have been kept singly, in individual cages during this period. Although this procedure would eliminate the possibility of the development of homosexual tendencies it is not improbable that it would have had at least as great an influence upon the results since it would affect the animals' responses to other animals of the same species regardless of sex.

C. Probably the ideal plan would be to leave the sexes unsegregated throughout, using only those females which happened to be non-pregnant when they reached the age of approximately 185 days. The first difficulty that this procedure would present would be that of detecting pregnancies. It is very likely that a number of animals would be used which would later be found to have been, at the time, in an early stage of pregnancy. Thus a large amount of data would have to be discarded so far as the present investigation is concerned. Furthermore, the percentage of non-pregnant females at any given time would be low since rats customarily mate shortly after parturition. In order, therefore, to use a large proportion of the animals most of them would have to be held over until one or probably several litters had been born before they could be used in a non-pregnant state. This would introduce a wide age difference, a factor which we desired to keep uniform since it may be related to sex behavior. Of course, if it were possible to secure animals far in excess of the numbers actually needed with the intention of discarding the bulk of them, data could be then secured on a sufficient number of unsegregated, non-pregnant females. Existing facilities rendered this method impractical. If this plan were followed it would be interesting to note whether the drive during the first oestrum following parturition is noticeably stronger than that during subsequent oestra. Observations of the animal's behavior have suggested this possibility.

4. Unpublished data (from this laboratory) of Marion P. Jenkins seem to indicate that the influence of the five weeks' segregation period upon the results is negligible. This investigator succeeded in securing a few unsegregated, non-pregnant females which were about 185 days of age. These she tested exactly as was done in the present investigation, with results almost identical to those reported here.

Sufficient and suitable food had been provided the animals throughout their development. Although the exact content of the diet given them at he Wistar Institute cannot be reported long and successful experience of this organization with the breeding of rats is a guarantee of its adequacy. During the five weeks spent in our laboratory the diet consisted of "McCollum's mixture" plus weekly rations of greens. Although McCollum's standard diet is described in several available sources it may be inserted here for the convenience of the reader:

	Per cent
Whole wheat flour	70.0
Whole milk powder	16.6
Casein	11.0
Sodium chloride	1.4
Calcium carbonate	1.0

This was mixed in an electric mixer for ten minutes or more and fed dry. There was at all times a supply available, i.e., food containers were refilled before they were entirely emptied. Water was supplied by means of inverted bottles and nozzles. Fouling of the water supply was impossible. For description of this device and for other valuable information concerning the care of rats see Greenman and Duhring (5).

Our purpose in keeping more than a sufficient amount of food continually before the animals was to keep uniform and constant the hunger drive, or food-seeking activity, at the time when the animal was used in the experiment. While a rat, even under these circumstances does not eat as uniformly throughout the twenty-four hours as does a guinea pig, for example, still we found that eating was much more generally indulged in throughout the day and night than is the case when rats are fed but a single day's ration at a time and are fed at a constant time, day after day, as is usually done in animal laboratories. We may assume that the rats were not in an active state of hunger when experimented upon since they were taken directly from a cage provided with food.

The animals were housed in wire mesh cages of about 420 square inches floor area and 5,900 cubic inches capacity. There were from five to ten animals in a cage. Bedding of coarse saw dust was provided.

B. The apparatus

The Columbia Obstruction apparatus (11) was employed in this study. As noted, the box for the control of the animals consisted of three compartments which may be called the entrance compartment, the obstruction compartment and the incentive compartment. A small section of the latter was separated off by an automatic door. The inside dimensions of the various sections were as follows:

	WIDTH	LENGTH	HEIGHT
	inches	*inches*	*inches*
Entrance compartment	10	10	10
Obstruction compartment	4	10	4
Incentive compartment:			
Main section	10	10	10
Small section	4	4	4

The entire floor of the obstruction compartment consisted of an electric grid. The grid which was described in the article cited above was found to possess one serious disadvantage. A drop or two of water could easily short circuit it, interfering with the electrical stimulation value of the grid. A micturition response of the animal occasionally had this effect, necessitating interruption and delay. To avoid such possible irregularities a second grid was built. Unlike the first, which consisted of copper wires wound around bakelite, this grid was formed of copper strips, one-eighth inch thick and three-eighths inch in width, running diagonally across the floor of the compartment. The strips were separated from each other by a uniform gap of one-eighth inch. This gap was of such width that it was impossible for a drop to hang suspended between two strips. The strips were so wired that those connected with the positive and the negative poles alternated. The mounting and wiring of the strips was from beneath in such a way as to secure insulation and protection from dirt and moisture. This grid proved to be perfectly satisfactory. Apparently the animals never failed to make simultaneous contact with strips from opposite poles and thus to receive the shock. They experienced no difficulty in walking across the grid, their habit of walking with outstretched toes preventing their catching their feet in the narrow gaps.

Except for the grid, the celluloid door and the glass covers, the entire apparatus was painted flat black. It was set up on a table in an improvised dark room. Since the experiments were all done at night we were not faced with the difficulty of eliminating daylight. The only source of light was an illumination hood 11 inches above the center of the apparatus. This cast uniform light in all parts of the apparatus. Since the illuminating surface extended without break over the whole length of the apparatus the partitions and the doors of the apparatus cast no shadows. The hood was so adjusted that light was thrown only into the box and not upon the other objects in the room nor upon the experimenter. Since the animal stood in the light and the experimenter in the shadow it did not seem necessary to use a screen and peephole as is usually done in well controlled animal experimentation.

Only one degree of shock was used in the present investigation. Alternating current of 60 cycles was used. The terminal pressure was 475 volts. The external resistance in the circuit totalled 10,-100,000 ohms and the current was thus 0.047 milliamperes.

The determination of a suitable degree of shock was no small problem. It was our purpose to adopt a degree which could be used not only in the study of the sex drive but also in the study of other drives, such as the hunger drive. Dr. Holden's study of the hunger drive in the white rat by means of the obstruction method (9) gives clear indication of the effect upon the drive of variation in the degree of shock. Of the three degrees she used the lowest gave the best results. Although the specific results of Dr. Holden's work could not be directly applied to the present study because of important differences in apparatus and technique her experience was helpful in guiding the preliminary work done for the purpose of determining the degree of shock, to be used. In this preliminary work data on both the hunger and sex drives were taken using various degrees of shock. The shock finally settled upon as the most satisfactory is less than the lowest shock used by Dr. Holden. The present work indicates that it is suitable for the study of the sex drive under the conditions here employed and further work from the Columbia laboratory, soon to be reported, proves its suitability in the study of the hunger drive. For Figures 1 and 2 of this monograph showing appartus, see Part I, 1, of this volume.

C. Detailed procedure

The male. Variation in the physiological condition of the male was obtained by varying the interval separating the test and the last previous mating. In every case the male was placed in a cage with a female in heat for two hours and the mating behavior observed. Distribution of the number of matings occurring during this period is shown in Table 2. The accuracy of these data cannot be vouched for since the experimenter was usually engaged in staining vaginal smears and in other laboratory duties while one or more such mating periods were in progress. It is unlikely, however, that the data seriously misrepresents the facts. The chief purpose of keeping a record of the mating was to determine whether the male in question behaved normally toward the female and to ascertain whether full opportunity for "satiation" was afforded.

After this two-hour period the male was placed in a cage with other males for the specified deprivation interval, at the end of which it was tested in the obstruction apparatus. Animals in the zero group were transferred directly from the cage with the receptive female into the apparatus and tested. The deprivation intervals studied were 0, 6 and 12 hours, 1, 4, 7, and 28 days. There were twenty animals in each of these groups. There was also tested

TABLE 2

Distribution table covering number of matings during the two-hour period prior to segregation of males

NUMBER OF COPULATIONS	FREQUENCY
6-10	7
11-15	5
16-20	11
21-25	10
26-30	13
31-35	16
36-40	17
41-45	18
46-50	16
51-55	19
56-60	11
61-65	6
66-70	4
71-75	7
76-80	3
81-85	2
	160

Average 40.6.
Standard deviation 18.0.

a control group of twenty animals. The procedure in the case of this group was exactly like that of the one-day group except that no stimulus animal or other incentive was used. The stimulus compartment was empty. In the case of all the other groups the stimulus was a female rat in oestrum, i.e., in the cornified stage. The stimulus animal was placed in the smaller section of the incentive compartment and the automatic door shut, preventing her from crossing the obstruction to the male. The male to be tested was placed in the entrance compartment, the door leading into the obstruction compartment being closed. The glass slides above all compartments were closed except when an animal was being introduced or removed. The male being restrained in the entrance compartment and the female in heat being restrained in the smaller section of the stimulus compartment the experiment proper could begin. This consisted of a short preliminary training period immediately followed by a twenty-minute test period. The purposes of the perliminary training were

(1) To accommodate the animals to the apparatus
(2) To permit the male to associate the stimulus compartment with the female in heat
(3) To permit the male to associate shock with the obstruction compartment, and with the obstruction compartment only

A considerable amount of data were taken in an effort to determine what type of preliminary training best served these ends. It was found that the final results were profoundly affected by the nature of the preliminary training. The procedure finally adopted is believed to fulfill the three purposes as adequately as this can be done.

Ten seconds after a male was placed in the entrance compartment the door leading into the obstruction compartment was opened. There was at this time no current in the grid. The test animal usually passed through the obstruction compartment and into the stimulus compartment within the next quarter minute or less. After he entered the stimulus compartment the door separating it from the obstruction compartment was quietly closed so that he could not retrace his steps. Meanwhile the automatic door had opened and the two animals were together in the stimulus compartment. At this moment the experimenter was always alert. The male was to be removed before he mated and yet not until he had had opportunity to identify the stimulus animal as a female. Since the most usual response of the male was the nosing of the female, and in particular of the genital region this behavior was taken as the criterion. Not until the male had done this was it removed and replaced in the entrance compartment. At the same time the stimulus animal was replaced behind the automatic door, and this closed. The animals and the doors were now in the positions they had been before. This procedure was repeated, four times in all. In each successive crossing the male tended to show less hesitation, to cross more promptly to the female, and, once in the stimulus compartment, to devote less time to exploration of the compartment and more to attention to the female.

It was not until the fifth time that the animal was allowed to cross to the female that electric shock was introduced. In this fifth crossing, the last of the preliminary series, the switch controlling the current to the grid was closed after the animal had entered the obstruction compartment. At the same instant the door was lowered behind him, cutting off possible retreat into the entrance compartment. The only way for the animal to end the electrical stimulation was by continuing on into the stimulus compartment. Again, as before, he was left there until his interest in the female crystallized

#	SX	WT			DR	GR	DATE			
							//			

	1	2	3	4	5	6	7	8	9	10	T
A											
T											
C											
	11	12	13	14	15	16	17	18	19	20	T
A											
T											
C											

Fig. 3. Reproduction of Data Card Used in the Present Investigation
Numerals represent the minute in the twenty-minute period during which the activity took place. *A* stands for approaches; *T*, for contacts; *C*, for crossings.

into the specific behavior of nosing the genital region. He was then replaced in the entrance compartment and the test period was immediately begun.

The procedure in the test period was like that in the preliminary period with the following exceptions:

1. The grid was electrified throughout the test period.

2. The test period was twenty minutes in length, irrespective of the behavior of the animals.

3. The behavior of the test animal (the male in this case) was observed and recorded in terms of approaches, contacts and crossings together with the time during the test period when each such act accurred. These data were recorded on a card such as shown in figure 3.

The three types of behavior recorded are defined as follows:

An approach: Some part of the animal (actually, the tip of its nose) crosses the threshold between the entrance and the obstruction compartment, but instead of proceeding and thus coming into contact with the grid, the animal withdraws.

A contact: Some part of the animal (its feet, occasionally its nose) touches the grid, but instead of proceeding across the grid, the animal retires into the entrance compartment again. The withdrawal is usually more abrupt in this case since it is aroused by the electrical stimulation.

A crossing: The animal passes through the entrance compartment and into the stimulus compartment, the door between the obstruction and the stimulus compartment being dropped behind him.

4. The final difference between the preliminary and the test period procedure lies in the criterion which determines that the male is to be returned to the entrance compartment after a crossing. In the preliminary trials he was not returned until he had nosed the genital region of the female even though this occassionally involved a delay of several minutes during which the male explored the stimulus compartment. During the test period such general exploratory activity was not so often observed and the attention of the male appeared fixed primarily upon the female. Nevertheless the definite nosing response did not always occur promptly and the criterion was broadened to include any specific contact between the two animals so that one or more of the following acts was sometimes substituted for the nosing act on the part of the male; nosing of other than genital regions; nipping or biting, most often of the back and neck regions. Such behavior on the part of the male was usually accompanied by similar behavior on the part of the female. In the case of a very few males who were extremely sluggish and, in most cases, crossed but a few times such behavior on the part of the females, only, was required since the males failed to make any

appropriate response within a reasonable amount of time. In general the test animal was returned after from five to fifteen seconds in the stimulus compartment.

The female. The physiological phenomenon most intimately related to the female sex drive is, of course, the oestrous cycle. The female is said to be in oestrum when it will accept the male. The other periods of the cycle are defined as related to this period. Mere observation of the behavior of a female in the presence of a mature male will indicate only that it will or that it will not mate. The obstruction method of measuring drives supplements such rather meager data since it permits us to rank those females which will mate in the order of the strength of the mating impulse. Just as the vaginal smear method developed by Stockard and Papanicolaou yields a much more exact and detailed picture of the physiological changes during the oestrous cycle than was theretofore obtainable so we feel that the obstruction method gives a more sensitive index of the behavior of the animal during the cycle than can be obtained by other behavior observation. We have used the vaginal smear method and the obstruction method in conjunction enabling us to follow the rise and fall of the strength of the sex drive during the cycle, thus correlating physiological condition and behavior in a more detailed way than has been possible in the past.

In testing the female the procedure was exactly like that for the male with the following exception:

1. The previous two-hour mating period and period of deprivation were omitted, the females being transferred directly from the living cage to the apparatus.

2. The stimulus animal was always a male (except in the control group, with which there was no stimulus animal used).

3. Vaginal smear records were taken for each female immediately following the test period.

Females for testing were selected at random from the living cage except that whenever a female was noticed, which, judging from its behavior and from macroscopic observation (non-manipulative), was in oestrum that animal was tested at once. If the animals had been selected entirely at random the majority of them would have been in the various phases of the dioestrum. Since we wished to test at least forty animals which were in heat, twenty of these to form the control group, selection with this in mind was necessary. In no case, however, was a vaginal smear taken prior to the testing of an animal.

A vaginal smear record was taken immediately after the removal of the female from the apparatus at the end of the test period. These were stained with haemotoxylin and eosin and permanently mounted in balsam. The classification of the females as to oestrous condition, and thus their grouping, was wholly on the basis of these histological records. After the slides had been examined and classified they were taken to Dr. G. N Papanicolaou who repeated this process (with no knowledge of the behavior data). Dr. Papanicolaou's grouping of the slides according to the stage of the cycle represented checked the previous classification but went further than it had in the way of subdividing the cycle. He found that he could distinguish eight groups, representing as many phases of the complete cycle. The behavior data was grouped accordingly as will be seen by referring to the results. Discarding the rather loosely defined terms such as Metoestrum, Postoestrum and so on, Dr. Papanicolaou has suggested the descriptive terms which we have used for the eight groups.

These microscopic records are open to examination at the Animal Laboratory, Columbia University.

The chief point in which our own observations do not verify those of Long and Evans concerns that stage of the cycle termed by them stage IV. The smears which most nearly resemble this stage are those placed by us in the post ovulative group. These do not show the clearcut picture described by Long and Evans. Both leucocytes and epithelial cells seem to have returned while a few cornified cells still remain. This would appear to be a rather mixed stage suggesting a gradual transition from the cornified stage to the stage of epithelial cells and leucocytes. Thus it contrasts with the more abrupt and clearcut onset of oestrum when the epithelial cells give place rather suddenly for the cornified, if we may judge from the fact that the two are not ordinarily found in the same smear. As will be seen below the behavior data likewise show a sudden onset and a gradual cessation of the characteristics of oestrum.

An important part of the technique, but one which baffles description, consists in the handling of the animals. The animals being tested must be lifted from the incentive compartment after crossing and replaced in the entrance compartment without arousing a fear response. Considerable preliminary work was required not merely for the determination of a suitable degree of shock, appropriate preliminary training, and so on but also to familiarize the experimenter with the apparatus, procedure and method of handling the animals. Careful observation of the animals during the experiment proper led to the impression that they were not seriously disturbed by the necessary handling. On the several occasions when such a disturbance was noted the animal was discarded. The chief points noted in the handling of animals when using the obstruction method were the following:

1. Strict silence was observed at all times.
2. All movements of the experimenter were slow.
3. A rhythm of removing and replacing the animals was acquired so that the process occupied very nearly the same amount of time in every case.
4. An animal was never lifted in such a way that, being insufficiently supported in the hand it would reach out for other support to which to cling. The animal never touched any object outside the apparatus except the hand of the experimenter.
5. When released from the hand the animal was not dropped into the entrance compartment but was lowered in the hand to the floor of the compartment and there gently released.
6. Manually operated doors and slides were manipulated with due regard for the position of the rat and in particular of its tail.

IV. Results

A. The male sex drive

Table 3 indicates the grouping of the one hundred and sixty male rats tested. The distribution of approaches, contacts, and crossings for the eight groups is given in Table 4. The medians, ranges, averages, standard deviations, and coefficient of variability are found in Table 5. Table 6 shows the reliability of the differences between the averages. The approaches, contacts, and crossings as distrib-

TABLE 3

Showing male groups

NUMBER OF ANIMALS	PERIOD OF SEX DEPRIVATION	OBJECT IN INCENTIVE COMPARTMENT
20	24 hours	None (control group)
20	0	Female in heat
20	6 hours	Female in heat
20	12 hours	Female in heat
20	1 day	Female in heat
20	4 days	Female in heat
20	7 days	Female in heat
20	28 days	Female in heat

TABLE 4

Distribution table covering approaches, contacts and crossings for the various peri of sex deprivation in the male rat

INTERVAL	CONTROL*			0			6 HOURS			12 HOURS			1 DAY			4 DAYS			7 DAYS			28 DAYS		
	Approaches	Contacts	Crossings	Approaches	Contacts	Crossings	Approaches	Contacts	Crossings	Approaches	Contacts	Crossings	Approaches	Contacts	Crossings	Approaches	Contacts	Crossings	Approaches	Contacts	Crossings	Approaches	Contacts	Crossings
0	5	8	2	8	15	6	3	2	3	5	4	1	4	6		7	5	1	5	3	2	6	6	2
1	9	7	4	6	4	2	2	1	1	2	3		6	6		3	6		3	3	2	5	4	2
2	2	3	2	2		3	2	2		3	3		3	4		5	6		2	3		1	4	1
3	4		4	3	1	3	1	2	1	3	3		1	2		1	2		5	4		2	1	
4		2	3				3	2	1	4	2	1	4	1	1	2	1	2	4	3	1	3	2	1
5			4	1		1	5	5		2	3		1						1	1			1	1
6							1	3	2		2			1	1	2						1	2	
7			1			1				1		1								1				
8						1	1	1										1		1				
9								1	2			2			2					1		1		
10						1			2				1									1		1
11							1	1	3			3			2			1			2			1
12						1	1		1			3			1			2						1
13						1			2			1			2						1			1
14												1						3			2			3
15									1			3			4			3			2			
16												1			3						4			2
17															1			1			1			1
18									1			1			2			3						
19																					1			1
20												2			1									
21																		1						1
22																		1			1			1
23																								
24																					1			1

* Control group, twenty-four hours' sex deprivation; no female in the incentive compartment.

TABLE 5

Showing results of the male groups

GROUP	APPROACHES					CONTACTS					CROSSINGS				
	Median	Range	Average	Standard deviation	Coefficient of variability	Median	Range	Average	Standard deviation	Coefficient of variability	Median	Range	Average	Standard deviation	Coefficient of variability
Control	1	0–3	1.2	1.03	85.8	1	0–4	1.05	1.18	112.3	3	0–7	2.95	1.8	61.0
0	1	0–5	1.2	1.36	113.3	0	0–3	0.35	0.7	200.0	2	0–13	3.6	4.08	113.5
6 hours	4	0–12	4.15	3.26	78.5	5	0–11	4.5	2.77	61.5	9.5	0–18	8.1	5.2	64.2
12 hours	2.5	0–7	2.7	2.03	75.2	2.5	0–6	2.65	1.98	74.7	12	0–20	12.2	4.8	39.3
1 day	1.5	0–10	2.3	2.35	102.2	1	0–6	1.5	1.5	100.0	15	4–20	13.45	4.03	29.9
4 days	1.5	0–6	1.8	1.9	105.5	1	0–4	1.85	1.56	84.3	14	0–22	12.6	6.24	49.5
7 days	2.5	0–5	2.25	1.5	66.6	3	0–9	3.1	2.5	80.6	14.5	0–24	12.25	7.07	57.7
28 days	1	0–10	2.5	2.9	116.0	1.5	0–6	2.0	1.97	98.5	12.5	0–22	10.55	7.1	66.4

TABLE 6

Showing reliability on the difference between the average crossings of the male groups

GROUPS	THE DIFFERENCE	STANDARD DEVIATION OF THE DIFFERENCE	DIFFERENCE / S.D. OF THE DIFF.	CHANCES IN 100 OF A TRUE DIFFERENCE GREATER THAN 0
0 hours and 6 hours.......	4.5	1.47	3.06	100.0
6 hours and 12 hours.......	4.1	1.58	2.59	99.5
12 hours and 1 day.........	1.25	1.4	0.89	81.0
1 day and 4 days..........	0.85	1.65	0.51	69.0
4 days and 7 days.........	0.5	2.1	0.24	60.0
7 days and 28 days.........	1.55	2.23	0.695	76.0
1 day and 28 days..........	2.9	1.8	1.6	94.0
1 day and control..........	10.50	1.14	9.2	100.0

uted during the twenty-minute test period (in one-minute intervals) is given in Table 7. In Table 8 these data are reduced to the percentages of such activity occurring during successive five minutes of the twenty-minute test period. Figure 4 shows graphically the effect of variation in the sex deprivation interval upon the approaches, contacts and crossings. Figure 5 shows graphically their distribution during the test period in terms of the percentage falling within each successive five minutes.

We will first take up the male groups in order from the shortest

TABLE 7

Showing temporal distribution of approaches, contacts and crossings of the male during the twenty-minute test period in one minute intervals

GROUP	ACTIVITY	1ST MINUTE	2ND MINUTE	3RD MINUTE	4TH MINUTE	5TH MINUTE	6TH MINUTE	7TH MINUTE	8TH MINUTE	9TH MINUTE	10TH MINUTE	11TH MINUTE	12TH MINUTE	13TH MINUTE	14TH MINUTE	15TH MINUTE	16TH MINUTE	17TH MINUTE	18TH MINUTE	19TH MINUTE	20TH MINUTE
Control	Approaches	0	0	2	0	1	0	0	3	1	0	1	1	3	1	3	3	2	0	3	1
	Contacts	0	4	2	1	1	0	3	2	1	1	0	0	1	0	2	0	1	1	0	1
	Crossings	4	9	5	6	4	4	2	2	5	4	1	1	2	1	4	1	1	0	2	1
0	Approaches	1	0	1	2	0	1	1	0	1	0	3	4	2	2	1	1	2	1	0	1
	Contacts	0	0	1	1	0	0	1	1	0	0	0	0	1	1	0	1	0	0	0	0
	Crossings	3	1	2	1	3	3	4	4	5	3	4	2	7	2	3	3	5	7	5	5
6 hours	Approaches	6	5	5	6	4	4	5	4	2	2	4	5	6	3	6	5	4	3	2	2
	Contacts	4	4	7	10	6	6	5	3	4	4	2	6	3	2	6	7	4	3	3	1
	Crossings	7	10	13	12	10	21	9	5	15	11	10	9	6	2	1	2	4	2	8	5
12 hours	Approaches	2	0	1	6	3	3	6	9	3	1	3	8	3	1	0	1	2	0	2	2
	Contacts	3	3	3	3	1	3	2	1	4	3	1	0	2	1	4	5	4	3	5	2
	Crossings	10	8	17	22	16	19	24	19	10	5	8	11	7	7	14	8	9	10	9	11
1 day	Approaches	7	2	1	3	6	6	5	1	2	2	1	1	2	0	2	1	0	1	2	1
	Contacts	3	1	1	2	2	3	3	1	1	1	2	1	0	3	2	2	0	1	1	0
	Crossings	12	17	13	16	19	17	15	9	11	10	10	13	16	13	10	7	12	17	15	17
4 days	Approaches	6	3	3	2	1	1	3	1	2	1	3	1	2	1	2	1	0	1	1	1
	Contacts	6	1	2	2	1	2	1	1	0	1	4	1	2	1	0	1	0	1	1	0
	Crossings	14	9	12	15	9	11	16	14	7	7	11	10	13	15	16	17	11	13	16	14
7 days	Approaches	4	1	2	3	2	1	4	1	3	2	2	0	3	3	3	3	1	2	2	1
	Contacts	3	4	5	3	2	2	0	1	4	2	7	4	3	6	4	5	2	2	1	2
	Crossings	11	7	16	10	3	6	10	8	5	4	11	7	11	13	16	21	20	21	23	22
28 days	Approaches	4	2	0	3	0	0	1	2	4	1	1	1	3	6	4	5	2	4	4	3
	Contacts	3	0	1	0	0	1	2	0	2	4	4	6	5	4	3	4	0	0	1	0
	Crossings	2	7	3	1	2	4	6	7	11	13	7	11	11	9	12	21	19	24	22	19

TABLE 8

Showing temporal distribution of approaches, contacts and crossings of the male groups during the twenty-minute test period in five-minute intervals

GROUP	ACTIVITY	0 TO 5TH MINUTE		6TH TO 10TH MINUTE		11TH TO 15TH MINUTE		16TH TO 20TH MINUTE		TOTAL
		Number	Per cent	Number	Per cent	Number	Per cent	Number	Per cent	
Control	Approaches	3	12 0	4	16.0	9	36 0	9	36.0	25
	Contacts	8	38 1	7	33.0	3	14.0	3	14.0	21
	Crossings	28	47.6	17	28.9	9	15.3	5	8.5	59
0	Approaches	4	16 6	3	12.5	12	50.0	5	20 9	24
	Contacts	2	28 6	2	28.6	2	28.6	1	14 3	7
	Crossings	10	13 6	19	26.4	18	25.0	25	34 7	72
6 hours	Approaches	26	31.2	17	20.4	24	28.8	16	19 2	83
	Contacts	31	34.4	22	24.4	19	21.1	18	19 9	90
	Crossings	52	32.1	61	37 6	28	17.3	21	13.0	162
12 hours	Approaches	6	12.0	22	44.0	15	30.0	7	14.0	50
	Contacts	13	24.7	13	24 7	8	15.2	19	36.1	53
	Crossings	74	30.3	76	31.0	44	18 0	50	20.5	244
1 day	Approaches	19	41.8	16	35 2	6	13.2	5	11.0	46
	Contacts	9	29 9	9	29 9	8	26 6	4	13.3	30
	Crossing	76	28.2	63	23 4	60	22.3	70	26.0	269
4 days	Approaches	15	42 0	8	22 4	9	25.2	4	11.2	36
	Contact	12	43.2	5	18 0	8	28.8	3	10.8	28
	Crossings	62	24 6	55	21 8	65	25.7	70	27 7	252
7 days	Approaches	12	27 6	11	25 3	11	25 3	9	20.7	43
	Contacts	17	27 2	9	14 4	24	38.4	12	19.2	62
	Crossings	47	19 2	33	13.5	58	23 7	107	43 7	245
28 days	Approaches	9	18	8	16.0	15	30.0	18	36.0	50
	Contacts	4	10 0	9	22.0	22	55.0	5	12.5	40
	Crossings	16	7 6	40	18 9	54	25.6	101	47.9	211

to the longest deprivation interval, noting the influence upon the behavior of varying this factor.

Animals of the 0 group, that is, those which had mated freely for two hours immediately preceding the test period, were not inclined — either to approach or to make contact with the grid at first. During the first five minutes a total of but four approaches, two contacts and ten crossings were recorded for the twenty animals in the group. The animals became more and more active as the test proceeded, a total of twenty-five crossings being recorded for the last five minutes of the period. Even so, the average number of approaches, of contacts and of crossings were lower for this group than for any other but the control. Six of the animals did

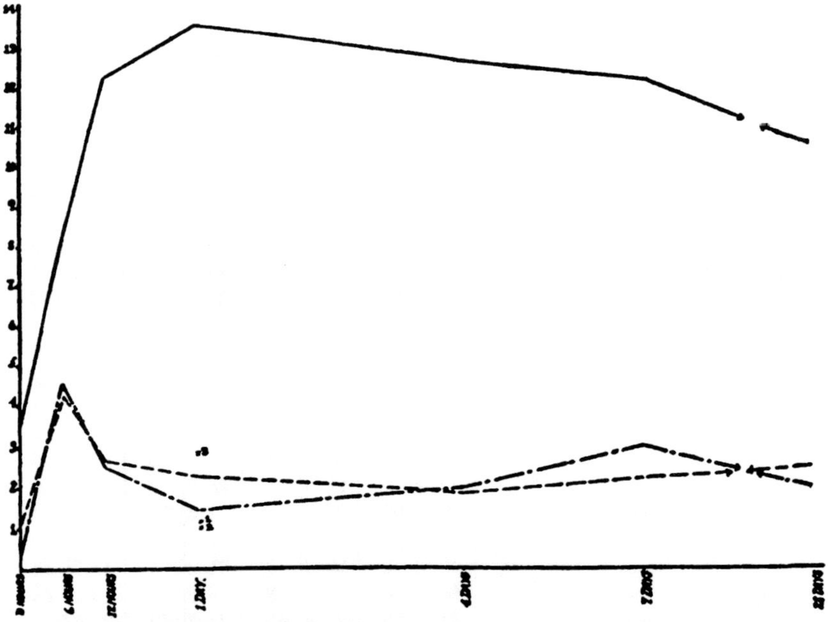

FIG. 4. Showing average number of crossings (solid line), contacts (dot-dash line) and approaches (dash line) for each interval of sex deprivation in the male rat. Results for the control group are shown on the one day abscissa since animals of this group were tested after a one day sex deprivation period, the average number of approaches being represented by the point labeled *1*, contacts by *2*, crossings by *3*.

not cross at all, while fourteen crossed three times or less. Notes taken during the course of the experiment indicate that the characteristic behavior of the animals of this group, and especially of the fourteen least active animals was, apparently, sleep. At any rate these animals spent considerable time lying practically motionless in a corner of the entrance compartment. Five of the six did not cross at all during the test period, apparently slept the entire twenty minutes and were sleeping when removed.

The preliminary training period for this group involved considerable temporal irregularity due to the extreme inactivity of many of the subjects. Although all of them crossed at least once and all but three crossed at least twice during the first fifteen minutes devoted to the preliminary training, it is doubtful whether certain of the animals would have crossed the five times (as required by the technique adopted) in less than two hours or more.

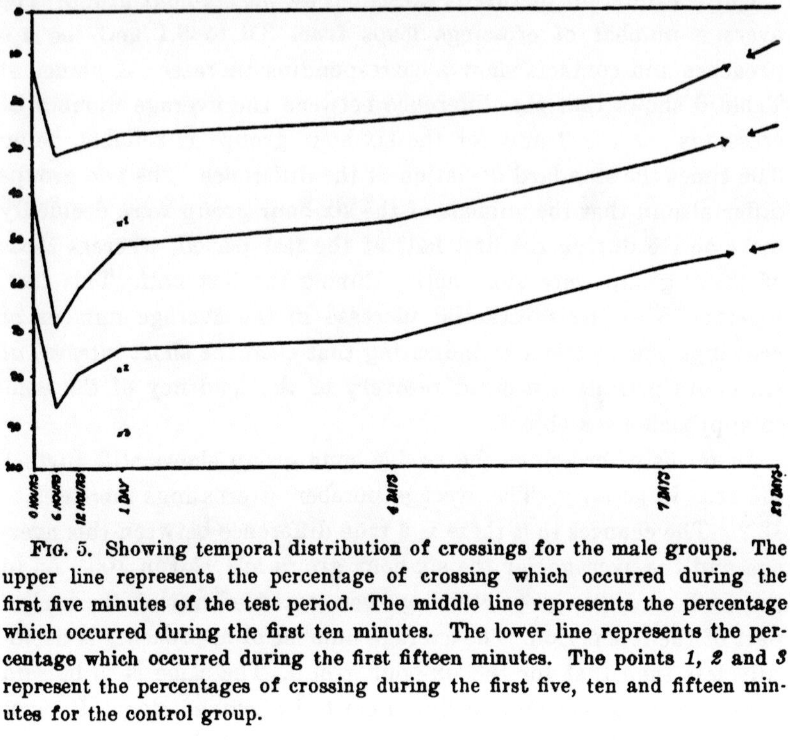

FIG. 5. Showing temporal distribution of crossings for the male groups. The upper line represents the percentage of crossing which occurred during the first five minutes of the test period. The middle line represents the percentage which occurred during the first ten minutes. The lower line represents the percentage which occurred during the first fifteen minutes. The points *1*, *2* and *3* represent the percentages of crossing during the first five, ten and fifteen minutes for the control group.

For practical purposes it was necessary to make the following exception to the usual procedure. If an animal had crossed but two or three times during the first thirty minutes of the preliminary training period the electric shock was introduced during the next crossing (as was done normally during the fifth crossing) and immediatly following this the test period was begun. It seems unlikely that this irregularity should vitiate the results. If an animal was so lethargic as not to cross to the female even though no shock was introduced it is quite unlikely that it would cross when the electric obstruction was present. Had we not made this exception the preliminary training period would have required, in certain cases, sufficient time to permit a partial recovery from the inactive state which normally follows a period of mating. An examination of the data for the six-hour group leads to the supposition that this recovery is relatively rapid.

Quite a different picture is presented by the six-hour group. The average number of crossings leaps from 3.6 to 8.1 and the approaches and contacts show a corresponding increase. A glance at Table 6 shows that the difference between the average number of crossings for the 0 and for the six-hour groups is reliable, being 3.06 times the standard deviation of the difference. The two groups differ also in that the animals of the six-hour group were decidedly more active during the first half of the test period, whereas those of the 0 group were more active during the last half. This fact, together with the substantial increase in the average number of crossings can be taken as indicating that even the short interval of six hours permits a decided recovery of the tendency of the male to approach a sex object.

In terms of crossings the twelve-hour group shows still further increase in activity. The average number of crossings increases to 12.2. The chances that there is a true difference between this average and the average for the six-hour group are 99.5 in 100. As in the six-hour group more crossings wer erecorded during the earlier part of the test period. The average number of approaches is decidedly less than that for the six-hour group. The same is to be said for contacts. This would at first seem to be contradictory. It must be remembered, however, that by an "approach" is meant "an approach and withdrawal" and that by a "contact" is meant "a contact and a withdrawal." It is then quite to be expected that as the animals cross more frequently they will approach or make contact without crossing less frequently.

A note should be made with respect to the twelve-hour group. On the average these animals were tested later at night than were the animals of the other groups. Most of them were tested between 2 and 4 A.M. Any effect that this difference in conditions might have had would probably be in the direction of decreasing the activity of the animals of this group, since the rats become less active with the approach of dawn.

Animals of the one-day group did not react very differently from those of the twelve-hour group. The average number of crossings, 13.45, is but slightly larger, the difference not being reliable. The approaches and touches show a further, but hardly significant decrease. The chief way in which this group differs from the previous relates to the time during the test period in which the crossing occurred. Whereas the crossings were most frequent during the first half in the six- and the twelve-hour groups, in the twenty-four group this advantage was eliminated and the crossing was remarkably uniform throughout the twenty-minute period. A definite and increasing tendency for the animals to cross more often in the second half of the period was found in all groups of more than twenty-four hours deprivation. Table 8 and Figure 5 show this increasing tendency clearly.

In other respects than that of the temporal distribution of activity just mentioned, the four and the seven-day groups differ but slightly from the twenty-four hour group and from each other. The slight drop in the average number of crossings from the twenty-four-hour to the four-day group and from the latter to the seven-day group is not reliable. When taken in conjunction with the more pronounced drop found in the twenty-eight-day group this tendency may be considered to have some significance. The average number of crossings in the twenty-eight-day group is only 10.55. The difference between this average and that for the twenty-four-hour group is 1.6 times the standard deviation of the difference, i.e., the chances that this difference is a true difference are 94 in a 100. It may be that there is an optimum deprivation interval (optimum from the standpoint of the tendency of the animal to cross the shock to a sex object) and that beyond this interval further deprivation interferes with the operation of the normal sex drive. Perhaps sex activity is dependent to a slight extent upon habit elements which are weaned by a long period of inactivity. It might be that homosexual tendencies began to develop during the longer deprivation intervals since the males were kept with other males during these intervals, and that these interfered with the response to a heterosexual stimulus. Examination of tables 7 and 8 shows that the animals of the twenty-eight-day group crossed more and more as the test proceeded, crossing more often in the last six minutes than during the first fourteen minutes. This might be taken to indicate a gradual reawakening of interest in a heterosexual stimulus.

The sex deprivation interval used in the male control group was twenty-four hours since it had been found that of the intervals used this one had resulted in the maximum activity. The results for the control group are therefore compared with those for the twenty-four-hour group. These two groups differed in but one respect. In the case of the control group there was no object in the reward compartment while in the case of the twenty-four-hour group there was always a female in heat in this compartment. The results for the two groups are strikingly and reliably different. The average number of crossings for the twenty-four-hour group was 13.45, for the control 2.95. The difference between these averages is 9.2 times the standard deviation of this difference. This proves conclusively that the presence of the female in heat was the dominating factor in determining the activity of the males of the twenty-four-hour group. Twenty-eight of the fifty-nine crossings recorded for the control group occurred during the first five minutes of the period and forty-five during the first ten minutes. In the twenty-four-hour group, on the other hand almost as many crossings occurred during the second half as during the first half of the period. An interpretation might be that animals of the control group underwent the punishment in response to a comparatively less powerful and more transient drive related to exploring activity whereas those of the twenty-four-hour group showed by their much more frequent crossing and by the fact that they crossed with practically undiminished vigor through the test period that they were influenced by a drive that was both stronger and more lasting. Obviously it should not be inferred from these remarks that it is supposed that this experiment gives a fair and comparable measure of the exploring drive, if we may use the term, since the region was too limited and simple to provide suitable "reward" for such a drive.

To summarize: The male sex drive appeared to be at its lowest point immediately after a two-hour period during which mating had taken place freely. Recovery was rapid during the first six hours and almost as rapid during the succeeding six. By twenty-four hours the tendency of the males to cross the grid to the female had reached its high point. During the following six days there was apparently little change, but what change there was was in the direction of a reduction in the strength of this tendency. Three more weeks of deprivation resulted in a more certain indication of such reduction. If the reduction in the strength of the drive is a true one it is in all probablity capable of rapid restoration as indicated by the increasing number of crosses during the final minutes of the test period in the animals sex deprived for twenty-eight days.

TABLE 9

Showing female groups

NUMBER OF ANIMALS	OESTROUS CONDITION	OBJECT IN INCENTIVE COMPARTMENT
22	Cornified	None (control group)
12	Early inactive	Male
6	Inactive	Male
4	Late inactive	Male
7	Early congestive	Male
21	Cornified	Male
3	Late cornified	Male
6	Post-ovulative	Male
10	Recuperative	Male

B. The female sex drive

Table 9 indicates the grouping of the 91 female animals used. The distribution of approaches, contacts and crossings for the nine female groups is shown in table 10. The medians, ranges, averages and average deviations are given in table 11. In table 12, four of the female groups (recuperative, early inactive, inactive, late inactive) have been combined. This composite group, containing 32 animals, represents the general dioestrous or inactive period of the oestrous cycle. This composite group is compared with the cornified group (21 animals) and the control group (22 animals) in this table (table 12). The medians, ranges, averages, standard deviations and coefficients of variability are given. Table 13 deals with the reliability of the difference between the average number of crossings for the cornified and control groups and for the cornified and composite dioestrum group. The approaches, contacts and crossings for all the female groups as distributed during the twenty

TABLE 10

*Distribution table covering approaches, contacts and crossings of the females as correlated with the histological character of the vaginal secretions**

INTERVAL	CONTROL (22) (NO SEX OBJECT)			EARLY INACTIVE (12)			INACTIVE (6)			LATE INACTIVE (4)			EARLY CONGESTIVE (7)			CORNIFIED (21)			LATE CORNIFIED (3)			POST-OVULATIVE (6)			RECUPERATIVE (10)		
	Approaches	Contacts	Crossings	Approaches	Contacts	Crossings	Approaches	Contacts	Crossings	Approaches	Contacts	Crossings	Approaches	Contacts	Crossings	Approaches	Contacts	Crossings	Approaches	Contacts	Crossings	Approaches	Contacts	Crossings	Approaches	Contacts	Crossings
0	5	6		9	3	5	2	1	4	1	1	1	1		1	2	3					3	1	1	3	2	5
1	3	3	3	1	5	2	2	3	1		1		1	1		1	5		1			1		1	4	3	2
2	2	3	1	2	2	2		2		1	2	2				2	1	1	1	1	1	1	1	1	1	2	
3	3	3	3		2	1	1			1		1			1	3	5						1	1	1	1	1
4	3	5	3			1	1		1				1	3			3	1	1	1		1	1		1		2
5			2			1				1				1		3		1		1						2	
6	2	2	1													2	1						1				
7	3		5										1	1		3											
8			2												1	1	1										
9			2											1		2	2	1					1	1			
10	1												1								1						
11													1					1									
12																1											
13																								1			
14													1		1	1		3			1						
15															1			2									
16															1			3									
17																		2									
18																		1									
19																		2									
20																											
21																		2									
22															1												
23																											
24																		1									
25																											

* Grouping is based upon histological characteristics of the vaginal secretion.

minute test period (in one-minute units) is given in table 14. In table 15 these data are given in terms of the percentage of each activity occurring during the four successive five minutes of the twenty-minute test period. Figure 6 shows graphically the effect of variation in oestrous condition upon the average number of approaches, contacts and crossings. Figure 7 shows the temporal

TABLE 11

Showing results of the female groups

Group	Approaches				Contacts				Crossings			
	Median	Range	Average	Average deviation	Median	Range	Average	Average deviation	Median	Range	Average	Average deviation
Control	3.0	0–10	3.23	2.34	2.0	0–6	2.27	1.66	5.0	1–9	5.05	2.2
Early inactive	0.0	0–2	0.42	0.63	1.0	0–3	1.25	0.75	1.0	0–5	1.5	1.42
Inactive	1.0	0–4	1.5	1.3	1	0–2	1.17	0.56	0	0–4	0.83	1.11
Late inactive	2.5	0–5	2.5	1.5	1.5	0–2	1.25	0.75	2.0	0–3	1.75	0.87
Early congestive	7.0	0–14	6.71	4.3	4	1–9	4.86	1.84	14.0	0–22	11.14	4.85
Cornified	5.0	0–14	5.43	2.93	3.0	0–9	3.14	2.1	16.0	2–24	14.61	4.21
Late cornified	2.0	1–4	2.33	1.11	4.0	2–5	3.66	1.11	10.0	2–14	8.66	4.43
Post-ovulative	0.5	0–4	1.17	0.83	3.5	0–9	4.0	2.33	2.5	0–12	4.66	4.21
Recuperative	1.0	0–4	1.3	0.99	1.5	0–5	2.0	1.4	0.5	0–4	0.9	1.3

TABLE 12

Showing results of the cornified group, the control group and a composite dioestrous group composed of the recuperative, the early inactive, the inactive and the late inactive groups

Number of animals	Approaches					Contacts					Crossings				
	Median	Range	Average	Standard deviation	Coefficient of variability	Median	Range	Average	Standard deviation	Coefficient of variability	Median	Range	Average	Standard deviation	Coefficient of variability
Cornified (21)	5	0–14	5.43	3.63	66.9	3	0–9	3.14	2.73	86.8	16	2–24	14.61	5.14	36.4
Composite (dioestrum) (32)	1	0–4	1.16	1.42	117.3	1	0–5	1.47	1.27	86.3	1	0–5	1.34	1.57	117.1
Control (cornified, no sex object) (22)	3	0–10	3.23	2.7	83.6	2	0–6	2.27	1.91	84.1	5	1–9	5.05	2.55	50.5

distribution of such activity in terms of the per cent occuring during successive five minutes.

The data on the behavior of the female animals were grouped according to the interpretation of the vaginal smears taken immediately after the test periods. Of the 69 animals tested with a male in the reward compartment 21 were found to have been in the cornified stage at the time of testing. It has been commonly observed

TABLE 13

Showing reliability of the difference between the average crossings of the control, the cornified and the composite dioestrous group

GROUP	STANDARD DEVIATION OF THE DIFFERENCE	THE DIFFERENCE	DIFFERENCE / S.D. OF DIFF.	CHANCES IN 100 OF A TRUE DIFFERENCE GREATER THAN 0
Cornified and control (1)	1.35	9.1	6.74	100
Cornified and composite (dioestrum) (2)	1.16	12.8	11.0	100
Composite and control	0.61	3.71	6.08	100

(1) These two groups differ only in that the incentive compartment contained a male during the testing of the cornified group and was empty during the testing of the control group. Animals of both groups were in the cornified stage.

(2) These two groups differ only with respect to oestrous condition, the first representing oestrum, the second dioestrum. A male was in the incentive compartment when both groups were tested.

that the most active sex behavior on the part of the female occurs at this time. Our results are in accordance with such observation. Animals in this group crossd, on the average, more than ten times as often as those in the four groups which together may be called the composite dioestrous group. The difference between the averages is a true one being over five times the standard deviation of that difference. Animals in dioestrum not only showed little tendency to cross (13 of the 28 did not cross at all) but were not inclined to approach or make contact with the grid more than once. The average number of approaches was 1.16, of contacts, 1.47. For the cornified group the figures are 5.43 and 3.14. The differences are reliable though not so great as in the case of the crossings. With respect to the time during the testing period when the animals crossed most often the cornified and dioestrous differ. Animals in the former group tended to cross more and more as the test period proceeded. Over half of the crossings occurred during the final six minutes. Just the reverse is true for animals in dioestrum. Such animals not only crossed less often but were less and less inclined to cross as the test period progressed. Over half of the crossings occurred during the first six minutes and less than a tenth during the final six minutes.

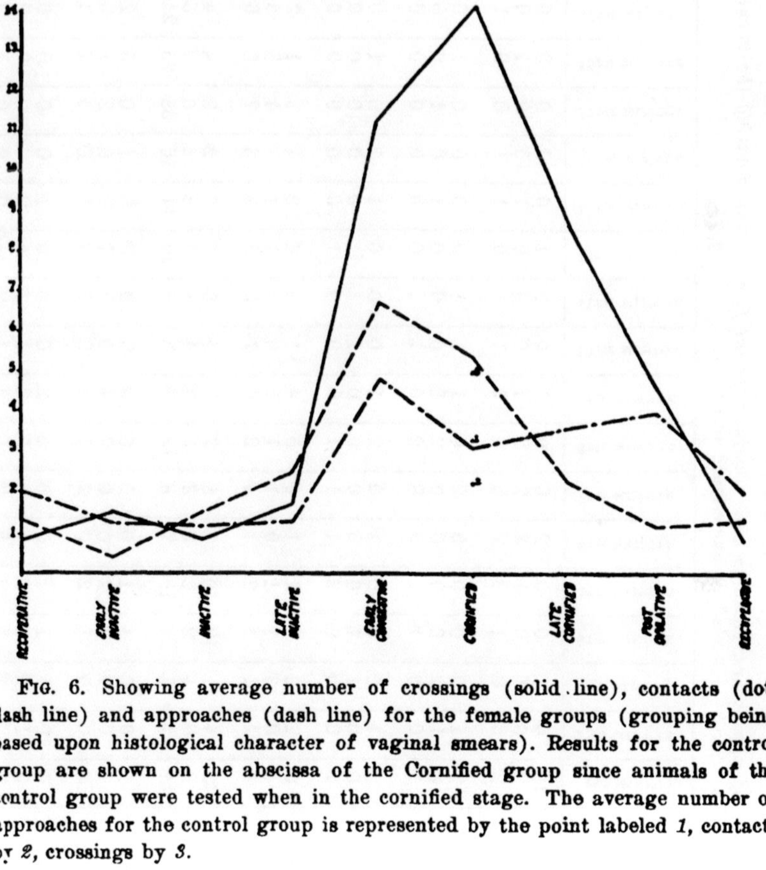

FIG. 6. Showing average number of crossings (solid line), contacts (dot-dash line) and approaches (dash line) for the female groups (grouping being based upon histological character of vaginal smears). Results for the control group are shown on the abscissa of the Cornified group since animals of the control group were tested when in the cornified stage. The average number of approaches for the control group is represented by the point labeled *1*, contacts by *2*, crossings by *3*.

Since we believe that the physiological condition related to the oestrous rhythm was the only factor which was different in the two groups we feel justified in the conclusion that the decided difference in behavior which has just been noted is related to the oestrous cycle.

The relationship between the cycle and the behavior recorded is demonstrated clearly by consideration of the progressive changes

TABLE 14

Showing temporal distribution of approaches, contacts and crossings for the female groups during the twenty-minute test period in one-minute intervals

GROUP	ACTIVITY	1ST MINUTE	2ND MINUTE	3RD MINUTE	4TH MINUTE	5TH MINUTE	6TH MINUTE	7TH MINUTE	8TH MINUTE	9TH MINUTE	10TH MINUTE	11TH MINUTE	12TH MINUTE	13TH MINUTE	14TH MINUTE	15TH MINUTE	16TH MINUTE	17TH MINUTE	18TH MINUTE	19TH MINUTE	20TH MINUTE
Early inactive	Approaches	10	0	1	0	0	2	0	0	1	0	0	1	0	0	0	0	0	0	0	0
	Contacts	2	0	1	0	1	3	2	2	1	0	0	0	1	0	0	1	0	0	0	1
	Crossings	2	0	0	1	3	2	0	1	2	1	1	0	1	1	0	0	1	0	1	1
Inactive	Approaches	2	1	0	0	0	0	2	0	1	1	1	0	0	0	0	1	0	0	0	0
	Contacts	0	1	2	0	0	0	0	2	0	0	0	0	1	0	1	0	0	0	0	0
	Crossings	1	0	0	1	1	0	0	0	0	1	1	0	0	0	0	0	0	0	0	0
Late inactive	Approaches	2	1	0	0	0	1	0	2	1	0	0	0	1	1	0	1	0	0	0	0
	Contacts	0	0	1	1	0	0	0	0	0	1	1	1	0	0	0	0	0	0	0	0
	Crossings	1	0	0	0	2	1	1	1	0	0	0	1	0	0	0	0	0	0	0	0
Early congestive	Approaches	3	2	3	1	2	1	6	5	4	1	3	4	2	1	3	1	2	2	0	0
	Contacts	1	0	3	1	1	0	1	1	1	0	4	1	1	1	1	3	5	3	3	1
	Crossings	4	1	3	1	2	1	4	2	4	7	5	3	3	3	7	8	6	7	2	3
Cornified	Approaches	12	4	11	12	10	7	8	7	2	5	6	7	6	2	2	4	4	2	2	5
	Contacts	3	0	1	2	2	3	3	3	3	4	3	0	3	4	6	9	5	8	2	1
	Crossings	14	11	12	12	10	7	8	12	9	9	9	10	11	16	35	26	24	30	28	14
Late cornified	Approaches	1	0	0	1	1	0	1	0	0	0	0	0	0	1	0	0	2	0	0	14
	Contacts	1	0	1	0	2	0	0	0	1	0	0	1	0	0	2	1	0	0	1	0
	Crossings	2	0	0	1	2	0	2	1	0	0	1	0	1	2	3	4	2	1	3	1
Post-ovulative	Approaches	0	0	1	1	1	0	0	0	2	0	0	0	1	0	0	0	0	0	0	0
	Contacts	0	1	2	1	1	1	2	3	2	1	3	4	2	0	0	0	0	0	0	1
	Crossings	2	2	2	1	2	5	2	0	2	1	2	1	0	0	0	1	3	1	1	0

TABLE 14 — *Continued*

Showing temporal distribution of approaches, contacts and crossings for the female groups during the twenty-minute test period in five-minute intervals

GROUP	ACTIVITY	1ST MINUTE	2ND MINUTE	3RD MINUTE	4TH MINUTE	5TH MINUTE	6TH MINUTE	7TH MINUTE	8TH MINUTE	9TH MINUTE	10TH MINUTE	11TH MINUTE	12TH MINUTE	13TH MINUTE	14TH MINUTE	15TH MINUTE	16TH MINUTE	17TH MINUTE	18TH MINUTE	19TH MINUTE	20TH MINUTE
Recuperative	Approaches	2	0	1	0	2	2	1	1	1	1	0	0	0	1	0	0	1	0	0	0
	Contacts	3	0	0	2	1	0	2	2	2	1	2	0	1	0	1	1	1	0	1	0
	Crossings	1	0	2	1	2	2	0	0	0	1	1	0	1	1	1	0	0	0	0	0
Control	Approaches	2	0	5	5	7	9	7	1	6	9	6	5	2	1	0	5	0	0	1	0
	Contacts	3	1	5	6	2	0	5	5	3	0	5	7	0	0	3	1	1	1	1	1
	Crossings	1	10	10	7	10	23	7	1	3	2	7	3	2	2	3	4	1	1	1	3
Composite (dioestrum)	Approaches	5	2	2	0	2	5	3	3	4	2	1	1	1	2	0	2	1	0	0	0
	Contacts	5	1	4	3	2	3	4	1	3	2	3	1	3	0	2	2	1	0	1	1
	Crossings	5	0	2	3	8	5	1	2	2	3	3	1	2	2	1	0	1	0	1	1

TABLE 15

Showing temporal distribution of approaches, contacts and crossings of the female groups during the twenty-minute test period in five-minute intervals

GROUP	ACTIVITY	0 TO 5TH MINUTE		6TH TO 10TH MINUTE		11TH TO 15TH MINUTE		16TH TO 20TH MINUTE		TOTAL
		Number	Per cent	Number	Per cent	Number	Per cent	Number	Per cent	
Control	Approaches	19	26.8	32	45.1	14	19.7	6	8.5	71
	Contacts	17	34.0	13	26.0	15	30.0	5	10.0	50
	Crossings	48	43.2	36	32.4	17	15.3	10	9.9	111
Early inactive	Approaches	1	20.0	3	60.0	1	20.0	0	0.0	5
	Contacts	4	26.6	8	53.3	1	6.6	2	13.0	15
	Crossings	6	33.3	6	33.3	3	16.6	3	16.6	18
Inactive	Approaches	3	33.3	4	44 4	1	11.1	1	11.1	9
	Contacts	3	42.9	2	28.6	2	28.6	0	0 0	7
	Crossings	3	60.0	1	20.0	1	20.0	0	0.0	5
Late inactive	Approaches	3	30.0	4	40.0	2	20.0	1	10.0	10
	Contacts	2	40.0	1	20.0	2	40.0	0	0.0	5
	Crossings	3	42.9	3	42.9	1	14.3	0	0.0	7
Early congestive	Approaches	11	23.4	17	36.2	13	27.7	6	12.8	47
	Contacts	6	17.7	3	8.8	8	23.5	17	50.0	34
	Crossings	11	14.1	18	23.1	21	26.9	28	35.9	78
Cornified	Approaches	49	43.0	29	25.4	23	20.2	13	11.4	114
	Contacts	8	12.1	16	24.2	16	24.2	26	39.4	66
	Crossings	59	19.2	45	14.7	81	26.4	122	39.7	307
Late cornified	Approaches	3	42.9	1	14.3	1	14.3	2	28.6	7
	Contacts	4	36.4	1	10.0	3	27.3	3	27.3	11
	Crossings	5	19.2	3	11.5	7	26.9	11	42.3	26
Post-ovulative	Approaches	3	42.9	2	28.6	1	14.3	1	14.3	7
	Contacts	5	20.8	9	37.5	9	37.5	1	4.2	24
	Crossings	9	32.1	10	35.7	3	10.7	5	21.4	28
Recuperative	Approaches	5	38.5	6	46.2	1	7.7	1	7.7	13
	Contacts	6	30.0	7	33.0	4	20.0	3	15.0	20
	Crossings	6	46.2	3	23.1	4	30.7	0	0.0	13
Composite	Approaches	12	32.4	17	45.9	5	13.5	3	8.1	37
	Contacts	15	31.9	18	38.3	9	19.1	5	10.7	47
	Crossings	18	41.9	13	30.2	9	20.9	3	7.0	43

in behavior accompanying the physiological change. As noted above, the interpretation of the vaginal smear record has led to the classification of the animals into eight groups. Four of these have been treated together (table 11), as representing the general dioestrous period and have been compared with a fifth, the cornified group representing oestrum, or heat. These groups represent the extremes, the periods of the least and the greatest sex activity. The

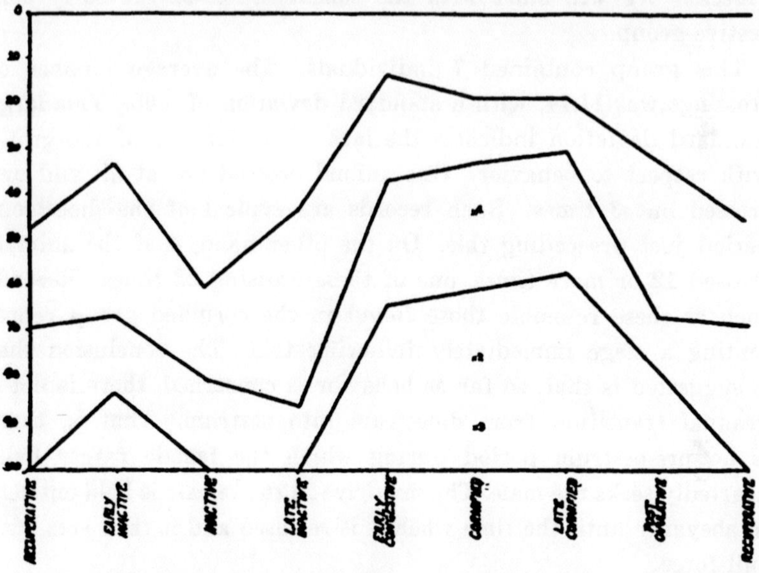

FIG. 7. Showing temporal distribution of crossings for the female groups. The upper line represents the percentage of crossing which occurred during the first five minutes of the test period. The middle line represents the percentage which occurred during the first ten minutes. The lower line represents the percentage which occurred during the first fifteen minutes. The points *1*, *2*, and *3* represent the percentages of crossing during the first five, ten and fifteen minutes for the control group.

remaining three groups represent transitional stages. The early congestive group contains animals which were no longer in dioestrum but had not yet reached the stage when only cornified cells were present in the smear. This period could be described as one of preparation. The two remaining groups, the late cornified and the post-ovulative, represent the transition from the condition of the greatest sex activity to that of the least. These groups are small and so might well have been combined into a single post-oestrous group. But since they can be definitely separated into two groups, the one slightly more closely related to the cornified period and the other to the inactive period, there seems no harm in letting them stand as originally classified.

Since the complete oestrous cycle is a continuously recurring phenomenon a description might be started at any point in the process. We will start with the consideration of the early congestive group.

This group contained 7 individuals. The average number of crossings was 11.14, with a standard deviation of 4.85. This large standard deviation indicates the lack of uniformity of the group with respect to behavior. One animal crossed not at all and one crossed but 3 times. Such records are typical of the dioestrous period just preceeding this. On the other hand, 4 of the animals crossed 12 or more times, one of these crossing 22 times. Records such as these resemble those found in the cornified group representing a stage immediately following this. The conclusion that is suggested is that, so far as behavior is concerned, there is not a gradual transition from dioestrum into oestrum. That is, there is no pre-oestrum period during which the female rather half-heartedly seeks the male. The sex drive in the female is held entirely in abeyance until the time when it is released and it then acts with full force.

Results for the cornified group have been mentioned above where they have been contrasted with those for the dioestrous groups. Of the 21 animals in the cornified group 16 crossed 14 or more times. The average number of crossings, 14.14 is the highest group average, verifying the assumption that the maximum sex activity accompanies the early cornified stage. The average number of approaches, 5.43 and of contacts, 3.14 are higher than the corresponding averages for any of the dioestrous groups, though not quite so high as found in certain transitional groups. That the average number of approaches and contacts is not decidedly greater than in the case of the transitional groups might be explained as follows. Whereas approaches, contacts and crossings are all measures of general activity and are all, therefore, likely to show higher values for the general oestrous period (including transitional stages) than for the general inactive period, the crossings, rather more than the approaches and contacts, measure also the specific orientation to this activity toward the sex object. Such orientation is, of course, characteristic of the cornified stage, or oestrum proper.

The late cornified group contains but 3 animals. Thus the averages are not particularly significant (approaches, 2.33; contacts, 3.66; crossings, 8.66). As far as the data go, the indication is that the drive is less intense or specific than earlier during the cornified stage.

The post-ovulative group is also rather small (6 animals). Since, however, the average number of crossings for the late cornified, the post-ovulate and the recuperative groups shows a consistent decline it seems reasonable to suppose that the downward trend is a true one and, moreover, a gradual one especially when compared with the more sudden onset of heat.

The recuperative group, containing ten animals, shows a low degree of activity, as 5 animals refused to cross and 2 crossed but once. The average is 0.9. Approaches and contacts are also low. This stage may be said, then, to mark the beginning of the dioestrum.

The early inactive group contains 12 animals. The average number of approaches and of contacts are lower for this group than for the preceding one, but there is a slight increase in the average number of crossings. While this increase is not great, there is a suggestion that it may possibly represent something other than a chance irregularity in the data regarding the temporal distribution of crossings. (Fig. 7). It is seen that in the other three groups representing the dioestrum (recuperative, inactive and late inactive) about half of the crossings occurred during the first five minutes, about a quarter during the second five minutes, the remaining quarter during the third five minutes and none at all during the final five minutes. In the early inactive group, on the other hand, there is found crossing rather more uniformly distributed throughout the twenty minutes. In other words, this group resembles the cornified group more than does any of the other dioestrous groups.

The suggestion, and it is only that, is that there may be a slight recurrence of oestrum-like behavior occurring at about the middle of the dioestrous period.

The inactive group, containing 6 animals, represents the lowest point in the cycle, so far as average number of crossings are concerned, although it is but a shade lower than the recuperative; 4 of the 6 animals failed to cross. The average number of contacts is also at its lowest point in this group. The average number of approaches shows a slight increase, an increase which is continued in the two succeeding groups.

The late inactive group (4 animals) may be said to close the dioestrous period. All three forms of activity recorded show a moderate increase, but the group is clearly a dioestrous group, giving a very different behavior picture from the rather hybrid group which follows it (the early congestive) with which this description of the cycle commenced.

The control group differs from all the other female groups in that there was no male in the incentive compartment during either the preliminary or the test period. This compartment was entirely empty. This group resembles the cornified group in all other respects, i.e., the animals were in the cornified stage and any differences in the behavior of the two groups must have been due to the presence of the sex object in the one case and its absence in the other. There are decided and reliable differences. The control group was much less active in terms of approaches, contacts and crossings. The average number of crossings for the control group was 5.05, for the cornified group, 14.14. The difference between these averages is 6.74 times the standard deviation of that difference. In the cornified group 17 of the 21 animals crossed more frequently than did any of the 22 animals in the control. The conclusion is that the behavior observed in the cornified group was largely determined by the presence of the male.

A further comparison, however, shows that important though the presence of the male is in determining the behavior of the animals of the cornified group, the physiological condition of the animals also plays an important part. It should be noted that the average number of crossings for the control group is higher than that for any other group except the cornified. In other words, a female in oestrum is more likely to undergo the electric shock in crossing to an empty compartment than is a female not in oestrum in crossing to a compartment containing a male. The implication is that the internal stimulation accompanying oestrum furnishes more drive than the external stimulation of the presence of the male. No claim is made for the specificity of the drive furnished by such internal stimulation. Wang and others have noticed the great increase in "spontaneous" activity accompanying oestrum. Such activity, irrespective of its source and normal mode of "satisfaction," would be reflected in data taken using the obstruction method.

These data indicate a definite relationship between the histological characteristics of the vaginal smear and the behavior of the female rat in a situation involving crossing an electric grid to reach the male. The onset of activity directed toward a sex object is apparently rather abrupt and corresponds to the very early part of the cornified stage. The cessation of this activity appears to be more gradual. During the period when the vaginal secretion contains predominantly epithelial cells and leucocytes activity is much reduced. None of the 32 females which were in this condition when tested crossed more than 5 times whereas only 3 of the 21 females of the cornified group crossed as few as 5 times or less while 16 crossed 14 times or more.

C. Comparison of the male and the female sex drive

To obtain a fair comparison of the sex drive in the male and in the female we should compare the two groups which represent the drive at its maximum in each sex. These are the twenty-four hours sex deprivation group for the male and the cornified group for the female. In terms of crossings there is not much difference. The average number for the females, 14.14 in twenty minutes is slightly but unreliably greater than that for the males, 13.45. In terms of approaches and contacts the female group has a more decided advantage, indicating, perhaps, that the behavior in the females was more definitely oriented toward the incentive compartment even though they did not cross much more frequently than did the males.

An important influence operated in the favor of the male drive.

This consisted of the behavior of the stimulus animal. When the females were tested the stimulus animals were, of course, males. These stimulus males were selcted for their docility and activity. No male that showed pugnacious tendencies or that was lethargic was used. Nevertheless, as stimulus objects they were hardly comparable to the female stimulus animals used in the testing of the males. These female stimulus animals were in the cornified stage when used. They were much more active than were the male stimulus animals and their activity was of the type which arouses and encourages sex behavior in the male. Were these stimulus females as passive as the male stimulus animals it seems not unlikely that the number of crossings of the male test animals would have been less and that there might even have been a reliable difference between the average number of crossings for the two sexes in the favor of the female.

For a detailed description of the type of female behavior referred to above see C. P. Stone (32, a. e).

A further difference in the operation of the sex drive in the male and in the female is suggested by an examination of the results for the control groups for the two sexes. We have noted above that the female control group (cornified, empty incentive compartment) crossed much more than did the female dioestrous groups (inactive stages, male in incentive compartment) indicating that the internal stimulation accompanying oestrum plays an important part in the behavior. That this is not the case for the males is seen by glancing at the results for the male control group. Animals in this group, although in the physiological condition during which the maximum sex activity is apparently displayed (twenty-four hours sex deprivation) did not display as much activity in terms of either approaches, contacts or crossings as did the animals in the other male groups. Even the "satiated" males (0 group) showed more activity than did the control. The indication is, then, that the external stimulus situation, the presence of the female in heat, is the dominating factor in the determination of sex activity in the male.

V. Summary of Conclusions

A. Male

1. The tendency of a male rat to approach a female rat in oestrum may be measured in terms of the number of times it will cross an electrical obstruction to reach the female within a given period of time. That the behavior of such males was dominated by the presence of the female in the present experiment was indicated by the results of a control group tested under conditions exactly like those of one of the test groups except that there was no female present in the incentive compartment. The males in this control group crossed the obstruction to the incentive compartment decidedly and reliably less often than did the animals in the comparable test group.

2. The tendency of a male rat to cross an electrical obstruction to a female rat in oestrum is at its low point immediately after a period (two hours in this case) during which the male has had access to and mated frequently with a female in oestrum. Recovery of the tendency from this low point is rapid during the first six hours after such a period of mating, almost as rapid during the second six hours after which it has reached a point only slightly below the maximum manifestation of this tendency which is found in an animal twenty-four hours after a period of mating.

3. Intervals of sex deprivation longer than one day do not increase the tendency of the male to cross. The data at hand suggest that this tendency decreases slightly from this point on to a deprivation interval of twenty-eight days, the longest interval studied. This decrease is not statistically reliable, however, and the only conclusion upon this point is that there is no increase in the tendency after the first day.

B. Female

1. The tendency of a female rat to approach a male rate may be measured in terms of the number of times it will cross an electrical obstruction to reach the male within a given period of time. That the behavior of the females in oestrum was dominated by the presence of the male in the present experiment was indicated by the results of a control group run under conditions exactly like those

of one of the test groups except that there was no male present in the incentive compartment. The females in this control group crossed the obstruction to the incentive compartment decidedly and reliably less often than did the animals in the comparable test group.

2. There is a very definite relationship between the histological character of the vaginal secretion of a female rat and its tendency, at the moment, to cross an electrical obstruction to reach a male, i.e., between the oestrous rhythm and behavior toward a sex object. The group of animals, which judging by the vaginal smear, were tested during the early part of the stage during which cornified cells only are found in the secretion displayed a decidedly and reliably greater amount of crossing and other activity oriented toward the male than did the group whose vaginal secretion was characterized by the presence of epithelial cells and leucocytes. In other words, activity directed toward a male is confined rather rigidly to a single period during the oestrous cycle, the period usually known as oestrum.

3. From the standpoint of behavior the onset of oestrum is sudden while its cessation is relatively gradual. The behavior of the group representing the transition into oestrum indicates that the group is not homogeneous but contains certain animals whose behavior resembles that commonly seen in dioestrum, and others whose behavior resembles that commonly seen in oestrum. When oestrum sets in it is apparently with practically full force. The groups representing transition out of oestrum show, in general, behavior midway between that seen in oestrum and that seen in dioestrum. The transition out of oestrum appears to be gradual.

C. Comparison of the male and the female

1. Since the strength of the tendency of the rat to cross an electric obstruction to reach a sex object is related to quite different physiological factors in the two sexes, the only fair comparison of the sexes from this standpoint is found in the comparison of those groups in which the physiological condition is such that the tendency was at its maximum for each sex. For the male this group is that tested after one day of sex deprivation. For the female it is

the group of animals tested when in the early cornified stage. The behavior record shows that of these two groups the female group was the more active although the difference is not statistically reliable. When it is considered that the incentive stimulus in the case of the male (a female in oestrum) was in most cases extremely active, displaying that type of activity commonly preceding mating whereas the stimulus animal for the female (a male) was decidedly less active and less frequently made such advances, the assumption seems justified that were it possible to equate the activity of the incentive stimuli the female group would be found to show a reliably greater amount of activity directed toward a sex object.

2. The behavior data indicate that sex activity is initiated rather more by the external stimulus situation in the male than in the female and rather more by internal stimulation in the female than in the male. The female rat in oestrum displays more activity in the form of crossing the electrical obstruction to the incentive compartment *even though that compartment be empty* than does the female in dioestrum *even though the compartment contains a male.* A male rat which has not mated for twenty-four hours (the interval which, of those studied, found sex activity at its maximum) will cross electrical obstruction less often to an empty incentive compartment than will a male of any of the other sex deprivation intervals studied (0, 6, 12 hours, 4, 7, 28 days) to an incentive compartment containing a female in oestrum.

Bibliography

(1) Allen, E. 1922. The oestrous cycle in the mouse. *Amer. Jour. Anat.*, xxx, 297-348.

——— 1923. Racial and familial cyclic inheritance and other evidence from the mouse concerning the cause of oestrous phenomena. *Amer. Jour. Anat.*, xxxii, 293-304.

(2) Bischoff, Th. L. W. 1852. Entwicklungsgeschichte des Meerschweinchens. Giessen.

(3) Blair, E. W. 1922. Contraction rate of the uterine musculature of the rat with reference to the oestrous cycle. *Proc. Amer. Assoc. Anat., Anat. Rec.* xxiii, 9-10.

(4) Corner, G. W. 1921. A review of some recent work on the mammalian reproductive cycle. *Jour. Mammal,* ii, 227-31.

CORNER, G. W. 1921. Cyclic variation in uterine and tubal contraction waves. *Amer. Jour. Anat.*, xxxii, 345-51.

(5) GREENMAN, M. J., AND F. L. DUHRING. 1923. Breeding and care of the albino rat for research purposes. The Wistar Institute of Anatomy and Biology, Philadelphia.

(6) HARTMAN, C. 1923. The oestrous cycle in the opposum. *Amer. Jour. Anat.*, xxxii, 353-95.

(7) HEAPE, W. 1900. The sexual season of mammals and the relation of the "Pro-oestrum" to menstruation. *Quar. Jour. Micr. Sci.* xliv, 1-70.

(8) HENSEN, V. 1876. Beobachtungen über die Befruchtung und Entwickelungen des Kaninchens und Meerscheinchens. *Zeit. f. Anat. u. Entwick.*, i, 213-72; 353-423.

(9) HOLDEN, F. 1926. A study of the effect of starvation upon behavior by means of the obstruction method. *Comp. Psych. Mon.*, 3, No. 17, 45 p.

(10) ISHII, O. 1920. Observations on the sexual cycle of the guinea-pig. *Biol. Bull.*, xxxviii, 237.

——— 1922. Observations on the sexual cycle in the white rat. *Anat. Rec..*, xxviii, 311.

(11) JENKINS, T. N., L. H. WARNER, AND C. J. WARDEN. 1926. Standard apparatus for the study of animal motivation. *Jour. Comp. Psych.*, vi, 361.

(12) KEYE, J. D. 1923. Periodic variation in spontaneous contraction of uterine muscle, in relation to the oestrous cycle and early pregnancy. *Bull. Johns Hopkins Hosp.*, xxxiv.

(13) KÖNIGSTEIN, H. 1907. Die Veränderungen der Genital Schleimhaut während der Gravidität und Brunst bei einigen Nagern. *Arch. f. Physiol.*, cxix, 553-70.

(14) LATASTE, F. 1892. Transformation périodique de l'épithélium du vagin des rongeurs. *Mem. de la Soc. de Biol.*, xliv, 765-69.

——— 1893. Rhythme vaginal des Mammifères. *Ibid.*, xlv, 135-46.

(15) LOEB, L. a. 1911. The cyclic changes in the ovary of the guinea-pig. *Jour. Morph.*, xxii, 37-70.

———b. 1914. The correlation between the cyclic changes in the uterus and the ovaries of the guinea-pig. *Biol. Bull.*, xxvii, 1-44.

———c. 1923. The mechanism of the sexual cycle with special reference to the corpus luteum. *Amer. Jour. Anat.*, xxxii, 305-43.

(16) LONG, J. A., AND H. M. EVANS. 1922. The oestrous cycle in the rat and its related phenomena. *Mem. Univ. Calif.*, vi, 1-148.

(17) MARSHALL, F. H. A. 1922. Physiology of reproduction. London: Longmans, Green & Co.

(18) MORAU, H. 1889. Des transformations épithéliales de la muqueuse

du vagin de quelques rongeurs. *J. de l'Anat. et de la Physiol.*, xxv, 275-97.

(19) Moss, F. A. 1924. A study of animal drives. *Jour. Ex. Psych.*, vii, 165.

(20) Papanicolaou, G. N. 1923. Oestrus in mammals from the comparative point of view. *Amer. Jour. Anat.*, xxxii, 284-92.

(21) Rein, G. 1883. Beiträge zur Kenntniss der Reifungserscheinungen und Befruchtungsvorgänge am Säugethierei. *Arch. f. Mikr, Anat.*, xxii, 233-70.

(22) Retterer, E. a. 1892. Sur la morphologie et l'évolution de épithélium au vagin mammifères. *Mém. Compt. rend. Soc. de biol.*, xliv, 101-7.

———b. 1892. Evolution de l'épithélium du vagin. *Ibid*, 566-8.

———c. Sur les modifications de la muqueuse utérine l'époque du rut. *Ibid.*, 637-42.

(23) Rubaschkin, W. 1905. Uber die Reigungs- und Befruchtungsprozesse des Meerschweincheneies. *Anat. Hefte, Wiesbaden*, xxix, 507-53.

(24) Seckinger, D. D. 1923. Spontaneous contractions of the fallopian tube of the domestic pig with reference to the oestrous cycle. *Bull. Johns Hopkins Hosp.*, xxxiv.

(25) Selle, P. M. 1922. Changes in the vaginal epithelium of the guinea-pig during the oestrous cycle. *Amer. Jour. Anat.*, xxx, 429-49.

(26) Simmons, R. 1924. The relative effectiveness of certain incentives in animal learning. *Comp. Psych. Monographs*, 2, No. 7.

(27) Slonaker, J. R. 1924. The effect of pubescence, oestruation, and menopause on the voluntary activity in the albino rat. *Amer. Jour. Physiol.*, lxviii.

———1925. Analysis of the daily activity of the albino rat. *Ibid.*, lxxiii.

———1926. Long fluctuations in voluntary activity in the albino rat. *Ibid.*, lxxvii.

(28) Smith, H. P. 1917. The ovarian cycle in mice. *Jour. Roy. Mic. Soc.*, p. 252.

(29) Sobotta, J. 1895. Die Befruchtung und Furchung des Eies der Maus. *Arch. f. Mikr. Anat.*, xlv, 15-93.

(30) Stockard, C. R. 1923. The general morphological and physiological importance of the oestrous problem. *Amer. Jour. Anat.*, xxxii, 227-83.

(31) Stockard, C. R., and G. N. Papanicolaou. a. 1917. The existence of a typical oestrous cycle in the guinea-pig, with a study of its histological and physiological changes. *Amer. Jour. Anat.*, xxii, 225-83.

———b. 1919. The vaginal closure membrane, copulation, and the vaginal plug in the guinea-pig, with further consideration of the oestrous rhythm. *Biol. Bull.*, xxxvii, 222-43.

(32) STONE, C. P. a. 1922. The congenital sexual behavior of the young male albino rat. *Jour. Comp. Psych.*, ii.

———b. 1923. Further study of sensory function in activation of sexual behavior in the male rat. *Jour. Comp. Psych.*, iii.

———c. 1924. The awakening of copulation ability in the male albino rat. *Amer. Jour. Physiol.*, lxviii.

———d. 1924. Delay in the awakening of copulatory ability in the male albino rat incurred by defective diets. *Jour. Comp. Psych.*, iv and v.

———e. 1926. The initial copulatory response of female rats reared in isolation from the age of twenty days to the age of puberty. *Jour. Comp. Psych.*, vi, 73-83.

(33) TSAI, C. 1925. The relative strength of sex and hunger motives in the albino rat. *Jour. Comp. Psych.*, v, 407.

(34) WANG, G. H. 1923. Spontaneous activity and the oestrous cycle. *Comp. Psych. Monographs*, ii, No. 6.

2. THE EFFECT OF SEGREGATION ON THE SEX BEHAVIOR OF THE WHITE RAT AS MEASURED BY THE OBSTRUCTION METHOD [1]

Marion Jenkins

I. Introduction

Through various experimental conditions it is possible to prevent the normal expression of sex behavior. Environmental conditions, for example, may be imposed which interfere with the consummation of the normal sex drive. *Normal* is here used in the biological sense to designate that form of sex behavior which ordinarily culminates in reproduction; or more specifically, the response of a male to a receptive female, or the response of a receptive female to a male. Interference with the operation of the normal sex drive is commonly brought about by the use of two methods: namely, isolation and segregation. Of these two methods, isolation is probably by far the greatest variation from the normal condition. It has usually been found impracticable to isolate completely. In fact, isolation rarely means more than eliminating visual and tactual stimuli from other animals of the same species. Usually little or no control of odors, sounds, etc., is exercised. The usual condition of isolation is separation from other members of the same species. In such a case, it may be possible for objects in the cage or human beings who care for the animals to elicit the sex response. Sex segregation, as we use the term, differs from isolation in that, although the normal outlet for the consummation of the sex drive has been prevented, nevertheless segregated animals still have an opportunity to exhibit sexual behavior toward members of their own sex.

These two methods of environmental control have been used by

[1] Reprinted with modifications from *Genetic Psychology Monographs, 1928,* 3: 457-571.

various investigators, but, in most cases, the studies have been limited to cage observations without experimental measurement. Studies of *isolation* will be considered first.

Interesting among these are the cage studies of Craig (2), who observed the behavior of doves which had been isolated long before puberty. He states that, with his conditions of isolation, the doves could "sometimes hear other doves, but they could never touch or see them." The four cases which he discusses in some detail did not manifest "masculine display behavior" until socially stimulated by the presence of a female dove. Then it appeared suddenly, but "human beings seemed to be the symbol"; the sex response appeared to be elicited chiefly by the hand and shoe. Three of the four finally "gave up their intimate friendship for human beings. But they gave it up slowly and gradually, showing interesting division of attention between human and dove companions." The other one never entirely gave up cooing to human beings. Indeed Craig says, "when I came near his cage, he still showed a desire to get out to me, and jealousy of other doves in my presence."

More extensive and detailed observations of the effect of isolation have been made by Stone (13). He observed the initial copulatory response of 21 female rats reared in isolation from the age of twenty days to the age of puberty. Nineteen individuals copulated within a few seconds after being put with the males. He therefore concludes that "the immediacy of the copulatory response in these females would seem to justify the conclusion that no environmental influences or factors beyond those necessary to insure normal somatic development are required to bring about sexual maturity as manifested by ability to perform the copulatory act during the receptive stage of œstrus." While these results show that the normal response is possible after isolation, they do not disprove the possibility of the occurrence of an abnormal response when the normal stimulus is not present.

Loutitt (7) made a study of the effects of isolation on the guinea pig, using both receptive and non-receptive females as the sex object; that is, he put males with both receptive and non-receptive females. He isolated males and females at 70 days of age and tested them at 150 to 300 days of age. He does not state the number of animals which he used. His results indicate that the receptive female is a more effective stimulus to sex activity than the non-receptive female. The following characteristics in the behavior of the male to a receptive female were observed: "periods of quiet are less and shorter, copulatory strokes are long and relatively slow, the male succeeds in intromission, he makes the first mount sooner and subsequent mounts at shorter intervals." Although Loutitt finds more activity in response to a receptive female, he states that after a period of isolation mounting occurs in response to the homosexual as well as to the normal stimulus. In these instances, both males attempted to play the part of the male. Of course this means very little since he does not state the length of the period of isolation. Homosexual behavior among females, on the other hand, he finds to be limited chiefly to a "short period before and during the receptive stage . . . . She takes the part of aggressor." When put with a male, the female exhibits "the ordinary behavior of the receptive female."

Cage studies on *segregated* groups show that conditions of segregation lead to the development of forms of sex behavior which one would naturally be led to predict on the basis of observation of animals kept in isolation. In a study of the congenital sexual behavior of the young male albino rat, Stone (9) segregated fifteen males from weaning till about the onset of puberty, or longer, six of which were normal, the others having the function of certain receptors destroyed by operation. In addition, eight animals were isolated at 21 days. While he does not discusss these two groups separately, he states in his general conclusions that "on the whole, the evidence brought forward indicates that the chief elements of the primary reproductive act appear within a sort period of time at puberty." Stone makes no attempt to test for the presence of homosexual behavior. His investigations were limited entirely to heterosexual stimuli.

The presence of homosexuality is discussed, however, by most other investigators who employ methods of segregation. Loutitt (7) states that when six male guinea pigs were kept together "at different times the same pig might play the part of male or female . . . The male having the part of female did not submit willingly but usually ran or fought with his aggressor.'' He considers that the ''exhibition of mating behavior toward animals of the same sex, at certain times or under certain conditions, is a normal response.'' However, since he at no point states that he has used other than isolated and segregated animals, this conclusion does not seem to be justified by his data, for it appears that his conclusions are not based upon observations of unsegregated animals — a condition which probably most closely approximates the normal.

Homosexuality among segregated guinea pigs has been observed by Avery (1). ''Male guinea pigs,'' he says, ''ofttimes mount each other. This occurs most frequently, however, when several males are kept together or when small young males are introduced into the pens of sexually mature males.'' Like Loutitt (7), he finds homosexuality among females chiefly at œstrum. ''Experimental observation directed to some 60 pigs over a period of about three months showed that homosexual behavior occurred in no less than ten to fifteen per cent of animals coming into heat during that time.''

Homosexuality has also been noted among pigeons. Several cases are found among the scattered observations of Whitman (16, p. 9). In one case, when two males were paired, one ''attempted courting during an entire month. He then began the course of incubation on the floor of the cage.'' Whitman (16, pp. 98-99) has further emphasized the importance of environmental conditions in the development of abnormal sexual behavior. ''Young birds raised under foster-parents of a different species are very apt to prefer a mating not with their own kind, but with a member of the species among which they have been reared.'' Confinement of two animals alone has a similar effect (16, pp. 99-100). ''The continued confinement and isolation of two birds will secure a mating when otherwise it would not occur. Confinement alone is sometimes efficacious, while at other times isolation from the sight and sound of other birds is necessary. Even inter-species mating can be secured in this manner.'' Whitman does not give the previous history of his animals nor the length of isolation or segregation period.

A few observers have noted that even unsegregated animals may exhibit homosexual behavior (16, p. 35). Whitman noted an instance of this in the case of passenger-pigeons. "Two of my male passenger-pigeons," he states, "mated with each other, notwithstanding they were in a pen where there were several unmated females desirous of mating." Such behavior has also been observed by Hamilton and Kempf in the case of unsegregated monkeys. Hamilton (3) states his conclusions in the following passage: "Homosexual behavior is normally an expression of tendencies which come to expression even when opportunities for heterosexual intercourse are present. Sexually immature male monkeys appear to be normally impelled toward homosexual behavior by sexual hunger. The fact that homosexual tendencies come to less frequent expression in the mature than in the immature male suggests the possibility that in their native habitat these animals may wholly abandon homosexual behavior . . . on arriving at sexual maturity." He finds that homosexuality among females is "rarely manifested in response to sexual hunger." Kempf's (5) findings are substantially in agreement with Hamilton's although he had but six monkeys in contrast with Hamilton's group of 28. Hamilton specifically states that he used both segregated and unsegregated animals under varying conditions of confinement (18 situations in all).

Nothing on mammals beyond the simple method of cage study was developed until Moss (8) introduced an experimental method for the objective study of animal drives. He utilized the principle of separating the test animal from the incentive by an electric grill, the animal being required to cross the same in order to obtain access to the incentive. The apparatus, technique, and results of Moss have been criticized by Jenkins, Warner, and Warden (4), who, in the same connection, describe an apparatus, which they have named the Obstruction Apparatus. The technique for the measurement of the sex drive has been described in detail by Warner (14) whose study on the normal sex drive was the first of the series to appear.

Warner was interested in the effect of sex deprivation on the sexual behavior of the male white rat and in relating the various stages of the œstrus cycle to the sexual behavior of the females. His animals were raised to maturity (150 days) unsegregated and, after a short period of segregation (35 days) necessary to his experimental technique (for details see his account), were tested by the obstruction method. In his investigation Warner found that long periods of sex deprivation seemed to weaken the normal sex drive in the male — a fact which led him to suspect the development of homosexual tendencies. However, Warner's procedure did not enable him to determine definitely whether changes in the strength of the normal sex drive are due to a transfer of the sex response to the homosexual stimulus. Futhermore, since he did not employ both sexes in the incentive compartment, he was unable to determine whether the response to the normal sex stimulus is essentially sexual, or primarily social in character.

The present investigation is an attempt to determine the effect of sex segregation upon the strength of the sex drive, and the relative incentive value of various kinds of stimuli (males, and receptive and non-receptive females). Animals were segregated before and after puberty, in order to determine the importance of the disintegration of sex habits in the development of homosexuality[2] and its effect upon variations in the strength of the drive. To this end, a mating period was provided for some groups one day before the test, to ascertain whether an interval of sex activity will effectively reinstate old sex habits or develop new ones, or whether, on the contrary, it only will tend to intensify homosexual habits formed as a result of sex segregation.

The present investigation involves two sets of essentially different conditions. First, there is the variation in the conditions *preceding* the test period, which is represented by sex segregation and mating periods. Second, there is the variation of the conditions *in* the test period, which is achieved by using three types of incentive animals—namely, males, and receptive and non-receptive females. Varying the incentive stimuli in this manner enables one to deter-

[2] By homosexuality is meant merely that certain behavior definitely sexual in character is called out by an animal of the same sex, even though at the same time similar behavior may be exhibited toward an animal of the opposite sex. The term has not been employed in the extreme sense in which homosexuality is supposed to preclude all normal or heterosexual behavior. Perhaps, the term homosexual tendency would be, in general, more exact than homosexuality and we have used the former term exclusively in dealing with the male, since here the tendency was somewhat less clearly marked than in the female.

mine not only the strength but the relative incentive value of different types of stimuli. A determination of the incentive value of various stimuli helps to account for the variations in the intensity of the rsponse to the normal sex stimulus; it also helps to ascertain whether the drive is predominantly sexual or whether it is due primarily to social or to other factors.

II. Method and Procedure

A. *Age and care of the animals*

The laboratory conditions which Warner has described in his study of sex behavior were observed in this experiment (cf. Warner, pp. 24-27), inasmuch as both are part of a comprehensive investigation of animal drives as above mentioned. The "Experimental Colony Strain" of albino rats of the Wistar Institute of Anatomy, Philadelphia, was used throughout. The ages at which the animals were tested were identical in both researches: that is, approximately 185 days, the range being from 175 to 196 days. The diet consisted of McCollum's mixture, plus weekly rations of greens. In the case of the unsegregated groups, an attempt was made to achieve constancy of conditions by always keeping an equal number of males and females in the same cage. This, of course, does not insure an equal number of receptive females in each cage at all times, but if a long period of time is considered, the average number of receptive females in a cage would be about the same.

B. *The apparatus*

The apparatus used in this investigation has been fully described in a recent article by Jenkins, Warner, and Warden (4). The modified grill (see Warner, pp. 28-31), and an improved device designed by Dr. T. N. Jenkins for the better manipulatory control of the apparatus to be described later (see Figure 1), were used.

The general procedure as presented by Warner was followed and, for the convenience of the reader, will be briefly outlined here. In the preliminary training period the test animal was placed in the entrance compartment and allowed to cross the grill five times to the stimulus animal. In the first four trials no current was passing through the grill. On the fifth trial the electric shock was introduced. The test animal was returnd as usual to the entrance compartment after the fourth trial. At the end of ten seconds the door leading into the obstruction compartment was opened, as in the first four trials. However, a change in procedure occurred at this point as a result of the introduction of the electric shock for the first time. When the animal entered the obstruction compartment, the door was lowered behind him and the switch closed. In this way, retreat to the entrance compartment was prevented and electrical stimulation continued until the animal had crossed the grill into the incentive compartment. As soon as the rat had traversed the grill, the door to the incentive compartment was closed behind the rat.

Since the function of the fifth trial is to permit the animal to associate shock with the obstruction compartment, it is probably more important to have conditions uniform for all individuals in this trial than in any other. The technique of closing the door behind the animal and throwing in the switch should, of course, be well standardized. Without this automatic device, the operation of closing the two doors was affected by the use of one hand, the other hand being required in throwing the switch which electrified the grill. Much skill is necessary in order to perfect this movement with even a fair degree of uniformity from one test to another. It is very important that the switch and entrance door be simultaneously closed, since each animal should receive the electric stimulation immediately upon leaving the entrance compartment. Otherwise it would be quite possible to condition the animals unequally. That is, there would be more or less stimulation according to whether the switch were closed before or after the door. Divided attention on the part of the experimenter between observing the animal and operating the various devices would tend to introduce widely differing conditions of stimulation for different animals. The automatic device (see Figure 1) successfully eliminates the irregularities arising from the performance of three separate operations by modifying the manipulatory procedure so as to minimize the importance of the personal equation in the conduct of the experiment.

First, this device permitted the switch to be closed before the fifth trial began. Second, simultaneity of the electrification of the grill and the closure of the entrance door was assured, since the closing of the door automatically turned on the current. The actual procedure followed in the fifth trial in this investigation was as follows: Switch S_1 is first closed. After the animal has remained

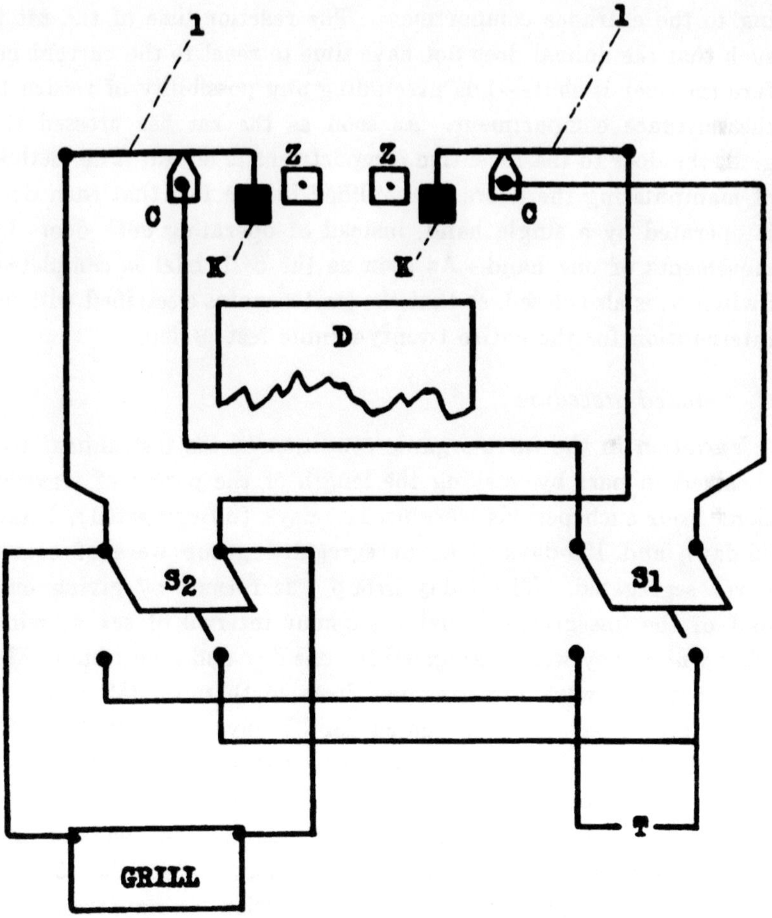

FIG. 1. Diagram of mechanism to facilitate the manipulatory control of the obstruction box. S_1 and S_2, switches through which current is conducted from the transformer T to the grill. D, manually operated door of the entrance compartment. l, l, contact springs. C, C, contacts. Z, Z, sponge rubber stops for manually operated door. K, K, sponge rubber tips of the contact springs.

in the entrance compartment for ten seconds, the door *D* is opened. When open, the door rests against the stops *Z, Z*. In this position the pressure of the door against the rubber tips *K, K*, breaks the electrical circuit at contact *C, C*, so that no voltage is impressed upon the grill. As soon as the animal is on the grill, the door is immediately lowered, and the initiation of the movement closes the circuit through contacts *C, C*. This prevents the rat from returning to the entrance compartment. The reaction-time of the rat is such that the animal does not have time to react to the current before the door is shut—thus precluding any possibility of return to the entrance compartment. As soon as the rat has crossed the grill, the door to the incentive compartment is closed. The method of manipulating the doors is simplified by the fact that each door is operated by a single hand, instead of operating *both* doors by movements of one hand. As soon as the fifth trial is completed, Switch S_2 is also closed, so that the grill remains electrified without interruption for the entire twenty-minute test period.

C. Detailed procedure

Variation in the intra-organic condition of the test animal was obtained in part by varying the length of the period of segregation. Four such periods were used: o-days (unsegregated), 1-day, 35-days, and 155-days. The unsegregated group were of course never segregated. The 1-day group was formed by giving one-half of the unsegregated males a 2-hour interval of sex activity, after which they were segregated for one day and then tested. The 35-day group were unsegregated from birth until 150 days, at which time they were segregated for 35 days and tested at 185 days. The 155-day group were unsegregated for the first 30 days of their lives. When they were 30 days of age they were weaned and segregated, being kept segregated until tested at 185 days of age.

It is important to note here that all groups except the 155-day group have had, up to the time of segregation, an opportunity to copulate freely. Copulatory activity of the 155-day group was precluded by segregating them at 30 days, since the earliest age at which copulation has ever been observed is 37 days. Stone's work is probably the most extensive study which has been done concerning the awakening of copulatory activity in the male albino rat. He used 56 males. He finds (11) "the mean age at which copulatory ability was first manifested was 48.06 days and the ages ranged from 37 to 72 days."

Segregation not only prevented copulatory activity, but precluded the possibility of any tactual stimulation by members of the opposite sex. No attempt was made to rule out olfactory stimulation. Inasmuch as the males were placed in the top row of cages and the females in the lower row, this tended to minimize visual stimulation. The cages were wire mesh, resting on heavy galvanized removable trays which extended out about two inches on all sides.

An endeavor was made to secure constancy of temperature in the laboratory, but seasonal changes could not be avoided. The unsegregated, 1-day, and 35-day groups were tested between January 20 and May 15, the 155-day animals from June 8 to July 19. In spite of this, the temperature conditions did not vary greatly. During the cool season the temperature was kept at approximately 75 degrees F. as detected by thermostat readings. In summer the laboratory was quite cool as it is located in the northeast corner of a modern brick building with cement floors. Owing to artificial heat in the building, the range of variation in humidity during the cool months may have varied somewhat from that of the summer months.

The male test animal

1. *Two-hour interval of sex activity.* At the outset of the investigation it was intended to have at least 20 animals in each group. However, in order to determine whether the effects of segregation are lasting, it was decided to give half the animals in each group a 2-hour interval of sex activity 24 hours before the test period. The 2-hour interval was placed 24 hours before the test period since Warner has shown (13) that this is approximately the point at which recovery from complete satiation (2-hour mating period) takes place. The writer had intended to combine the results of these two subgroups. It was found, however, that this dif-

ference in conditions so changed the results for these groups that it did not seem fair to combine them.

In this paper the data used include two groups of males which were not tested by the writer—namely, Warner's 1-day and 28-day groups. The 28-day group was the longest interval which he used, but was considered sufficiently near to 35 days to be used here. However, there is one very important difference between this group and those used by the writer. Warner's 28-day group was first segregated for 35 days, then given 2 hours of sex activity, and again segregated for 28 days. The writer's groups, on the other hand, were tested at the end of the 35-day period.

The 2-hour interval of sex activity requires further description. Half of the males for each segregation period was caged for two hours with an isolated receptive female. The males were chosen at random from among the group which happened to be 185 days of age at the time. Receptive females were chosen on the basis of behavior which is characteristic of oestrum, and the arousal of the copulatory response in indicator males.

It will be well, at this point, to explain the function of indicator males. These indicator males were selected because they showed exceptionally strong sex behavior, especially in the matter of frequent copulation. For example, when placed in a cage of females, an indicator male would manifest a minimum of general exploratory behavior; almost immediately he would begin to nose the genital region of the females and would copulate as soon as the female displayed characteristic mating behavior, such as the short run described by Stone (9). The use of indicator males for selecting receptive females for the 2-hour interval was found to have one rather serious disadvantage. Male sex behavior, such as nosing the genital region of the female, was often omitted. After a time, varying from two or three days to one month, the experienced indicator would no longer wait for the manifestation of the mating behavior on the part of the receptive female. When placed in a cage of females, he often mounted any female, and copulated. Consequently, if the arousal of the copulatory response in an indicator had been used exclusively as a criterion for the selection of receptive females, non-receptive, and even pregnant females would sometimes have been chosen. In order to overcome this difficulty, only those females were used which had aroused the copulatory response in at least two, and in the majority of cases three, indicator males.

To insure further the selection of a receptive female for the 2-hour interval, great care was used in observing their own active mating behavior. As the investigation proceeded the writer became more experienced in discriminating intensely active from fairly active receptive females. Since the 155-day group were the last tested, it is quite possible that the receptive females used in the 2-hour interval for the 155-day group were somewhat more intense in their mating behavior than those used with the 1-day and 35-day groups.

Having selected the receptive females to be used in the 2-hour interval, they were each placed in separate wire mesh mating cages (15″ x 8″ x 6″), the galvanized metal floor of which was covered with a thin layer of sawdust. The females were cage-adapted for at least 15 minutes before the male was admitted. The purpose of this period was to allow the female to explore thoroughly the cage so that she would display mating behavior as soon as the male was admitted. After being cage-adapted, any new stimulus, such as a male, tends to elicit exploratory behavior from the female.

These 2-hour intervals of sex activity usually occurred between 1:30 A.M. and 3:30 A.M., but a few were given as early as 11:30 P.M., or as late as 4:30 A.M. A close examination of the data showed that there was no correlation between the hour of sex activity and the number of copulations. Table 1 shows the distribution of the copulatory acts which occured in the 2-hour interval of sex activity. The writer was careful to secure accurate data for this interval.

It will be noted that, of the 79 animals given the 2-hour opportunity for sex activity, 58 did not copulate (Table 1). This should not be interpreted to mean that these animals did not manifest any sex activity whatever. In many cases they persistently nosed the genital region of the female and actively pursued her about the cage. They would sometimes stop for a moment and lick the penis, and then continue the pursuit. Often, this behavior very closely resembled that of males which were about to copulate. These males seemed on the verge of mounting and yet never actually mounted (although this described behavior lasted for some time). Meanwhile, the receptive female would usually display as intense mating behavior as those females which actually elicited the copulatory response. Indeed the receptive female would usually continue to display active mating behavior for some time (perhaps five min-

TABLE 1

Distribution table covering number of matings during the two-hour period prior to segregation of males

NUMBER OF COPULATIONS	FREQUENCY
No copulatons	58
1— 5	1
6—10	3
11—15	3
16—20	2
21—25	1
26—30	4
31—35	0
36—40	0
41—45	0
46—50	2
51—55	2
56—60	1
61—65	0
66—70	0
71—75	0
76—80	0
81—85	0
86—90	1
91—95	1
	79
Average for those which copulated	31.8
Average deviation	20.9

utes) after the male had relapsed again into a state of inactivity. After this, the female would continue to display activity, but of the random variety, as exploring the cage, washing her body, or nosing the male genitals.

The writer was astonished by the lack of copulatory activity on the part of the male in response to the aggressive mating behavior of the female. Two explanations may be advanced for this lack of activity. First, the male had not been cage-adapted before the 2-hour interval. Second, the mating behavior of the female may have lacked the persistence and specific pattern necessary to elicit copulation on the part of these males, although the females used in the mating period were apparently quite normal. In order to determine whether such conditions were of any great importance, a group of fourteen sluggish males, which showed no tendency to copulate, were again given an opportunity for sex activity after they had been tested in the obstruction box the next day. These males were given approximately one-half hour of cage-adaptation before admitting the receptive female. Furthermore, an attempt was made to elicit copulation from these males by placing them with an exceptionally excitable and persistent receptive female. In about half the cases the female used was one which was extremely exciting, having at one time elicited 88 copulations from a single male in two hours. Out of this group, only one copulated at all and after two copulations this male was quiescent for the rest of the period. The other males nosed the genitals for varying lengths of time and sometimes pursued the female, but never attempted to mount. The mating behavior of the female continued for some time after the male had ceased to respond.

The disparity between these results and Warner's as regards the number of males which did not copulate is somewhat puzzling. Much of Warner's work was done in a warm summer in a laboratory which had southeastern exposure and was therefore considerably warmer than the laboratory used in the present experiment. Weather conditions then may have been effective in causing a difference in the copulatory activity of the males of the two groups.

A second explanation should be considered. Dr. Warner has privately informed the writer that he only tested males which copulated on the ground that only males which copulate are normal. In this investigation, on the other hand, every male was tested which had had a 2-hour *opportunity* for sex activity. The reason for this difference in technique becomes apparent when the problems of the two investigators are considered. Warner, as he states in his monograph, was studying the sex behavior of the "normal" rat. The present investigation, on the other hand, was an attempt to determine the effect of segregation upon *all* rats regardless of whether they possessed the normal drive or not. Furthermore, we might expect that segregation, since it is an abnormal condition (that is, imposed by an external agency) would produce abnormal behavior. One of the important problems under investigation, therefore, was the development of abnormal behavior as the result of segregation. Obviously, then, it was impossible for the writer to ignore the behavior of rats which displayed a deficency in the type of behavior usually considered characteristic of the normal sex drive.

It is interesting to note in this connection that the distribution of copulations for the males which actually copulated is quite similar to that obtained by Warner. Furthermore, the present investigation is not unique in the observation that all males are not sexually aggressive. Avery (1) has clearly stated the same finding: "Extremes in sexual aggressiveness of adult males are readily observable in breeding experiments wherein one has an opportunity to study the behavior of many males. Some males will exhibit sexual aggressiveness day after day even though allowed to copulate almost daily, whereas others are sluggish for a day or two after repeated copulations with a female; *still others are never sexually aggressive although in good health and in other respects quite vigorous.* These points are of great importance in breeding experiments when an assumption is made by the experimenter that the male will copulate at the first opportunity given by the female. Such assumptions are precarious at all times and should be made only, if at all, when the male in question has been thoroughly and recently tested with receptive females." (Italics by the writer.)

2. *The test period.* The majority of the tests in the present investigation occurred between 11:30 P.M. and 3:30 A.M. However, some animals were tested as early as 9:00 P.M., and others as late as 5:00 A.M., but the range of hours was approximately constant for each group. The procedure of the test period has been discribed in considerable detail by Warner (p. 32-39).

However, it was not possible to follow this technique in every detail. It was necessary in the present work to quantify the temporal aspect of the criterion for sex behavior, although Warner had not done so. He states that "in the preliminary trials (the male) was not returned until he had nosed the genital region of the female, even though this occasionally involved a delay of several minutes." It will be recalled that, since he was studying the normal sex drive, he tested only animals which copulated in the 2-hour interval of sex activity. If the animals which did not copulate had been tested, it is quite probable that he would have discovered some extraordinary delays in nosing the genitals. It has previously been stated that the writer tested all animals regardless of whether they actually copulated or not during the 2-hour interval. Since many animals were tested which did not copulate it can be understood why the writer discovered in the preliminary experiment that animals very often failed to nose the genitals of a female stimulus even after a protracted period of delay. In one case a delay of 30 minutes was insufficient time for a stimulus animal to elicit a nosing response from the male test animal. For practical reasons a time limit was set for the period of delay in nosing the genitals. One reason for setting a definite limit upon the time spent by the test animal in the incentive compartment was to make it possible for the experimenter actually to carry out the investigation, since all animals must be tested between the age limits of 175 and 196 days. Furthermore, a technique should be broad enough to include all animals in the group. Finally, it is necessary to quantify the temporal aspect of the criterion in order that all tests might be comparable. A limit of three minutes in the incentive compartment was allowed for the nosing of the genitals. This interval of time was selected because, in the preliminary investigation, it was found that an animal which failed to nose in three minutes was not very likely to nose, even if given a much longer time.

In the actual test (20 minutes) Warner broadens his criterion to include "any specific contact between the two animals." It should be noted, however, that he always used a normal sex object as a stimulus animal whereas in the present investigation the incentive animal in some cases was an animal of the same sex. It was found that males which show marked homosexuality repeatedly nose the genital region of one another, whereas males which are definitely heterosexual do not continue such behavior long. The criterion in the preliminary period (nosing the genitals) was also employed in the test. A time limitation has the additional advantage of quantifying the temporal aspects of the criterion. A limit of 30 seconds was employed largely on the basis of preliminary work regarding the length of time at which nosing occurs. If neither animal nosed the genital region of the other within this interval of time, the test animal was removed and again placed in the entrance compartment.

The female test animal

With the exception of the 1-day series, the same segregation groups were used as in the case of the males — namely, unsegregated (0-day), 35-day and 155-day groups. It was found impractical to test a 1-day group of females because the œstrus cycle is such that it would involve an excessive waste of animals, for a female if given an interval of sex activity when in œstrum would usually not be in œstrum at the time of the test 24 hours later as the conditions of this group required.

The data presented in this study include only test animals which were in the cornified stage of œstrum. The criteria for the cornified stage were satisfied by means of the vaginal smear. Since the vaginal smear could not be taken until after the test (cf. Warner, p. 37), the female to be tested could not be chosen by this method. On the contrary, the experimenter relied on behavior criteria to determine whether the female was probably in the cornified stage of œstrum, in order not to waste an excessive number of females. With the exception of the unsegregated group, the behavior of the females toward indicator males could not be used as a criterion, since any association with a male would amount to interpolating an unsegregated interval into the segregation period. The only feasible method then was to choose test females on the basis of their general behavior. The writer would sometimes spend an entire hour in observing a single female to determine whether she was in œstrum. Various cues were found helpful in deciding which females were in œstrum, such as the gait of the animal, the intensity of her activity, jerky walk, nervousness when approached, etc. Under such circumstances, it was inevitable that an occasional female would be tested, and later found not to be in the cornified stage as indicated by the vaginal smear. All such animals were eliminated entirely.

1. *Copulatory interval.* On account of the short duration of the receptive period in the female, it was impossible to test animals which had had a 2-hour interval of sex activity 24 hours before the test period. Consequently, no female groups could be tested which were exactly comparable to the male groups which were given a

2-hour opportunity for sex activity. However, in the case of the 155-day group, an attempt was made to introduce an interval of sex activity before the test period. Since this group was segregated before puberty, there had never been any opportunity for the female to associate the male with the sex response. Consequently, the male would be a novel, but not necessarily a sex, stimulus for the female. In addition to a group which had had no sex activity, it was thought desirable to test also a group which had had an opportunity to become familiar with the male as a sex stimulus. When this group was formed, the intention was to combine the results of this group with the regular 155-day group. However, it was found that this interval of sex activity had such a profound effect upon the behavior of the females during the test period, that it was deemed inadvisable to combine the data for these two groups.

The procedure of giving the females an interval of sex activity prior to the test period is, in some respects, similar to that used in the case of the male groups. The female was placed in the regular mating cage and allowed at least fifteen minutes for cage-adaptation. Sex activity for each female was provided by one of two indicator males. In order to equalize as far as possible the opportunities for sex activity for all members of the group, provision for copulatory activity was restricted to the alternate use of the two indicator males.

Since this interval was given in order that the female might associate the male with the copulatory response, it became essential for actual copulation to take place. Five copulations were allowed each female. After the five copulations had taken place the indicator was removed and the female left undisturbed for one hour at the end of which she was removed and tested in the usual manner.

2. *The test period.* The conditions of testing the females were precisely the same as for the males. Immediately following the test period a vaginal smear was taken and a permanent slide made. These were later examined by Dr. Warner and are on file in the Animal Laboratory. The records of all animals whose smears did not definitely show the cornified stage were excluded from the data herein presented.

The incentive animals

For both the male and female groups, three types of stimulus animals were used in the present investigation — namely, male, receptive female, and non-receptive female. Throughout this paper, the term receptive is used interchangeably for female in the cornified stage, female in heat, and female in œstrum; the term non-receptive is used to designate a female in the recuperative stage of diœstrum. These stages were determined on the basis of mating behavior and the usual vaginal smear.

In order to keep the male stimulus as constant as possible, only three male stimulus animals were used throughout the entire investigation. These animals will be designated as No. 1, No. 2, and No. 3. In the case of the female groups, No. 3 was always used as the stimulus. This animal was selected because it possessed a very stereotyped form of behavior, and therefore the advantage of being a singularly constant stimulus. When the door of the restraining compartment opened, he would come out at once, and immediately approach the female and nose the genital region. As soon as the experimenter removed the female from the incentive compartment, he would return to the restraining compartment and remain there while the door was being closed.

The function of the incentive animal is to stimulate the test animal. In order to carry out this function to the greatest possible extent, the incentive animal was always allowed a period of adaptation in the obstruction box before the test period began. The object of introducing this period of adaptation was to reduce exploratory behavior to a minimum and hence increase its value as a stimulus. The incentive animal was placed in the box at least one-half hour before the test animal was admitted. The restraining door was kept open in order to permit the incentive animal to explore freely the incentive compartment.

The experience of testing over 300 rats has thoroughly convinced the writer that a period of adaptation to the apparatus itself (no incentive animal present) should also be allowed in the case of the test animal, in order to minimize the importance of exploratory behavior during the test period. The primary function of the preliminary trials is to permit the test animal to associate the stimulus with the incentive compartment. An attempt should be made to rule out exploratory behavior before the experiment begins, so that when the incentive animal is introduced, it will immediately become the dominant stimulus in the situation. The desirability of eliminating as far as possible exploratory behavior in the test period is supported by the following reasons: (1) Exploratory behavior probably tends to increase the dispersion around the averages — a fact which would decrease the reliability and introduce an element of heterogeneity into the results; (2) Exploratory behavior is likely to complicate the situation and render an analysis of the causes underlying variations in the strength of the sex drive less easy or valid; (3) In the case of small differences, the exploratory factor may even shift the relative positions of a set of averages to such an extent as to minimize the significance of the trends indicated on the face of the data.

The question might be raised as to the manner in which the three types of incentive animals could operate as differential incentives under our conditions. For, after all, the different types (receptive female, non-receptive female, and male) appear in general very much alike. From what we know of the visual capacity of the white rat it is hardly reasonable to suppose that one type could be distinguished from another on the basis of visual cues. It is true that the males were quite noticeably larger to the human eye than either type of female (receptive and non-receptive approximately the same size), but there is little or no reason to supposee that this difference served as a cue. Differences in odor seem to offer a more reasonable basis of explanation. Long and Evans (6) state that during the cornified stage of the œstrus cycle "there is usually a disagreeable odor attached to the vaginal secretion, which perhaps is a means of attracting and exciting the male." Also Watson (15) found that in the maze "adult rats showed preferences for entrances that contained the odor of the opposite sex." The obstruction apparatus is constructed so as to favor the passage of odor through the tunnel to the test animal — both incentive compartments (*C, D*) being closed by a glass cover during the test, and the door between perforated so as to permit the odor of the incentive animal to pass freely through. However, we have no evidence of differences in odor between male and non-receptive female sufficiently marked to serve as a cue. There may or may not be such. Whether, then, differences in odor between the three types of incentive animals operating *directly* upon the test animal through the 10-inch tunnel was an important factor in determining the nature of the response, cannot be certainly known from our results. We were interested primarily not in an analysis of the incentive animal as a stimulus (although this is a problem of importance in itself) but in the behavior of the test animal to these three types of incentive stimuli, assuming for the moment their differential character.

The specific manner in which the incentive operated as an excitant becomes a problem in this connection only when the fact is taken into account that the test animal is in the entrance compartment at the moment of response and is thus separated from the incentive animal. This does not mean, however, that we must suppose that the response of the test animal to the incentive must be based upon necessarily internal sensory cues operating at the moment. It is much more reasonable to believe that during the five preliminary crossings to the incentive (each animal was tested but once and to only one of the three types of incentive animals), the incentive compartment and the specific type of animal contained therein became associated with the act of crossing over. The incentive-drive value of the incentive animal is thus operating as an excitant to the test animal before the test proper is begun and continues throughout the test as the animal crosses over and comes in direct contact with the incentive animal again and again. The excitation so aroused by the incentive animal is thus cumulative, the nature and amount of the excitation depending upon the type of incentive animal used in testing any given animal or group. Under our conditions no animal was required to discriminate between incentive types in making a response since only a single type was present during a given test and for a given animal.

The writer believes that it is quite possible to furnish a still simpler interpretation of differences in incentive power of the various types of stimulus animals (as indicated by number of crossings, etc.). It is quite reasonable to suppose that the preliminary trials, as well as those that follow, cause changes in internal stimulation — an after-effect consequent to stimulation in the incentive compartment. An incentive which calls forth sex behavior on the part of the test animal probably sets up internal activity which continues or perseverates for a time after the animal has been returned to the entrance compartment. The effect of this increased excitement is greater activity. This would tend to elicit the crossing response, even if no association were formed between the incentive compartment and crossing over, because the obstruction apparatus is so constructed that crossing over to the incentive compartment is one of the simplest and most natural responses for the animal to make.

III. Results for Males

The main results for males will be presented in four general tables. Table 2 indicates the grouping of the 197 male rats tested. The distribution of approaches, contacts, and crossings for the eighteen groups is given in Table 3. The medians, ranges, averages, mean deviations, coefficients of variation, and temporal medians are shown in Table 6. The approaches, contacts, and crossings as distributed during the 20-minute test (in one-minute intervals) are given in Table 4. To these general tables, special tables will be added when necessary.

The four tables mentioned above give an inadequate picture of the situation. The many interrelated problems involved require the designation of the various trends and their subsequent evaluation through an analysis of the data by means of intercomparison and special statistical and graphic methods. Such an analysis will be made by treating the results under four headings:

A. Variations in the strength of the drive and the relative incentive values of the stimuli. This topic involves a study of the effects produced by two sets of variables. First, there are the changes in the *strength* of the drive which result from *varying the conditions (segregation periods) preceding the test.* Secondly, there are the effects produced by *varying the conditions (incentive stimuli) in the test*, due to the use of the three types of stimuli for each period of segregation. Varying the incentive stimulus, as was done in this experiment, furnishes data which are of great importance in the explanation of the drive for various segregation periods.

B. Individual differences.

C. Persistence of the drive as shown by the temporal distribution of drive behavior during the test.

TABLE 2

Showing male groups

Number of animals	Period of sex deprivation	Object in incentive compartment
13	0 hours	Female — cornified stage*
10	"	Female in diœstrum — recuperative stage†
10	"	Male
15	0 hours plus 2 hours sex activity plus 1 day deprivation	Female — cornified stage
11	"	Female in diœstrum — recuperative stage
10	"	Male
20	28 days‡	Female — cornified stage
10	35 days	Female in diœstrum — recuperative stage
10	"	Male
20	35 days plus 2 hours sex activity plus 1 day deprivation‡	Female — cornified stage
10	"	Female in diœstrum — recuperative stage
10	"	Male
5	155 days	Female — cornified stage
10	"	Female in diœstrum — recuperative stage
10	"	Male
5	155 days plus 2 hours sex activity plus 1 day deprivation	Female — cornified stage
8	"	Female in diœstrum — recuperative stage
10	"	Male

*Throughout this paper the writer has used "female in cornified stage" and "receptive female" interchangeably.

†Throughout this paper the writer has used "female in diœstrum" and "non-receptive female" interchangeably.

‡These groups were tested by L. H. Warner and are reported in his monograph on sex behavior.

D. The effect of the mating period upon the test 24 hours later.

The standard measures dealt with in these topical discussions are average number of crossings, indices of the strength and incentive value of the drive, variability, and persistence. In view of the fact that frequent reference will be made to these measures in different

TABLE 3

Frequency distributions covering approaches, contacts, and crossings for the various periods of sex deprivation in that male rat

DEPRIVATION IN THE MALE RAT

Group: Unsegregated

Stimulus animal	Female—cornified stage									Female in diœstrum—recuperative stage									Male								
Deprivation of test animal	0 days			1 day			Total			0 days			1 day			Total			0 days			1 day			Total		
Number of responses	Approaches	Contacts	Crossings	Approaches	Contacts	Crossings	Approaches	Contacts	Crossings	Approaches	Contacts	Crossings	Approaches	Contacts	Crossings	Approaches	Contacts	Crossings	Approaches	Contacts	Crossings	Approaches	Contacts	Crossings	Approaches	Contacts	Crossings
0	4	4	2	8	10		12	14	2	8	6		4	5		12	11		7	7		6	4		13	11	
1	1	2	1	3	2		4	4	1	1	2		5	3		6	5		1	3	1	2	2		3	5	1
2	2	1		3	2		5	3		1	1		1	1		2	2		1			1	1	1	2	1	1
3			1						1		1	1		1			2	1				1		2	1		2
4	1	2		1	1		2	3				1						1			1			2			3
5	1	3					1	3				1						1	1		1		1		1	1	1
6	2	1	1				2	1	1			2	1	1		1	1	2			1		1	1		1	2
7			3			1			4			1						1			1		1	1		1	2
8	1		1			1	1		2			2						2			1						1
9															3			3			1						1
10						2			2						2			2						1			1
11						1			1						1			1			1			1			2
12			2			1			3						3			3			1			1			2
13						1			1						2			2									
14						3			3																		
15						1			1			2						2									
16						1			1																		
17	1					1	1		1																		
18						1			1												1						1
19			2						2																		
20																											
21																											
22																											
23						1			1																		
24																											
Number rats in group	13			15			28			10			11			21			10			10			20		

Group: Segregated 35 days

Stimulus animal	Female—cornified stage						Female in diœstrum—recuperative stage									Male								
Deprivation of test animal	28 days			1 day			35 days			1 day			Total			35 days			1 day			Total		
Number of responses	Approaches	Contacts	Crossings	Approaches	Contacts	Crossings	Approaches	Contacts	Crossings	Approaches	Contacts	Crossings	Approaches	Contacts	Crossings	Approaches	Contacts	Crossings	Approaches	Contacts	Crossings	Approaches	Contacts	Crossings
0	6	6	2	4	6		3	4	1	6	8		9	12	1	4	8	2	4	4		8	12	2
1	5	4	2	6	6		2	3		3	2		5	5		3	2	1	2	2		5	4	1
2	1	4	1	3	4		1	2	1	1		1	2	2	2			1		2			2	1
3	2	1		1	2		4	1					4	1					1			1		
4	3	2	1	4	1	1										2		1				2		1
5		1	1	1					1						1	1		1	1	1	1	2	1	2
6	1	2			1	1			2			1			3				1		1	1		1
7									1						1						1			1
8												1			1				1		2	1		2
9	1					2			1			1			2						1			1
10	1		1						1						1						2			2
11			1			2						1			1			1						1
12			1			1						2			2			1						1
13			1			2			1						1									
14			3						1			1			2			1		1			1	1
15						4						2			2			1						1
16			2			3															1			1
17			1			1																		
18						2																		
19			1																					
20						1																		
21			1																					
22			1																					
23																								
24																					1			1
Number rats in group	20			20			10			10			20			10			10			20		

Group: Segregated 155 days

Stimulus animal	Female—cornified stage									Female in diœstrum—recuperative stage									Male								
Deprivation of test animal	155 days			1 day			Total			155 days			1 day			Total			155 days			1 day			Total		
Number of responses	Approaches	Contacts	Crossings	Approaches	Contacts	Crossings	Approaches	Contacts	Crossings	Approaches	Contacts	Crossings	Approaches	Contacts	Crossings	Approaches	Contacts	Crossings	Approaches	Contacts	Crossings	Approaches	Contacts	Crossings	Approaches	Contacts	Crossings
0	4	1		4	2		8	3		5	2		3	1		8	3		6	2		7	4	2	13	6	2
1	1			1	1		2	1		2	3		3	3		5	6		1	3		2	3		3	6	
2		3			2			5		2	1	2	1	1	1	3	2	3	1				2		1	2	
3		1						1		1	2	1	1	2	2	2	4	3	1	1			1		1	2	
4											1	1			1		1	2		2	1			3		2	4
5			1						1			1			1			2	1	2		1		1	2	2	1
6			1						1					1	1			1			2			1			3
7			1			1			2			2			1		1	3			2			1			3
8			1			1			2			1						1			2			2			4
9						2			2			1						1			1						1
10																											
11											1				1		1	1									
12			1						1																		
13						1			1												1						1
14																					1						1
15																											
16												1						1									
17																											
18																											
19																											
20																											
21																											
22																											
23																											
24																											
Number rats in group	5			5			10			10			8			18			10			10			20		

TABLE 4

Showing temporal distribution of approaches, contacts, and crossings of the male groups during the twenty-minute test period in one-minute intervals

Group	Stimulus animal	Deprivation of test animal	Activity	1st minute	2nd minute	3rd minute	4th minute	5th minute	6th minute	7th minute	8th minute	9th minute	10th minute	11th minute	12th minute	13th minute	14th minute	15th minute	16th minute	17th minute	18th minute	19th minute	20th minute
Unsegregated	Female—cornified stage	0 days	Approaches	0	2	2	1	1	2	2	2	1	1	0	2	3	3	6	4	5	6	3	5
			Contacts	0	4	1	1	1	0	0	3	2	0	0	0	0	3	3	6	3	3	1	2
			Crossings	11	7	8	7	3	4	6	5	4	5	3	7	3	4	6	5	3	4	4	2
		1 day	Approaches	1	0	0	1	0	2	1	0	1	0	0	1	0	2	2	0	0	1	0	1
			Contacts	0	1	0	1	1	0	2	0	1	0	2	0	0	1	0	0	1	0	0	0
			Crossings	22	16	15	12	11	14	10	13	12	5	11	9	8	8	8	9	5	5	5	4
	Female in diœstrum—recuperative stage	0 days	Approaches	0	0	2	0	0	0	0	0	0	0	0	0	0	0	0	0	0	0	0	1
			Contacts	2	0	0	0	0	1	1	0	0	1	0	0	0	0	1	0	0	1	0	0
			Crossings	8	7	6	5	5	5	5	2	2	2	3	3	5	2	5	3	2	2	3	2
		1 day	Approaches	1	1	1	1	0	1	1	1	1	0	1	1	0	1	0	1	1	0	0	0
			Contacts	1	0	2	0	0	0	0	0	2	1	4	2	0	1	0	1	0	0	0	0
			Crossings	11	11	11	6	7	8	4	8	6	6	3	6	5	3	4	4	4	5	3	5
	Male	0 days	Approaches	0	0	2	1	0	0	1	0	0	0	0	0	0	0	0	0	0	4	0	0
			Contacts	1	0	1	0	0	0	0	0	0	0	0	1	0	0	0	0	0	0	0	0
			Crossings	7	7	6	1	5	2	6	2	6	1	6	4	3	3	4	1	4	3	4	6
		1 day	Approaches	2	0	0	0	0	1	1	1	1	0	0	0	0	0	0	0	0	1	0	0
			Contacts	1	0	1	0	1	1	3	1	1	1	0	0	1	2	1	0	4	2	2	0
			Crossings	9	8	7	3	6	4	1	2	3	4	1	4	1	2	1	2	0	2	1	1
Segregated 35 days	Female—cornified stage	28 days	Approaches	4	2	0	3	0	0	1	2	4	1	1	1	3	6	4	5	2	4	4	3
			Contacts	3	0	1	0	0	1	2	0	2	4	4	6	5	4	3	4	0	0	1	0
			Crossings	2	7	3	1	2	4	6	7	11	13	7	11	11	9	12	21	19	24	22	19
		1 day	Approaches	7	2	1	3	6	6	5	1	2	2	1	1	2	0	2	1	0	1	2	1
			Contacts	3	1	1	2	2	3	3	1	1	1	2	1	0	3	2	2	0	1	1	0
			Crossings	12	17	13	16	19	17	15	9	11	10	10	13	16	13	10	7	12	17	15	17
	Female in diœstrum—recuperative stage	35 days	Approaches	2	0	0	0	2	2	0	0	3	2	0	2	0	0	1	0	0	1	1	0
			Contacts	0	1	0	0	2	0	0	0	1	0	0	1	0	1	1	0	1	0	2	0
			Crossings	7	8	6	4	3	4	5	6	5	1	5	3	2	2	2	2	2	3	0	2
		1 day	Approaches	1	0	0	0	0	0	0	1	0	0	1	0	0	0	2	0	0	2	0	1
			Contacts	0	0	0	0	0	0	0	1	0	0	0	0	0	0	0	1	0	0	0	0
			Crossings	9	8	8	5	5	7	9	4	3	2	7	6	4	3	6	5	2	3	3	5
	Male	35 days	Approaches	4	1	0	2	1	0	0	0	0	0	0	2	2	0	2	1	1	0	0	0
			Contacts	0	1	0	0	0	0	0	0	0	1	0	0	0	0	0	0	0	0	0	0
			Crossings	7	4	6	4	5	1	3	4	3	2	2	2	5	2	2	2	3	3	3	1
		1 day	Approaches	2	0	0	1	1	0	2	2	3	2	2	1	3	1	1	1	0	0	0	2
			Contacts	0	1	1	4	2	3	0	2	2	2	0	1	1	0	1	2	0	0	0	3
			Crossings	11	7	8	3	9	7	7	4	3	2	2	3	4	6	5	4	6	3	4	5
Segregated 155 days	Female—cornified stage	155 days	Approaches	0	0	0	0	0	0	0	0	0	0	0	0	0	0	0	1	0	0	0	0
			Contacts	0	0	0	0	0	0	2	0	1	0	0	1	0	1	1	0	2	1	0	0
			Crossings	5	3	4	5	2	2	0	3	1	0	4	3	1	0	0	2	1	1	1	0
		1 day	Approaches	0	0	0	0	1	0	0	0	0	0	0	0	0	0	0	0	0	0	0	0
			Contacts	1	0	1	0	1	0	0	0	0	0	0	0	0	1	0	0	1	0	0	0
			Crossings	5	4	3	1	2	3	4	3	3	3	1	3	1	1	3	2	0	1	3	0
	Female in diœstrum—recuperative stage	155 days	Approaches	0	0	0	1	0	0	0	2	0	1	1	0	1	0	0	0	1	0	0	2
			Contacts	2	0	0	0	0	2	1	1	1	2	2	2	0	1	1	0	1	0	4	6
			Crossings	6	7	6	4	4	5	4	4	3	1	2	1	2	3	3	1	2	2	1	2
		1 day	Approaches	2	0	0	1	1	0	1	0	0	0	0	0	0	1	0	0	1	0	0	1
			Contacts	1	1	2	0	1	2	0	0	1	2	0	0	0	1	1	0	2	3	0	1
			Crossings	6	3	2	4	2	4	3	3	2	0	1	1	3	1	0	2	0	1	2	1
	Male	155 days	Approaches	0	1	0	0	1	0	0	2	0	1	2	0	1	0	0	1	0	0	1	1
			Contacts	0	2	3	1	1	2	0	0	0	3	0	0	4	1	0	0	1	2	1	3
			Crossings	9	6	5	6	5	4	6	2	4	3	6	3	5	3	4	2	1	2	3	3
		1 day	Approaches	2	0	0	0	0	0	0	0	0	1	0	0	0	1	0	0	0	0	1	0
			Contacts	3	1	1	0	0	0	0	0	0	1	0	1	1	1	0	0	0	0	0	1
			Crossings	6	4	4	5	2	2	1	1	1	2	1	3	2	2	0	3	0	1	4	2

TABLE 5

Showing temporal distribution of approaches, contacts, and crossings of the male groups during the twenty-minute test period in five-minute intervals

Group	Stimulus animal	Deprivation of test animal	Activity	0 to 5th Minute		6th to 10th Minute		11th to 15th Minute		16th to 20th Minute		Total
				Number	Per Cent	Number	Per Cent	Number	Per Cent	Number	Per Cent	
Unsegregated	Female—cornified stage	0 days	Approaches	6	11.7	8	15.7	14	27.4	23	45.1	51
			Contacts	7	21.2	5	15.1	6	18.1	15	45.4	33
			Crossings	36	35.6	24	23.7	23	22.8	18	17.8	101
		1 day	Approaches	2	15.4	4	30.8	5	38.4	2	15.4	13
			Contacts	3	30.0	3	30.0	3	30.0	1	10.0	10
			Crossings	76	37.6	54	26.7	44	21.8	28	13.8	202
	Female in dioestrum—recuperative stage	0 days	Approaches	2	66.6	0	0	0	0	1	33.3	3
			Contacts	2	28.6	3	42.8	1	14.3	1	14.3	7
			Crossings	31	40.3	16	20.8	18	23.4	12	15.6	77
		1 day	Approaches	4	30.1	4	30.1	3	23.1	2	15.4	13
			Contacts	3	21.4	3	21.4	7	50.0	1	7.0	14
			Crossings	46	38.3	32	26.6	21	17.5	21	17.5	120
	Male	0 days	Approaches	3	37.5	1	12.5	0	0	4	50.0	8
			Contacts	2	66.6	0	0	1	33.3	0	0	3
			Crossings	26	32.1	17	21.0	20	24.7	18	22.2	81
		1 day	Approaches	2	28.6	4	57.1	0	0	1	14.3	7
			Contacts	3	13.6	7	31.8	4	18.2	8	36.5	22
			Crossings	33	53.2	14	22.6	9	14.5	6	9.7	62
Segregated 35 days	Female—cornified stage	28 days	Approaches	9	18.0	8	16.0	15	30.0	18	36.0	50
			Contacts	4	10.0	9	22.0	22	55.0	5	12.5	40
			Crossings	16	7.6	40	18.9	54	25.6	101	47.9	211
		1 day	Approaches	19	41.8	16	35.2	6	13.2	5	11.0	46
			Contacts	9	29.9	9	29.9	8	26.6	4	13.3	30
			Crossings	76	28.2	63	23.4	60	22.3	70	26.0	269
	Female in dioestrum—recuperative stage	35 days	Approaches	4	25.0	7	43.7	3	18.7	2	12.5	16
			Contacts	3	30.0	1	10.0	3	30.0	3	30.0	10
			Crossings	28	38.9	21	29.2	14	19.4	9	12.5	72
		1 day	Approaches	1	12.5	1	12.5	3	37.5	3	37.5	8
			Contacts	0	0	1	50.0	0	0	1	50.0	2
			Crossings	35	30.1	25	21.5	26	22.4	18	15.5	104
	Male	35 days	Approaches	8	50.0	0	0	6	37.5	2	12.5	16
			Contacts	1	50.0	1	50.0	0	0	0	0	2
			Crossings	26	40.6	13	20.3	13	20.3	12	18.7	64
		1 day	Approaches	4	16.7	9	37.5	8	33.3	3	12.5	24
			Contacts	8	32.0	9	36.0	3	12.0	5	20.0	25
			Crossings	38	36.8	23	22.3	20	19.4	22	21.3	103
Segregated 155 days	Female—cornified stage	155 days	Approaches	0	0	0	0	0	0	1	100.0	1
			Contacts	0	0	3	33.3	3	33.3	3	33.3	9
			Crossings	19	50.0	6	15.8	8	21.0	5	13.2	38
		1 day	Approaches	1	100.0	0	0	0	0	0	0	1
			Contacts	3	60.0	0	0	1	20.0	1	20.0	5
			Crossings	15	32.6	16	34.8	9	19.6	6	13.0	46
	Female in dioestrum—recuperative stage	155 days	Approaches	1	11.1	3	33.3	2	22.2	3	33.3	9
			Contacts	2	7.7	7	26.9	6	23.1	11	42.3	26
			Crossings	27	42.8	17	27.0	11	17.5	8	12.7	63
		1 day	Approaches	4	50.0	1	12.5	1	12.5	2	25.0	8
			Contacts	5	27.8	5	27.8	2	11.1	6	33.3	18
			Crossings	17	41.5	12	29.3	6	14.6	6	14.6	41
	Male	155 days	Approaches	2	18.1	3	27.3	3	27.3	3	27.3	11
			Contacts	7	29.2	5	20.8	5	20.8	7	29.2	24
			Crossings	31	37.8	19	23.2	21	25.6	11	13.4	82
		1 day	Approaches	2	28.6	1	14.3	3	42.8	1	14.3	7
			Contacts	5	50.0	1	10.0	3	30.0	1	10.0	10
			Crossings	21	45.6	7	15.2	8	17.4	10	21.7	46

TABLE 6

Showing results of the male groups

Groups	Stimulus animal	Deprivation of test animal	Approaches: Median	Approaches: Range	Approaches: Average	Approaches: Mean deviation	Approaches: Coefficient of variation	Contacts: Median	Contacts: Range	Contacts: Average	Contacts: Mean deviation	Contacts: Coefficient of variation	Crossings: Median	Crossings: Range	Crossings: Average	Crossings: Mean deviation	Crossings: Coefficient of variation	Crossings: Temporal median†
Unsegregated	Female—cornified stage	0 hours	2.7	0—17	3.9	3.4	109.2	2.5	0—6	2.5	2.1	105.2	7.5	0—19	7.8	4.8	77.0	7.9
		1 day	.9	0—4	.85	.9	132.9	.7	0—4	.7	.9	161.4	14.1	7—23	13.5	3.1	28.7	7.1
		Total	1.5	0—17	2.3	2.4	130.4	1.0	0—6	1.5	1.7	142.0	12.0	0—23	10.8	4.7	54.5	
	Female in diœstrum—recuperative stage	0 hours	.6	0—2	.3	.5	210.0	.8	0—3	.7	.8	142.8	7.0	3—15	7.7	3.0	48.8	6.5
		1 day	1.3	0—6	1.2	1.0	104.1	1.2	0—6	1.3	1.3	125.3	11.5	9—13	10.9	1.3	14.9	7.2
		Total	.9	0—6	.8	.9	141.2	1.0	0—6	1.0	1.0	125.0	9.8	3—15	9.4	2.8	37.3	
	Male	0 hours	.7	0—5	.8	1.1	172.5	.7	0—1	.3	.4	167.0	8.0	1—18	8.1	3.5	54.0	8.7
		1 day	.8	0—3	.7	.7	125.7	1.5	0—7	2.2	2.3	130.9	5.5	2—12	6.2	3.0	60.6	4.7
		Total	.7	0—5	.7	.9	161.4	.9	0—7	1.2	1.4	145.8	7.0	1—18	7.1	3.3	58.1	
Segregated 35 days	Female—* cornified stage	28 days	1.0	0—10	2.5	2.3	116.0	1.5	0—6	2.0	1.6	100.0	12.5	0—22	10.6	6.1	71.6	14.9
		1 day	1.5	0—10	2.3	1.7	91.3	1.0	0—6	1.5	1.2	100.0	15.0	4—20	13.4	3.3	29.9	9.6
	Female in diœstrum recuperative stage	35 days	2.0	0—3	1.6	1.2	93.7	1.3	0—3	1.0	.8	100.0	7.0	0—14	7.2	3.4	59.1	6.8
		1 day	.8	0—2	.5	.6	150.0	.6	0—1	.2	.3	190.0	12.0	2—15	10.4	3.3	39.7	7.2
		Total	1.2	0—3	1.0	1.0	125.0	.8	0—3	.6	.7	146.6	9.5	0—15	8.8	3.7	52.7	
	Male	35 days	1.3	0—5	1.6	1.6	125.0	.6	0—1	.2	.3	188.0	5.0	0—15	6.4	5.3	103.7	7.5
		1 day	1.5	0—8	2.4	2.5	130.4	1.5	0—14	2.5	2.8	140.4	9.0	5—24	10.3	3.9	47.4	6.9
		Total	1.4	0—8	2.0	2.1	131.5	.8	0—14	1.3	1.7	163.8	7.5	0—24	8.3	4.5	67.4	
Segregated 155 days	Female—cornified stage	155 days	.6	0—1	.2	.3	190.0	2.5	0—3	1.8	.7	48.8	7.5	5—12	7.6	1.9	25.0	5.0
		1 day	.6	0—1	.2	.3	190.0	1.5	0—2	1.0	.8	100.0	9.2	7—13	9.2	1.5	20.4	7.3
		Total	.6	0—1	.2	.3	190.0	2.2	0—3	1.4	.9	80.7	8.5	5—13	8.4	1.9	28.3	
	Female in diœstrum—recuperative stage	155 days	1.0	0—3	.9	.9	125.5	2.0	0—11	2.6	2.1	101.1	6.5	2—16	6.3	3.1	61.5	5.9
		1 day	1.3	0—3	1.0	.7	88.0	2.0	0—7	2.2	1.5	85.4	5.0	2—11	5.1	2.1	51.5	5.9
		Total	1.2	0—3	.9	.9	125.5	2.0	0—11	2.4	1.8	94.1	5.5	2—16	5.6	2.5	55.8	
	Male	155 days	.8	0—5	1.1	1.3	148.1	2.5	0—5	2.4	1.8	94.1	8.0	4—14	8.2	2.2	33.5	7.0
		1 day	.7	0—5	.7	1.0	178.5	1.3	0—3	1.0	.8	100.0	5.0	0—8	4.6	2.6	70.8	6.0
		Total	.8	0—5	.9	1.2	166.6	1.6	0—5	1.7	1.4	102.9	7.0	0—14	6.4	2.5	48.4	

*Temporal medians were computed by the writer and also mean deviations (not given in Dr. Warner's Monograph).

†The temporal median is that time of the twenty-minute test period at which fifty per cent of the total crossings have occurred (See Table 4).

parts of this paper, a little space will be given at this point to their definition and elucidation.

1. Discussion of the number of crossings will involve two kinds of averages which will be designated respectively as "total average" and "partial average." The former is the average number of crossings in the entire 20-minute test; the latter includes only the crossings that occurred during the last fifteen minutes of the test. One object for introducing the partial average is to furnish an additional check upon the trends indicated by the total averages. The primary reason, however, is that it tends to eliminate the disturbing effects of other forms of dynamic behavior, such as exploratory, which may be present in the early part of the test period. If we assume for the moment that the exploratory drive has not been completely exhausted during the preliminary period, more exploratory behavior would likely be exhibited during the early part of the test period than later. One might argue that this would tend to vitiate the results for the sex drive. The writer, however, believes that this is not the case. In the first place, since the conditions preceding the test period were the same for all the groups, the surplus exploratory behavior remaining after the preliminary period ought to be about the same for all the groups. The effect of exploratory behavior would probably be to increase the size of the average for each group; the relative differences between the groups should remain approximately the same. However, in order to furnish some statistical evidence on this point, the "partial average" was computed. The first five minutes were excluded on the assumption that exploratory behavior would rapidly diminish in the early part of the test period so that the sex drive would become progressively more dominant. A comparison of the total and partial averages shows quite strikingly that exploratory behavior, *even if present to some extent, does not affect the order of the differences between the various groups* (Table 8). In brief, there is a fairly high correlation between the two sets of averages.

2. The indices used in this paper are designed to depict two different aspects of behavior. First, there are the changes in behavior due to variations in conditions of segregation. Secondly, there are the changes that are due to variations in the incentive

stimulus. The relative changes occurring as a result of varying the segregation period (incentive stimulus constant) will be indicated by a *drive* or *motivation index.* Those that occur as a result of varying the incentive stimuli (segregation period constant) will be designated by an *incentive index.*[3] The general formula for the calculation of these indices is:

$$\text{Index} = \frac{Y - X}{X + Y}$$

where X is the standard, and Y is the comparison value. In other words, it is the difference between the averages for these two sets of conditions divided by their sum. The reason for using the sum of the standard and comparison values in the denominator instead of taking the standard as the unit was to keep the index from fluctuating between infinite values. By using the sum, the maximum numerical value is unity, so that the index can fluctuate only between the extreme values of + 1.0 and — 1.0. Positive values of the index indicate that conditions (segregation or incentive) are modified in such a way as to intensify the behavior represented by the standard; negative values indicate a decrease in the form of behavior represented by the standard. One advantage implied by this index is that the *value* of the index is in a sense independent of the averages, so that given *relative* differences appear equally significant. In other terms, the greater the magnitudes compared, the less significant the same absolute difference. Using absolute values, on the other hand, would tend to give too much weight to differences when the averages are large, and insufficient weight to differences when the averages are very small.

In the case of the motivation index, the average for the unsegregated group was always taken as the standard. This step was taken on the assumption that the unsegregated state approximates most closely to the normal condition for the animals in their natural habitat. In some respects, a better standard would have been an unsegregated group which had been given a short period of sex deprivation before the test, on the ground that sex deprivation

[3] The use and limitations of these indices and the formulæ for their calculation were suggested by Dr. T. N. Jenkins.

would more nearly equalize the intra-organic factors involved (cf. p. 485). Since optimal sex activity occurs after 24 hours of sex deprivation (Warner), the 1-day group, according to this criterion, would be the best norm for the males. However, the writer wished to compare the results for males and females . . . a fact which stressed the desirability of a norm common to both sexes. In the present case, this purpose is best answered by the unsegregated group, for no 1-day group of females was tested. In view of the temporal characteristics of the œstrus cycle of the female, it would be impractical to test a 1-day group of females on account of the excessive waste of animals as discussed in the previous section.

As the numerical value of the motivation index approaches nearer and nearer to unity, the change in the intra-organic factors, as represented by the form of behavior measured, becomes more and more complete. In the course of most work on drives by the obstruction method, it is not very likely that an index of + 1.0 or — 1.0 would occur, since this would indicate a complete change of the intra-organic state of the organism. However, in an extreme case of negative conditioning, it may be quite possible to obtain a motivation index of + 1.0 or — 1.0. When the motivation index is positive, it indicates that the intra-organic state of the animal has been modified so as to intensify the behavior represented by the average for the unsegregated group.[4]

In the case of the incentive index, the behavior toward the normal sex stimulus was always taken as the standard.

Then, for males,

X = average for the normal sex stimulus (receptive female), and

Y = average for the abnormal sex stimulus (male or non-receptive female).

For receptive females,

X = average for the normal sex stimulus (male), and

Y = average for the abnormal sex stimulus (receptive or non-receptive female).

An incentive index of zero indicates that the two stimuli compared are equally effective in eliciting a response. As the incentive

[4] To generalize beyond the data presented in this paper, these indices would require a statement of their probable errors. The writer relied, instead, upon the reliability of the difference as presented in Tables 7 and 10.

index approaches nearer and nearer to unity, the change in the incentive value of the stimulus, as represented by the behavior measured, becomes greater and greater. If there are no crossings to the abnormal stimulus, the index is — 1.0. A positive index points towards homosexuality. In fact it would be possible to define homosexuality in terms of a positive incentive index . . . the greater the index, the more specific the homosexual drive. *Small* negative values, on the other hand, would serve to indicate the development of homosexual tendencies. An index of + 1.0 would indicate that the heterosexual stimulus is wholly ineffectual in eliciting the measured response.[5] Other data, of course, have been employed throughout to determine the presence of homosexual tendencies (cf. sections on Individual Differences and the Effect of the Mating Period).

3. The magnitude and nature of the dispersion or variability is indicated by four measures. The magnitude is indicated by the average deviation, the coefficient of variation, and the range; and the nature of the scatter is indicated by the relative density. The first three measures are well known and need no further comment. The relative density[6] for each quartile of the distribution is calculated as follows: The first, second, and third quartiles are computed. From these values and the range for the entire distribution considered as a continuous series the range for each quartile is computed. Then the relative density is computed by finding the reciprocal of the quartile range and multiplying by 100. It may be expressed in the following form:

$$\text{Relative density} = \frac{100}{\text{quartile range (continuous series)}}$$

A table of relative densities is convenient in pointing out deviations

[5] The discussion of motivation and incentive indices does not imply that such indices have such a limited application as to apply only to problems in sex behavior. It is believed that they could be used equally well in connection with other drives, such as the hunger and the thirst drive. In the case of hunger, different periods of food deprivation would furnish one basis for a motivation index. On the other hand, the incentive index could be used in a study of the incentive value of various foods by using one kind of food, such as McCollum's diet, as the standard of comparison.

[6] Suggested by Dr. T. N. Jenkins.

from the normal form of the distribution, such as cases of bimodality.

4. The data for determining the persistence of the drive during the test period are furnished by the temporal distributions (Table 5 and Figure 6). If the number of crossings for a given animal falls off toward the end of the period, the drive may be said to be less persistent, as judged by behavior criteria, than when the rate of crossing is relatively constant throughout the test period, or actually increases.

An attempt was also made to determine the general or average effect of different periods of segregation by means of a combined score, obtained by pooling the results for the three incentive conditions. Such scores, however, involve several factors which tend to nullify their significance. These factors in certain cases tend to lower the average independently of changes in the average that can be assigned to variations in the strength of the sex drive. First, there is the factor of the physiological limit to the number of crossings which a rat can make in a test period of given length. In other words, we might say that the number of crossings is not a linear function of the strength of the drive. As the drive becomes stronger, the number of crossings approaches the physiological limit so that each additional increment to the strength of the drive is paralleled by smaller and smaller increments in the number of crossings. Therefore the absolute number of crossings as a measure of the strength of the sex drive tends to minimize the significance of the drive in intensely motivated animals. A second factor is negative adaptation to certain incentives used. For example, the 35-day females failed to cross as frequently to a receptive female as to an empty incentive compartment. In fact, it is the writer's opinion that the avoidance of the females in this case was evidence of a strong heterosexual sex drive in the animals of this group. Here it is quite plain that negative adaptation would tend to lower the value of the pooled result even though the animals may actually be quite active. Consequently, the method of pooling gives results which are spurious as measures of activity.

A. Variations in the strength of the drive and the relative incentive values of the stimuli

One day of segregation nearly doubles the strength of the normal sex drive,[7] as measured by the average number of crossings to a *receptive* female (Tables 7 and 8, and Figure 2). Unsegregated males, on the average, cross only 7.8 times, but when segregated for one day their average rises to 13.5. After one day the sex drive seems to diminish slowly in strength as the segregation period lengthens until at 155 days the activity is no greater than that of the unsegregated group. In fact, the motivation index shows a slight negative value for the 155-day group (Table 9 and Figure

TABLE 7

Reliability of the difference between the average number of crossings of males for various deprivation periods

GROUPS (deprivation periods)	STIMULUS ANIMAL	THE DIFFERENCE	STANDARD DEVIATION OF THE DIFFERENCE	DIFFERENCE / S. D. OF THE DIFFERENCE	CHANCES IN 100 OF A TRUE DIFFERENCE GREATER THAN 0
Unsegregated and 1 day	Cornified	5.7	1.94	2.93	99.8
	Diœstrum	3.2	1.27	2.51	99.4
	Male	1.9	1.81	1.04	85
1 day and 35 days	Cornified	2.9	1.97	1.47	93
	Diœstrum	3.7	1.43	2.58	99.5
	Male	.2	2.40	.08	53
35 days and 155 days	Cornified	3.0	2.00	1.50	93
	Diœstrum	.9	1.81	.49	69
	Male	1.8	2.27	.79	79
Unsegregated and 35 days	Cornified	2.8	2.37	1.18	88
	Diœstrum	.5	1.78	.29	62
	Male	1.7	2.51	.68	75
Unsegregated and 155 days	Cornified	.2	1.97	.10	54
	Diœstrum	1.4	1.69	.83	79
	Male	.1	1.63	.06	52
1 day and 155 days	Cornified	5.9	1.45	4.07	100
	Diœstrum	4.6	1.31	3.51	100
	Male	2.0	1.46	1.36	91

[7] By *normal sex drive* the writer means the sex drive to the *normal stimulus* or incentive animal. In the case of a male the normal incentive stimulus is the receptive female; in the case of the female the normal stimulus is the male.

Total and partial averages and average deviations for crossings of the male groups

Stimulus animal		Unsegregated		1 day		35 days		35 days—Sex activity		155 days		155 days—Sex activity	
		Total	Partial	Total	Partial	Total	Partial	Total	Partial	Total	Partial	Total	Partial
Cornified	*Av.*	7.8	5.0	13.5	8.4	10.6	9.7	13.4	9.6	7.6	3.8	9.2	6.2
	A.D.	4.8	3.5	3.1	2.6	6.1	•	3.3	•	1.9	1.8	1.5	1.4
Diœstrum	*Av.*	7.7	4.6	10.9	6.7	7.2	4.4	10.4	6.9	6.3	3.6	5.1	3.0
	A.D.	3.0	2.4	1.3	1.2	3.4	2.9	3.3	2.9	3.1	2.2	2.1	1.5
Male	*Av.*	8.1	5.5	6.2	2.9	6.4	3.8	10.3	6.5	8.2	5.1	4.6	2.5
	A.D.	3.5	2.6	3.0	2.3	5.3	3.7	3.9	3.1	2.2	2.1	2.6	1.4

TABLE 9

Motivation indices for the male groups

(Average for unsegregated group as the standard of comparison)

Stimulus	Cornified					Diœstrum					Male				
Deprivation Period	1 day	28 days	35 days—Sex activity	155 days	155 days—Sex activity	1 day	35 days	35 days—Sex activity	155 days	155 days—Sex activity	1 day	35 days	35 days Sex activity	155 days	155 days—Sex activity
Total	.267	.152	.264	—.012	.082	.172	—.033	.149	—.100	—.203	—.132	—.117	.119	.006	—.275
Partial	.253	.319	.312	—.136	.107	.185	—.022	.200	—.121	—.210	—.309	—.182	.080	—.037	—.375

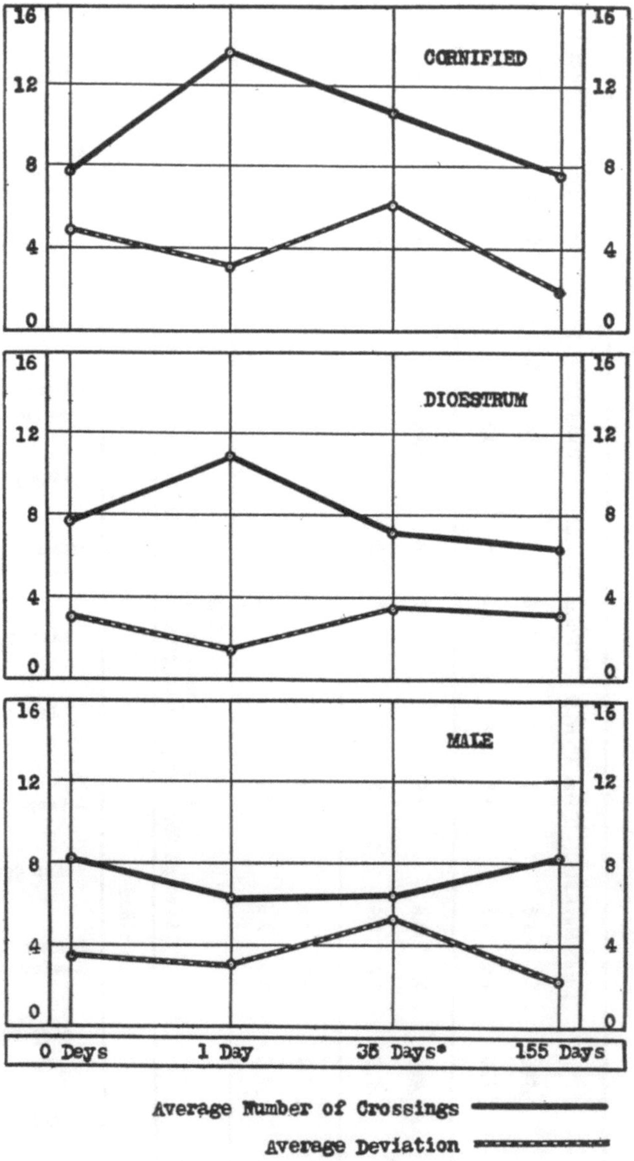

FIG. 2. Graphs of the average number of crossings and average deviations for males for various deprivation periods. Incentive stimulus indicated in upper right-hand corner of each diagram.

3). This becomes of more significance when it is considered that the plane of activity is low as compared with that of the unsegregated group. The number of approaches and contacts indicates that there is a preponderance of activity in the case of the unsegregated group, although the small number of crossings would suggest that they are quite effectually deterred by the grill.

The response to a non-receptive female shows somewhat similar relationships. The motivation index takes on a substantial positive value after one day of segregation; for longer periods it becomes increasingly negative. The rise in the number of crossings to a non-receptive female, which occurs after one day of segregation is practically as great as the drop from the first to the thirty-fifth day, the average for the unsegregated group being 7.7, the 1-day group 10.9, and the 35-day 7.2. After 35 days, the decline in the strength of the drive seems to be very slow, for the average falls only to 6.3 as the result of 155 days of segregation.

The response to a *male* incentive shows an almost complete reversal of trend from the results obtained in the case of the normal incentive. One day of segregation substantially decreases the intensity of the response to a male incentive animal. In fact, the motivation index shows a fairly large negative value, whereas those for one day of segregation in the case of the female incentives are strongly positive. Thereafter, the strength of the drive to a male gradually increases until at 155 days, the motivation index takes on a slight positive value. In other words, as the segregation period lengthens beyond one day, there is an increase in the stimulation value of the male.

A comparison of the responses to the three types of incentive stimuli shows some pronounced differences in the incentive values of the stimuli for various periods of segregation. In the case of the unsegregated group, all three types of incentive stimuli are about equally effective—a fact which indicates that there is not a very specific sex drive. One day of segregation, on the other hand, increases enormously the stimulus value of the receptive female as compared with the male (Table 10 and Figure 4). An inspection of the incentive indices shows that there is also a substantial increase in the incentive value of a non-receptive female (Table 11

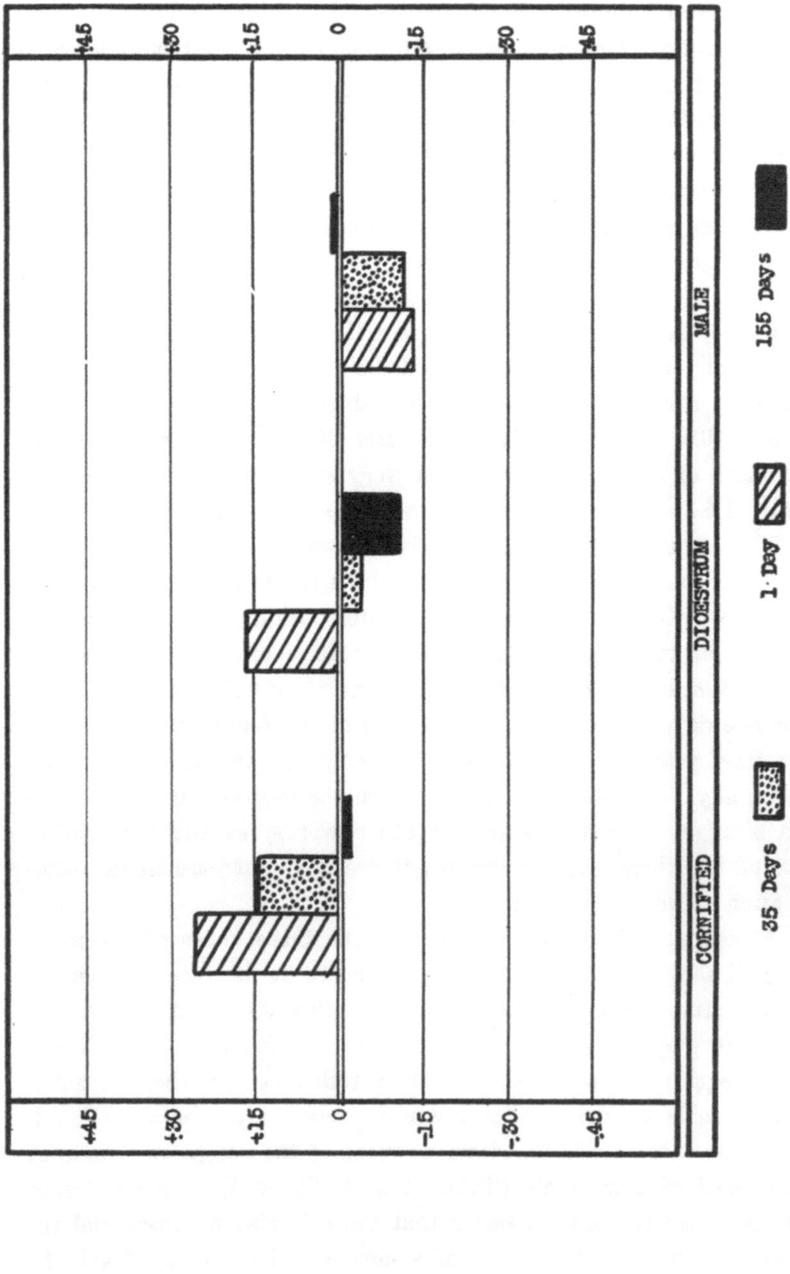

FIG. 3. Column diagram of the motivation indices for the male groups. The average for the unsegregated group is used as the standard of comparison.

and Figure 5). Although the three types of incentive animals rank in the same order after 35 days of segregation, the effectiveness of both types of female has diminished, the loss being greatest for the non-receptive female. The increase in the incentive value of the male, however, is somewhat obscured by the bimodality of the dis-

TABLE 10

Reliability of the difference between the average number of crossings of males for various stimulus animals

GROUPS (stimulus animals)	DEPRIVATION PERIOD	THE DIFFERENCE	STANDARD DEVIATION OF THE DIFFERENCE	DIFFERENCE / S. D. OF THE DIFFERENCE	CHANCES IN 100 OF A DIFFERENCE GREATER THAN 0
Cornified and Diœstrum	0 days	.1	2.03	.04	52
	1 day	2.6	1.11	1.34	91
	35 days	3.4	2.17	1.56	94
	155 days	1.3	1.61	.80	79
Diœstrum and Male	0 days	.4	1.80	.22	58
	1 day	4.7	1.27	3.70	100
	35 days	.8	2.49	.32	62
	155 days	1.9	1.50	1.26	89
Cornified and Male	0 days	.3	2.15	.14	56
	1 day	7.3	1.54	4.74	100
	35 days	4.2	2.70	1.55	93
	155 days	.6	1.37	.43	66

tribution. In the case of the 155-day group, differences in incentive value are not so great as in the 1-day and 35-day groups. In fact, the high degree of specificity to the receptive female has practically disappeared. Indeed the incentive index suggests homosexual tendencies.

The number of approaches and contacts (Table 6) for the 155-day group indicates that the degree of activity is practically the same and uniformly low for all incentive stimuli—a fact which shows among other things that the sex drive is rather weak as the result of this very long period of segregation.

An inspection of Table 8 shows that the trend of the partial averages parallels very closely that of the total averages. This fact, in conjunction with the reliability of the difference as presented in Tables 7 and 10, indicates that the directions of the observed differences are true.

It will be noted that the number of crossings is approximately the same for all three incentive conditions for the *unsegregated*

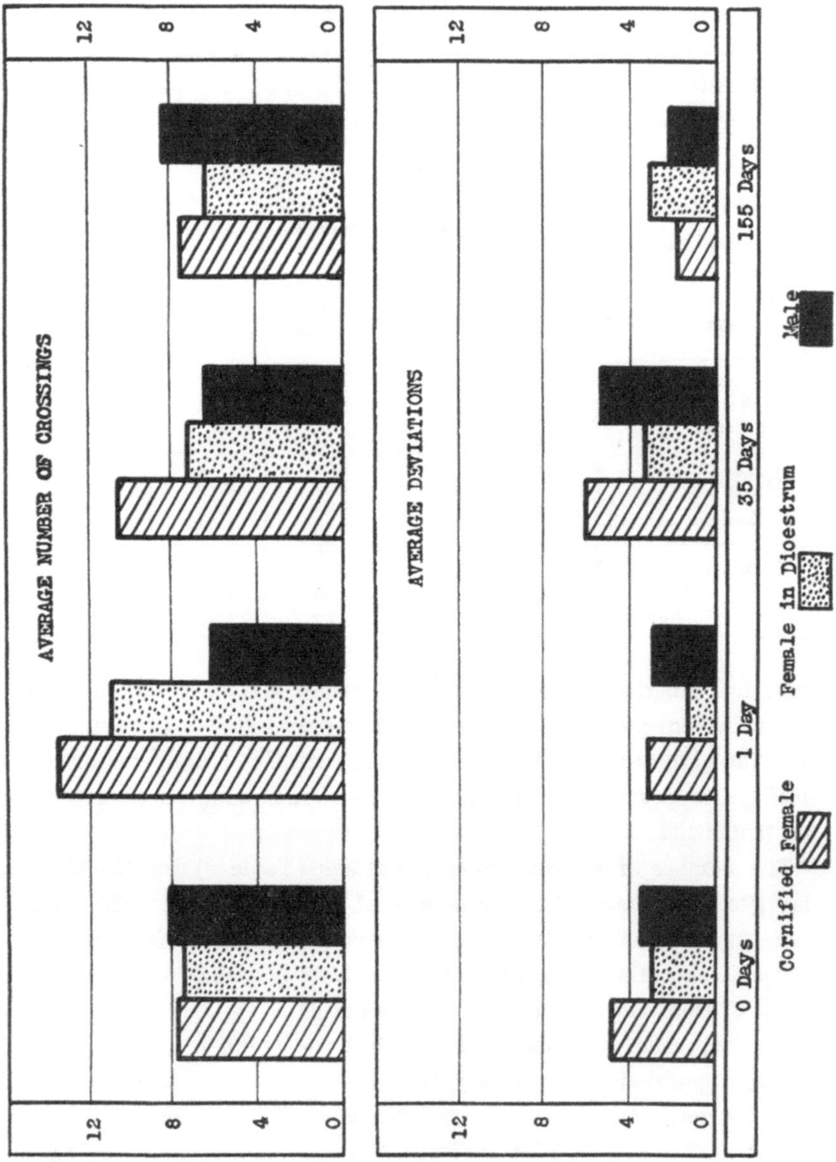

Fig. 4. Column diagram showing the average number of crossings and average deviations for males in re-

TABLE 11

Incentive indices for the male groups

(Average for the normal sex stimulus as the standard of comparison)

Deprivation period	Unsegregated		1 day		35 days		35 days—Sex activity		155 days		155 days—Sex activity	
Stimulus	Diœstrum	Male	Diœstrum	Male	Diœstrum	Male	Diœstrum	Male	Diœstrum	Male	Diœstrum	Male
Total	—.006	.018	—.106	—.370	—.191	—.247	—.126	—.130	—.093	.037	—.286	—.333
Partial	—.041	.047	—.112	—.486	—.375	—.437	—.163	—.192	—.027	.146	—.347	—.425

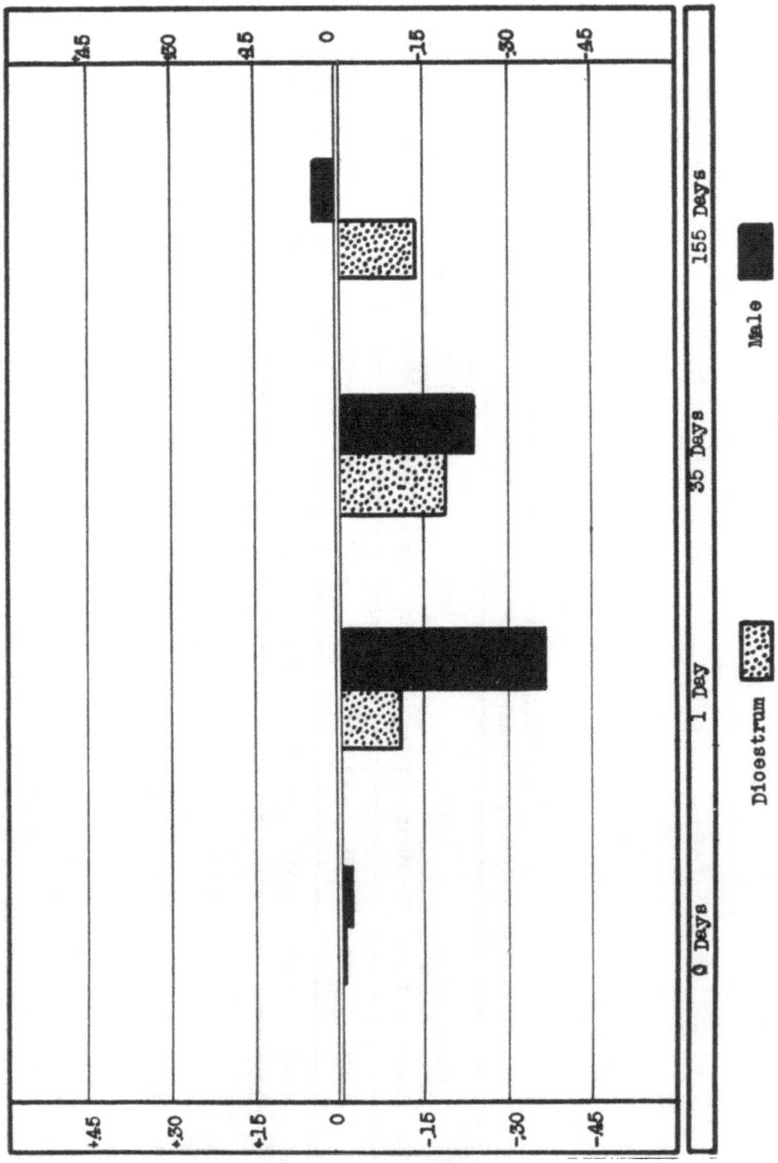

FIG. 5. Column diagram of the incentive indices for the male groups. The average for the normal sex stimulus is taken as the standard of comparison.

group, being low in all cases. The explanation seems to lie in the fact that the females are constantly present in the home cage, providing opportunity for frequent sex activity. That some copulatory activity had occurred recently is suggested by the fact that the average number of crossings to the receptive female is 7.8, whereas for satiated males it is only 3.6 and 12.2 after a 12-hour recovery period (Warner). Evidently the behavior of the unsegregated males is not wholly dominated by the sex drive, but involves exploratory and social factors; these males have had sufficient opportunity for sexual activity preceding the test to decrease materially the strength of the sex drive.

In the *1-day* series, the trend is distinctly upward for both receptive and non-receptive, and downward for male incentives. The conditions during the mating period were precisely the same for all three groups. In the case of the receptive female incentive, the rise was in all likelihood due to (1) the fact that a 24-hour recovery priod after mating was allowed, and (2) the fact that during this interval the males were segregated for the first time and thus deprived of the opportunity for stimulation by the females in the living cage. However, it should be noted that during the time previous to the mating period a *receptive* female may not have always been present in the living cage, since occasionally all females may have been in the dioestrus stage. The rise in the crossings to a non-receptive female offers greater difficulties since the male usually does not copulate with a non-receptive female. The fact that the rise is not so great in this case shows that the non-receptive female is less of a stimulus than the receptive and this is what would be naturally expected.

The behavior of the test animals suggests that the reaction to the non-receptive female might be looked upon as a case of substitute stimulus. Indicator males, as used by the writer, at first go through the entire repertory of sex reactions, but after a time (varying from two or three days to a month) omit such preliminary sex behavior as the exploratory nosing of the external genetalia of the female. The more experienced indicator males do not wait for the females to display such characteristic reactions as the short run, but mount immediately and finally come to mount both receptive and non-receptive females alike. The present writer has observed eight cases in which experienced indicator males mounted and appeared to copulate with pregnant females. In no case, however, were indicator males, after segregation for 24 hours, observed to mount males although nosing sometimes occurred .

The number of crossings to the male incentive is less after segregation for one day than for the unsegregated group. This may be due to negative adaptation to males during the segregation period.

The effect of increasing the segregation period from 1 day to *35 days* is to lessen the strength of the drive to receptive and non-receptive females in terms of the number of crossings, while the average is approximately the same to the male incentive. This decrease is evidently due in part to a waning of the sex drive. However, there is evidence of a disintegration of sex habits and a development of homosexual tendencies, as shown by the fact that the response to the male was almost as high (in absolute number of crossings) as to the non-receptive female and much closer to the value for the receptive female than in the 1-day group.

The value for the 155-day series is uniformly low as might have been expected since these males had been segregated before puberty. The female incentives (both receptive and non-receptive) appear to be merely a novel, rather than a sex stimulus. As will be noted, the crossings to the male incentive stimulus are greater than to the non-receptive, and measurably greater than to the normal stimulus. This suggests that segregation at so early an age brings about the development of a homosexual tendency. It is true that the drive toward the male incentive is, on the whole, weaker than that to the normal incentive. As indicated by the incentive index, however, it is relatively stronger after a very long (155 days) segregation period than a short one (1 day or 35 days). This may be accounted for in part on the assumption that after 155 days there is a development of definite homosexual tendencies. That these are in evidence will be more clearly indicated when Individual Differences and the Effect of the Mating Period are discussed.

It might be argued that the small differences between the various incentives, in terms of number of crossings, indicate not so much homosexuality as desexualization. Such a difference in interpretation becomes very largely a matter of definition of terms. If homosexuality be thought of as precluding normal sex behavior altogether, then desexualization might conceivably be the better term to use in this instance. But as stated in the introductory section we have used the term throughout this paper to include all specifically sexual behavior displayed toward an animal of the same sex even though at the same time similar behavior be exhibited toward an animal of the opposite sex. If this definition be allowed, the 155-day group did show a homosexual tendency, when averages, variability, etc., are all taken into account.

The observations of Stone (8) support the view that homosexual habits might thus be built up in the segregated male rat. He says: "Associated with scratching of the body is the nibbling of the external genitals which, in the case of the males, eventually comes to be chiefly licking of the penis. The stimulating effect of licking probably causes the act to be repeated and to become habitual. Whether or not this act may be considered a form of masturbation in the rat is an open question . . . No case of orgasm has been discovered as a result of the licking of the organ, but the difficulties of detecting the orgasm are such that one cannot be sure that it is not effected." Stone, however, has observed homosexual behavior among males. "Upon rare occasions adult male rats have been observed to mount one another under ordinary conditions of cage confinement. This occurs most frequently when strange males are introduced into a cage of males, or when several males still in a state of sexual excitement following prolonged copulation are brought together in their home cage."

Observations by the writer corroborate these facts and add further support to the suggestion that homosexuality may develop in the segregated male. Males of both the 35-day and 155-day groups were seen to go through the copulatory act as defined by Stone (11) — namely, pursuit and mounting of a male, palpation of the sides of the sex object, cessation of palpation and backward lunge, and licking of the penis. Furthermore, three indicator males which had been allowed to copulate rather freely with receptive females every night for over a month were observed to go through the copulatory act with each other on the second night after they had been confined to their cages and allowed no sex activity with females. Prior to this short segregation period they had been segregated during the day but never seen to display sexual behavior toward each other.

B. *Individual differences*

The magnitude and character of the scatter around the various averages are shown in Table 12 and Figure 2. These measures of variability confirm the conclusion that homosexuality has developed in certain individuals of the 35-day group and has become general in the 155-day group.

It will be noted that the relative variability is uniformly high in

TABLE 12

Measures of variability for males

Stimulus animal	Deprivation period	Average deviation	Quartile deviation	Coefficient of variation	Range	Relative Density* 1st quartile	2nd quartile	3rd quartile	4th quartile
Female—cornified stage	0 days	4.8	4.5	77.0	0—19	31	23	20	13
	1 day	3.1	2.6	28.7	7—23	34	31	47	12
	28 days	6.1	6.5	71.6	0—22	28	10	28	15
	35 days—sex activity	3.3	2.6	29.9	4—20	13	27	66	23
	155 days	1.9	1.2	25.0	5—12	83	76	83	23
	155 days—sex activity	1.5	2.8	20.4	7—13	83	100	166	23
Female in diœstrum—recuperative stage	0 days	3.0	1.6	48.8	3—15	40	66	55	13
	1 day	1.3	1.4	14.9	9—13	111	62	83	76
	35 days	3.4	2.5	59.1	0—14	18	66	28	22
	35 days—sex activity	3.3	3.0	39.7	2—15	15	28	40	66
	155 days	3.1	2.5	61.5	2—16	66	33	50	11
	155 days—sex activity	2.1	1.7	51.5	2—11	66	66	50	12
Male	0 days	3.5	3.0	54.0	1—18	22	40	28	13
	1 day	3.0	3.4	60.6	2—12	58	55	20	40
	35 days	5.3	5.5	103.7	0—15	66	28	13	28
	35 days—sex activity	3.9	1.6	47.4	5—24	40	66	58	7
	155 days	2.2	1.4	33.5	4—14	37	76	66	11
	155 days—sex activity	2.6	1.6	70.8	0—8	23	125	40	66

$$\text{* Relative density} = \frac{100}{\text{quartile range (continuous series)}}$$

the unsegregated series (0 days). This is very likely due to the fact that some of the animals have copulated more recently and more often than others previous to the test, inasmuch as the test animals were taken directly out of the living cages which included both sexes. The 1-day segregated series show marked homogeneity in comparison. This is not surprising in view of the fact that the same opportunity for copulation was given in the mating period, which was followed by the constant segregation interval, allowing precisely the same condition for recovery. It should be noted, however, that although the average number of crossings of the 1-day group to males is less than it is for the unsegregated group, nevertheless the variability continues to be relatively high. An examination of the relative densities discloses the presence of bimodality. It has previously been pointed out that the decrease in the number of crossings suggests negative adaptaion; the bimodality would indicate that part of the males react in this way. The greatest individual variation occurs in the 35-day series — in fact, the coefficient of variation is so tremendously high in the male incentive group that some explanation for this heterogeneity must be sought. As before mentioned, the distribution of this group is bimodal. When this fact is taken in conjunction with the low average, the only feasible explanation would seem to be that with certain animals the normal stimulus alone will elicit a sex response while in others the drive takes the homosexual direction. About 60 per cent of the group crossed to the male incentive stimulus between zero and five times, the modal value being two. The remaining 40 per cent crossed between eleven and fifteen times, the modal value being thirteen. The disproportionately large average deviation of this group when the incentive is a receptive female again suggests the disintegration of sex habits and the development of homosexuality.

The small amount of variability found in the 155-day group indicates the operation of causes which are common to all members of the group. One reason for this low degree of variability is that there has been no radical change in living conditions since puberty which would tend to induce different modes of adaptation, as has already been found in the case of the 35-day group. As previously stated, the number of approaches and contacts shows that activity is uniformly low for the 155-day group as compared with the unsegregated group, but the average number of crossings is about the same. Since the number of crossings is about the same, in spite of the fact that the approaches and contacts in the 155-day group are apparently much lower, it appears that the incentive stimulus must have been the effective factor in eliciting the crossing response. Variability is also greater in the case of the unsegregated group — a fact which corroborates the conclusion that general or random activity is high for the unsegregated group, and comparatively low in the 155-day group. Indeed, the very low variability in the case of the 155-day group indicates that one factor is relatively predominant in the drive to cross to the incentive compartment. Since the male is not a novel stimulus, the test animal must have reacted to it as a sex stimulus.

Thus we find that the average number of crossings to the homosexual stimulus is only slightly greater than that to the females. The relatively large number of crossings to the females is probably due to the fact that they are novel stimuli. The fact that the drive is apparently stronger to a receptive than to a non-receptive female is probably a consequence of the greater activity of the female in oestrum.

C. Persistence of the drive as shown by the temporal distribution of drive behavior during the test

Table 5 and Figure 6 indicate the persistence of the drive during the 20-minute test as shown by the per cent of crossings for the four five-minute intervals. According to this measure, the unsegregated groups are quite persistent in crossing to all three incentives. The 35-day series, on the other hand, show wide differences depending upon the incentive employed. The drive to the normal incentive seems to gather increasing force as the test proceeds, while that to the non-receptive female diminishes in persistence. The drive to the male is the most constant. With the 155-day group, the drive to the non-receptive female shows the greatest decrease in persistence during the last fifteen minutes of the test.

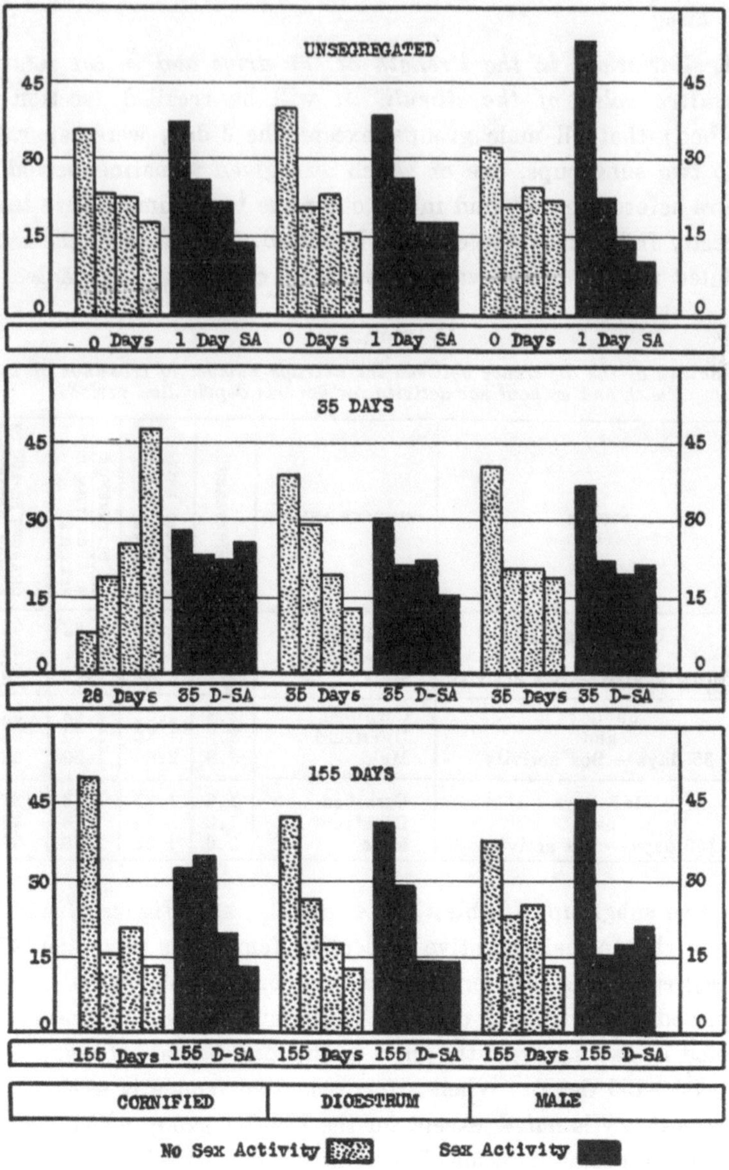

Fig. 6. Column diagram showing temporal distribution of crossings of the male groups during the twenty-minute test period in five-minute intervals.

D. The effect of the mating period upon the test twenty-four hours later

1. *Variations in the strength of the drive and in the relative incentive value of the stimuli.* It will be recalled (section on method) that all male groups, except the 1-day, were separated into two subgroups, one of which was given a mating period 24 hours before the test, and in the other the test animals were taken directly from the living cages to be tested. The effect of the interpolated mating period can be found by comparing the indices of

TABLE 13

Reliability of the difference between the average number of crossings of males with and without sex activity for various deprivation periods

GROUPS	STIMULUS ANIMAL	THE DIFFERENCE	STANDARD DEVIATION OF THE DIFFERENCE	DIFFERENCE / S. D. OF THE DIFFERENCE	CHANCES IN 100 OF A TRUE DIFFERENCE GREATER THAN 0
Unsegregated	Cornified	5.7	1.94	2.93	99.8
and	Diœstrum	3.2	1.27	2.51	99.4
Unsegregated — Sex activity	Male	1.9	1.81	1.04	85
35 days	Cornified	2.8	1.92	1.45	93
and	Diœstrum	3.2	1.88	1.70	96
35 days — Sex activity	Male	3.9	2.60	1.50	93
155 days	Cornified	1.6	1.35	1.18	88
and	Diœstrum	1.2	1.53	.78	78
155 days — Sex activity	Male	3.6	1.35	2.66	99.6

the two subgroups (Tables 13, 14, and 15, and Figures 7 and 8). When the normal incentive (receptive female) is used, the effect of segregation is counterbalanced to a considerable extent by such a period, there being a decrease in the effectiveness of the mating period in passing from the short to the long periods of segregation (0—35—155 days). When a non-receptive female is used, a similar tendency is noted, except for the 155-day group which shows a slightly opposite direction. When a male incentive is employed, the effect of the interpolated mating period does not seem consistent. As would be expected the mating period lowered the effectiveness of the male incentive for the zero group. A decided increase in crossings to the male in the 35-day group, apparently due to the mating period, would seem to mean that the opportunity to mate with a female stirred up greater sex activity even toward a male incentive animal 24 hours later. The mating period in the case of the longer segregation interval (the group which was segregated

TABLE 14

Reliability of the difference between the average number of crossings of males after sex activity

GROUPS	STIMULUS ANIMAL	THE DIFFERENCE	STANDARD DEVIATION OF THE DIFFERENCE	DIFFERENCE / S. D. OF THE DIFFERENCE	CHANCES IN 100 OF A TRUE DIFFERENCE GREATER THAN 0
Unsegregated — Sex activity	Cornified	.1	1.33	.07	53
and	Diœstrum	.5	1.40	.35	64
35 days — Sex activity	Male	4.1	1.94	2.11	98
35 days — Sex activity	Cornified	4.2	1.22	3.44	100
and	Diœstrum	5.3	1.60	3.31	100
155 days — Sex activity	Male	5.7	1.85	3.08	100
Unsegregated — Sex activity	Cornified	4.3	1.31	3.28	100
and	Diœstrum	5.8	1.05	5.52	100
155 days — Sex activity	Male	1.6	1.56	1.02	84

before puberty) seemed to arouse the normal sex drive, causing a partial breakdown of the homosexual tendency as indicated by the rather marked decrease in crossing to the male. Indeed, the fact that sex activity has operated to decrease the average number of crossings to a male from 8.2 to 4.6, points rather significantly to the conclusion that homosexual tendencies were present before the mating period was given.

2. *Individual differences.* The subgroups which were given the mating period showed less variability (except for the 155-day group crossing to a male in which case the variability is less when the last fifteen minutes are considered) regardless of the nature of the incentive, than the corresponding groups in which the mating period was omitted. The general effect of the 2-hour period of sex activity seems therefore to operate as a constant in reducing individual variation.

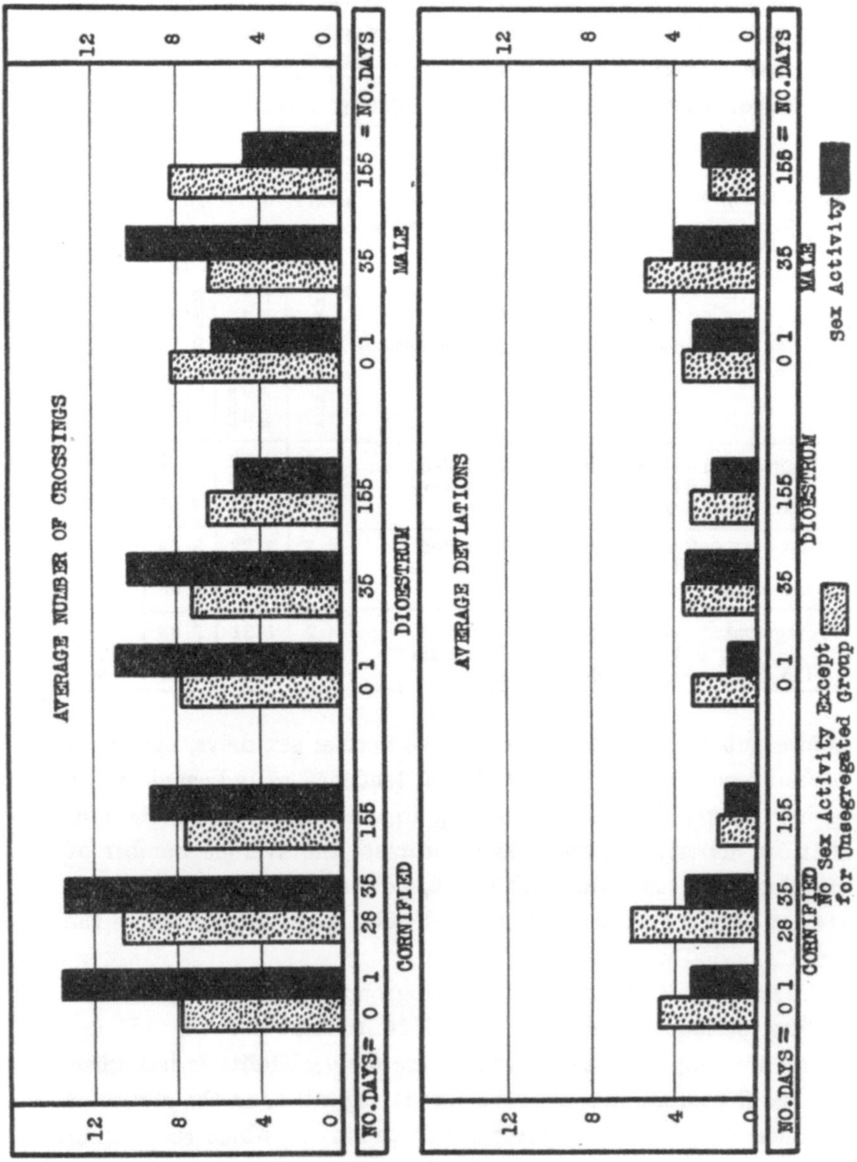

Fig. 7. Column diagram showing average number of crossings and average deviations for the male groups

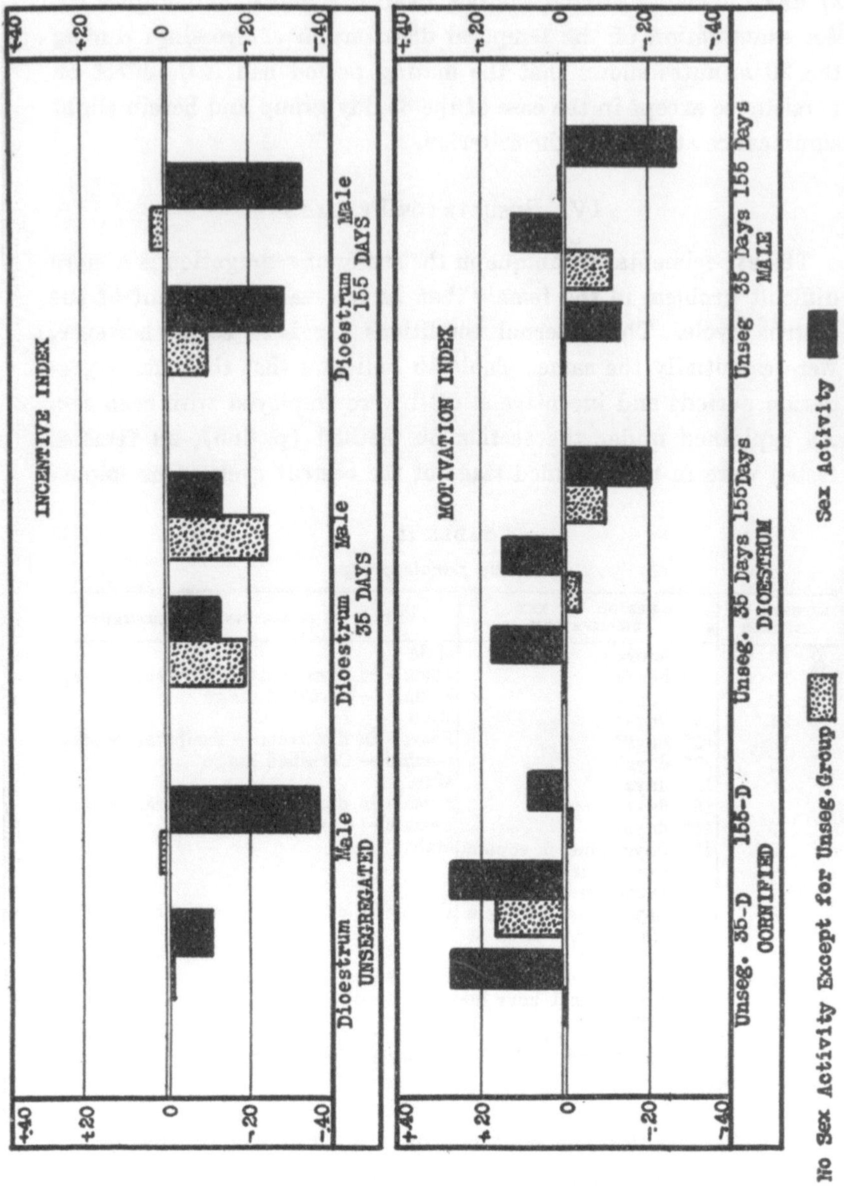

FIG. 8. Column diagram showing the motivation and incentive indices for the male groups to indicate the effect of sex activity.

3. *Persistence of the drive as shown by the temporal distribution of drive behavior during the test* (Tables 5 and 6, and Figure 6). An examination of the temporal distribution of crossings during the 20 minutes shows that the mating period had little effect on persistence except in the case of the 35-day group and herein slight significance attaches to the criterion.

IV. Results for Females

The experimental technique in the study of segregation is a more difficult problem in the female than in the male on account of the oestrus cycle. The external conditions for both sexes, however, were essentially the same. Table 15 indicates that the same segregation periods and incentive stimuli were employed with each sex. As explained under the section on method (p. 485), all females tested were in the cornified stage of the oestrus cycle. One important difference was necessitated by the character of the oestrus cycle, i.e., it was impossible to give the females the 2-hour mating period 24 hours before the test, as in the case of the males. The receptive stage is not long enough to allow a 24-hour lapse of time

TABLE 15

Showing female groups

NUMBER OF ANIMALS	PERIOD OF SEX DEPRIVATION	OBJECT IN INCENTIVE COMPARTMENT
10	0 hours	Male
15	0 hours	Female in diœstrum — recuperative stage
10	0 hours	Female — Cornified stage
21	35 days*	Male
12	35 days*	Female in diœstrum — recuperative stage
15	35 days*	Female — Cornified stage
10	155 days	Male
7	155 days	Female in diœstrum — recuperative stage
10	155 days	Female — Cornified stage
9	155 days plus 5 copulations plus 1 hour deprivation	Male
8	155 days plus 5 copulations plus 1 hour deprivation	Female in diœstrum — recuperative stage
6	155 days plus 5 copulations plus 1 hour deprivation	Female — Cornified stage

* This group was tested by L. H. Warner and is reported in his monograph on sex behavior.

between such a period and the test, since the test animal must be in the cornified stage. No 1-day group, therefore, could be tested. However, a variant to this condition was introduced by giving half of the females of the 155-day group a brief mating period one hour before the test (p. 196). In each case, the amount of sex activity allowed was five copulatory responses, which usually occurred within a few minutes, although fifteen minutes was required in some cases.

In order to facilitate a comparison, the results for the females will be treated under the same topical plan as that followed in the discussion of the results for males. The main topics are:

A. Variations in the strength of the drive and the relative incentive value of the stimuli.

B. Individual differences.

C. Persistence of the drive as shown by the temporal distributions of drive behavior during the test.

D. The effect of copulatory activity (five responses) upon the test one hour later.

A. Variations in the strength of the drive and the relative incentive values of the stimuli

The changes in the strength of the sex drive as a result of varying the segregation conditions preceding the test are best indicated in the graphs of Figures 9 and 10. The average number of crossings of the unsegregated females to the normal incentive stimulus *(male)* is greater than the number of crossings for any other group tested. However, there is practically no difference (14.2 and 14.1) between the unsegregated group and the 35-day groups, there being only 52 chances in 100 of a true difference. This is convincing evidence that the sex drive, after 35 days of segregation, in females raised unsegregated to maturity, showed no apparent loss to the appropriate stimulus. The data for the 155-day group furnish a reliable indication that segregation from before puberty has resulted in a substantial decrease in the strength of the normal sex drive.

The response to the *non-receptive female,* like that to the normal incentive, indicates that the drive is not altered appreciably as a

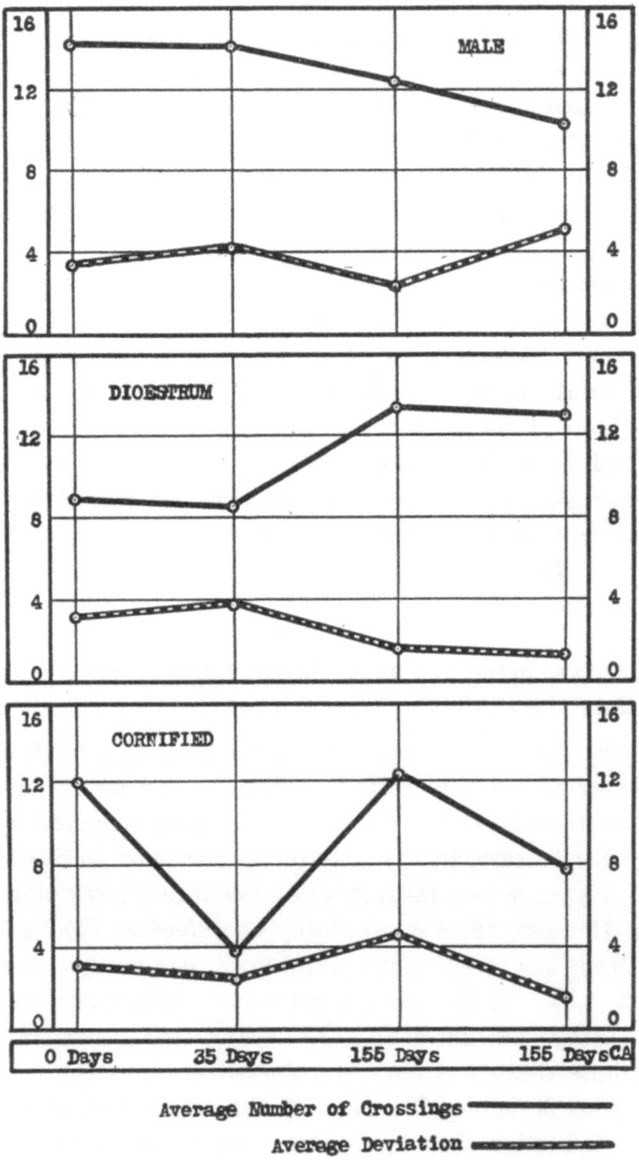

FIG. 9. Graphs of the average number of crossings and average deviations for females for various deprivation periods. Incentive stimulus indicated on each diagram.

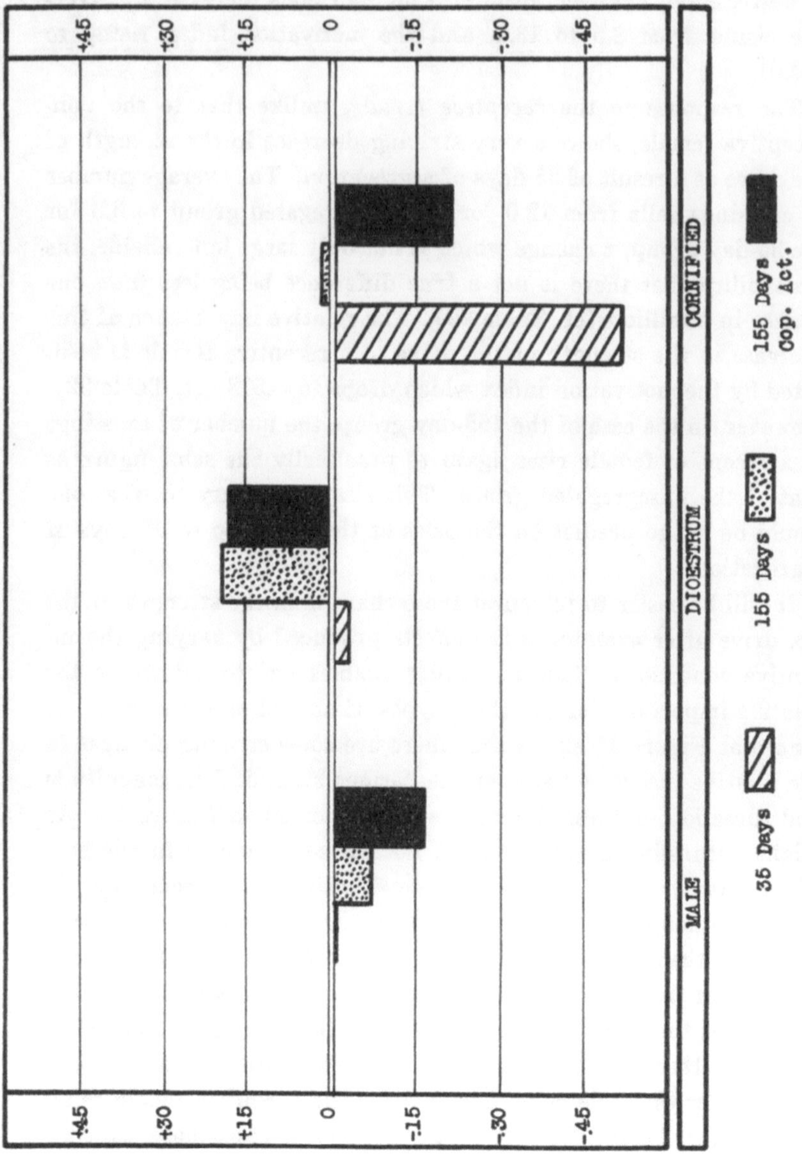

Fig. 10. Column diagram showing the motivation indices for the female groups. The average for the unsegregated group is used as the standard of comparison.

result of 35 days of segregation. One hundred and fifty-five days of segregation, however, results in an enormous increase, the average rising from 8.5 to 13.4, and the motivation index rising to +.201.

The response to the *receptive female*, unlike that to the non-receptive female, shows a very striking decrease in the strength of the drive as a result of 35 days of segregation. The average number of crossings falls from 12.0 for the unsegregated group to 3.8 for the 35-day group, a change which is not only large but reliable, the probability that there is not a true difference being less than one chance in a million (cf. Table 20). The relative importance of this decrease in the strength of the drive to a receptive female is indicated by the motivation index which drops to—.518 (cf. Table 22). However, in the case of the 155-day group, the number of crossings to a receptive female rises again to practically the same figure as that of the unsegregated group. This rise is contrary to what one would be led to predict on the basis of the trend up to 35 days of segregation.

It will be easier to interpret these changes in the strength of the sex drive after examining the effects produced by varying the incentive conditions. This procedure enables one to determine the relative importance of the three types of animal as sex stimuli. A glance at Figure 11 shows that there are some striking changes in the relative incentive value of the various stimuli. The magnitude and direction of these differences are indicated in Figure 12. As might naturally be expected, the normal sex stimulus in the case of the *unsegregated* group is more effective than either type of female, although the incentive value of the non-receptive female is less than that of the female in oestrum. An examination of the averages of the approaches and contacts for the unsegregated group shows that the activity is much higher in the case of the male than with the other incentive stimuli — the sum of the approaches and contacts for the male being 5.5, that for the receptive female being only 0.8 (cf. Table 19). The 35-day group also react more strongly to the normal stimulus than to the other incentive conditions. But there is a significant change in the specificity of the drive; in fact, differences in incentive value reach a maximum in this group,

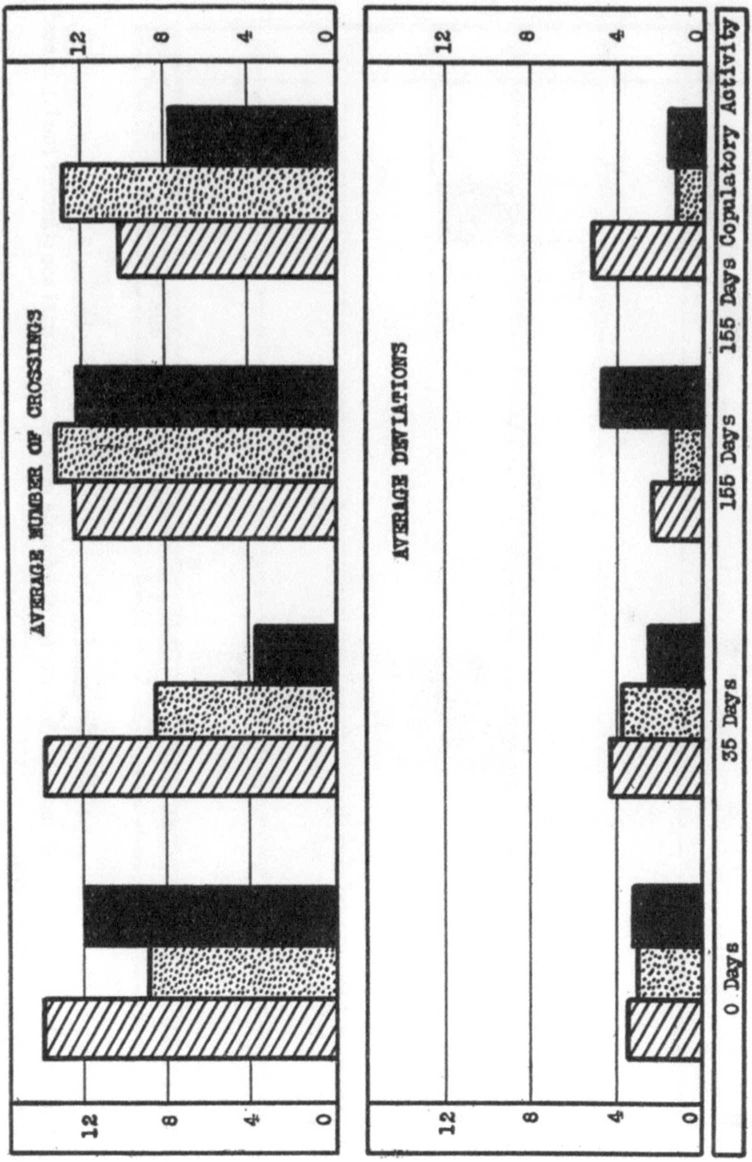

FIG. 11. Column diagram showing average number of crossings and average deviations for the female groups in response to various incentive stimuli.

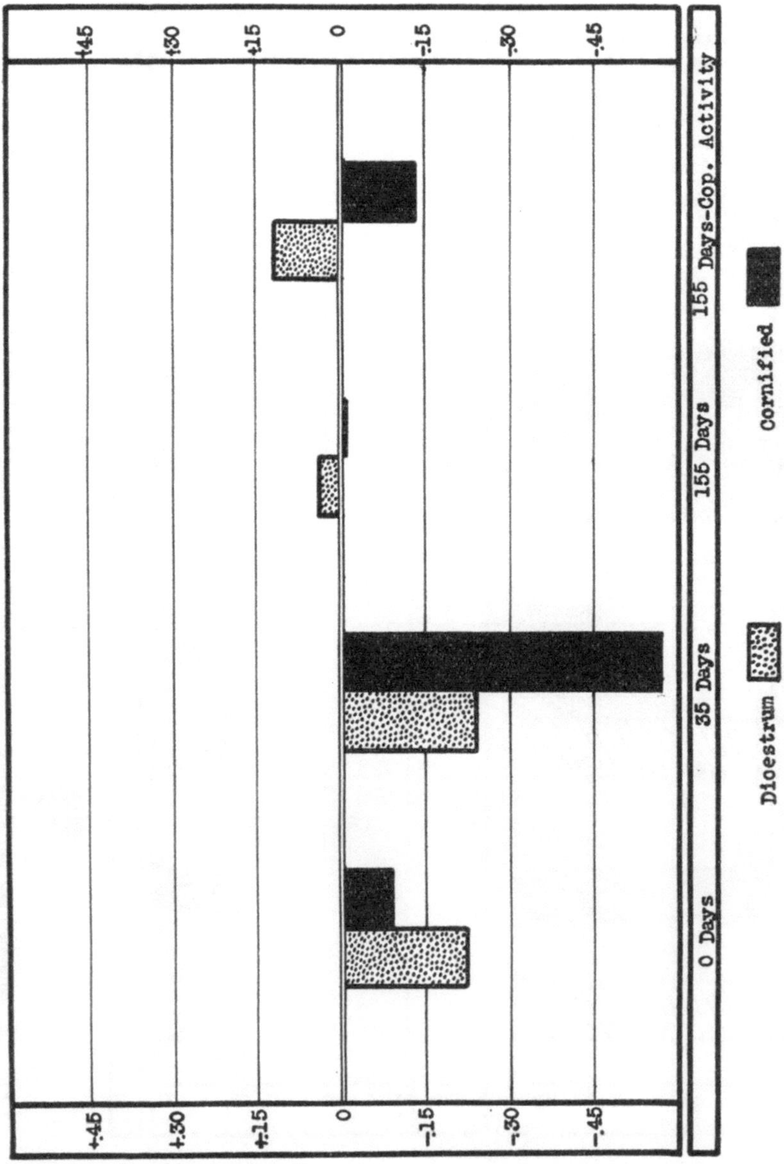

FIG. 12. Column diagram of incentive indices for the female groups. The normal sex stimulus (male) is used as the standard of comparison.

the incentive index dropping in one case (receptive females) to the low figure of —.575. This increase in value might lead one to expect a still greater degree of specificity as the segregation period is lengthened. But the data for the *155-day* group indicate that the contrary is true. Not only is the specificity very low, but what little there is seems to be directed toward a homosexual incentive (non-receptive female).

An examination of Table 21, shows that the order of magnitude and direction of the differences between the various partial averages follow the same general trend as those in the case of the total averages. This is additional evidence, other than that found in Tables 20 and 23, that the tendencies which have been pointed out above are warranted by the data.

In spite of the apparent reversals in trend, a careful interpretation of the data shows conclusively that most of these changes can be explained on the basis of the formation and disintegration of sex habits and that segregation from before puberty to 185 days is probably accompanied by the development of homosexual tendencies. One point which merits special examination is the striking drop in the number of crossings to a receptive female in the case of the 35-day group. This is all the more interesting when we note that the number of crossings for the non-receptive female is practically the same as for the unsegregated group. In fact, the average number of crossings to a receptive female is five less than that for any other group, the next lowest being 8.5. Seventy-five per cent of the animals cross less than five times and only one animal more than seven times. The number of crossings are even fewer than for Warner's control group, in which no incentive animal whatever was used in the incentive compartment. The average crossings for his control is 5.0, whereas the average for the present group is only 3.8. Since the same electrical obstruction was used in both groups, and since both groups had been segregated for the same length of time under the same conditions of care, the presence of the receptive female must have been the deterrent which reduced the number of crossings from 5.0 to 3.8.

This conclusion is further supported by the fact that there is a considerable increase in the number of approaches and contacts as

TABLE 16

Frequency distribution covering approaches, contacts, and crossings for the various period of sex deprivation in the female rat

Group	Unsegregated									Segregated 35 days					
Stimulus animal	Male			Female in diæstrum—recuperative stage			Female—cornified stage			Male			Female in diæstrum—recuperative stage		
Deprivation of test animal	0 days			0 days			0 days			35 days			35 days		
Number of responses	Approaches	Contacts	Crossings	Approaches	Contacts	Crossings	Approaches	Contacts	Crossings	Approaches	Contacts	Crossings	Approaches	Contacts	Crossings
0	1	5		8	10	1	8	6		2	3		2	9	1
1	2	1		4	3		1	3		1	5		6		
2	2	1		1	1		1	1		2	1	1	3	1	1
3		1		1						3	5		1	1	
4	1	1		1	1	1					3	1			
5	2	1				1				3		1			2
6	1		1			1			1	2	1				
7						1				3				1	1
8						1			2	1	1				
9						2				2	2	1			1
10						1			1						1
11			3			3						1			1
12						1			1	1					2
13									2						
14	1		1						1	1		3			1
15						2						2			1
16			1						1			3			
17			1									2			
18			1									1			
19			2									2			
20									1						
21												2			
22															
23															
24												1			
Number rats in group	10			15			10			21			12		

Group	Segregated 35 days			Segregated 155 days											
Stimulus animal	Female—cornified stage			Male			Male			Male			Female in diæstrum—recuperative stage		
Deprivation of test animal	35 days			155 days			1 hour			Total			155 days		
Number of responses	Approaches	Contacts	Crossings	Approaches	Contacts	Crossings	Approaches	Contacts	Crossings	Approaches	Contacts	Crossings	Approaches	Contacts	Crossings
0	2	4	3	6	3		5	4		11	7		5	2	
1	2	2		4	5		2	1		6	6		2	4	
2		1	3		1		1	1	1	1	2	1			
3	1	4	3		1						1	2		1	
4	2	1	2				1	2	2	1	2				
5	2	2	1					1			1				
6		1	1												
7	1		1												
8	2					1			1			2			
9						1						1			
10	1					1			1			2			1
11						1			1			2			
12						1						1			1
13	1					2						2			1
14	1														2
15			1												1
16						3						3			1
17									2			2			
18															
19									1			1			
20															
21															
22															
23															
24															
Number rats in group	15			10			9			19			7		

Group	Segregated 155 days														
Stimulus animal	Female in diæstrum—recuperative stage						Female—cornified stage								
Deprivation of test animal	1 hour			Total			155 days			1 hour			Total		
Number of responses	Contacts	Approaches	Crossings	Approaches	Contacts	Crossings	Approaches	Contacts	Crossings	Approaches	Contacts	Crossings	Approaches	Contacts	Crossings
0	4	4		9	6		8	3		3	4		11	7	
1	4	2		6	6			4			1			5	
2		1			1		1	1	1	2	1		3	2	1
3					1					1		1	2		1
4								1						1	
5		1			1			1						1	
6									1						1
7															
8									1			1			2
9												4			4
10						1			1						1
11			2			2									
12			1			2			1						1
13			2			3			1						1
14			1			3			1						1
15			2			3									
16						1									
17															
18															
19									2						2
20									1						1
21															
22															
23															
24															
Number rats in group	8			15			10			6			16		

compared with those of the other groups, indicating that, even though the number of crossings are few, nevertheless there is material evidence of considerable activity. These facts show conclusively that there has been no *positive* conditioning to the receptive female as a result of 35 days of segregation. On the other hand, certain facts about mating behavior support the hypothesis that *negative* conditioning may have been set up. When a female is in the receptive stage, sensitivity to tactual stimulation in the genital region seems to be increased. For example, Stone (8) observed that an application of acid to the base of the tail of a non-receptive female causes it to exhibit mating behavior, due to increased sensitivity. These facts suggest the possibility that, since these females are orientated toward the normal stimulus alone and yet are extremely sensitive to any tactual stimulation, the increased activity of the receptive female induces an avoiding reaction. In other words, the females tend to avoid sexual stimulation from all stimuli other than the normal one. In this manner, negative conditioning is perhaps developed to such an extent that a receptive female in the incentive compartment may become a definite deterrent.

It may seem strange that negative conditioning does not also occur in the unsegregated group when a receptive female is used as an incentive. This can probably be accounted for in either of two ways. The first explanation is a statistical one. Since the number of animals in the living cage is kept constant (the other half being males), there are only half as many females present in a cage as in the case of the 35-day group. Therefore, when a given female is in oestrum, the chances that another female will be in oestrum at the same time are correspondingly lowered. The chances of another female being in oestrum would be four times as great for the 35-day group as it would be for the unsegregated group. In the second place, if two receptive females should happen to be present in the cage at the same time, they would have little opportunity to stimulate each other because they are probably reacting entirely to the males. These two conditions seem able to account for the fact that there is no evidence of negative conditioning to receptive females in the unsegregated group.

A word should also be said concerning the apparent lack of negative conditioning in the case of the 155-day group. Since negative conditioning occurs in the 35-day group, still more would it naturally be expected to be found in the case of the 155-day group, but, as already noted, quite the contrary is true. The number of

TABLE 17

Showing temporal distribution of approaches, contacts, and crossings of the female groups during the twenty-minute test period in one-minute intervals

Group	Stimulus animal	Deprivation of test animal	Activity	1st minute	2nd minute	3rd minute	4th minute	5th minute	6th minute	7th minute	8th minute	9th minute	10th minute	11th minute	12th minute	13th minute	14th minute	15th minute	16th minute	17th minute	18th minute	19th minute	20th minute
Unsegregated	Male	0 days	Approaches	2	3	4	2	4	3	2	2	2	1	1	0	0	1	0	1	4	3	2	3
			Contacts	0	1	0	2	1	0	0	2	0	0	0	0	1	1	0	1	0	4	1	1
			Crossings	15	12	6	7	9	5	8	7	7	6	6	6	9	7	5	5	6	5	5	6
	Female in diœstrum—recuperative stage	0 days	Approaches	0	1	0	1	0	0	2	0	0	4	1	1	0	1	0	1	0	0	0	1
			Contacts	1	2	1	0	0	1	0	1	0	1	1	0	0	1	0	0	0	0	0	0
			Crossings	9	8	8	10	5	11	7	4	8	4	7	9	3	3	7	4	7	6	6	7
	Female—cornified stage	0 days	Approaches	0	0	0	0	0	0	1	0	0	1	0	0	0	1	0	0	1	0	0	0
			Contacts	1	0	0	0	2	0	0	0	1	0	0	0	0	0	0	0	1	0	0	0
			Crossings	10	8	5	7	6	7	3	7	6	7	4	7	5	5	7	5	4	5	6	6
Segregated 35 days	Male	35 days	Approaches	12	4	11	12	10	7	8	7	2	5	6	7	6	2	2	4	4	2	2	5
			Contacts	3	0	1	2	2	3	3	3	3	4	3	0	3	4	6	9	5	8	2	1
			Crossings	14	11	12	12	10	7	8	10	9	9	9	10	11	16	35	26	24	30	28	2
	Female in diœstrum—recuperative stage	35 days	Approaches	2	0	0	0	1	0	0	1	0	1	0	0	0	2	2	1	0	1	4	0
			Contacts	0	0	0	0	0	0	0	1	1	2	0	2	2	2	1	0	0	0	0	1
			Crossings	12	10	8	5	8	7	6	5	2	3	5	5	2	5	4	3	4	3	4	2
	Female—cornified stage	35 days	Approaches	6	4	3	3	5	3	8	8	4	5	5	2	5	0	2	3	8	7	0	2
			Contacts	4	3	1	3	4	0	3	0	4	3	0	0	1	2	0	2	1	3	1	1
			Crossings	14	10	6	1	3	4	2	2	2	1	1	1	1	1	2	0	1	1	2	2
Segregated 155 days	Male	155 days	Approaches	0	0	0	1	0	0	0	1	0	0	1	0	0	0	0	0	0	0	1	0
			Contacts	0	1	1	2	1	4	0	0	0	0	0	0	0	0	0	0	0	1	0	0
			Crossings	10	10	7	3	7	5	9	5	8	6	3	8	6	7	6	6	4	4	5	5
		1 hour	Approaches	2	0	0	0	0	0	2	1	0	2	0	0	0	0	0	0	0	0	1	0
			Contacts	0	2	1	1	3	0	0	1	0	2	1	2	2	0	1	0	0	0	0	0
			Crossings	7	7	7	7	6	5	4	5	5	4	5	2	4	4	4	4	1	3	4	4
	Female in diœstrum—recuperative stage	155 days	Approaches	0	1	0	0	0	0	0	0	0	0	0	0	0	0	0	1	0	0	0	0
			Contacts	0	0	1	0	1	2	1	1	0	1	0	0	0	0	0	0	0	0	0	0
			Crossings	7	7	5	7	5	4	5	2	6	5	5	3	3	5	6	0	4	4	4	7
		1 hour	Approaches	0	1	0	1	0	0	0	1	1	0	0	0	0	0	0	0	0	0	0	0
			Contacts	0	0	1	2	0	1	1	0	0	2	0	0	0	0	0	0	0	0	2	0
			Crossings	8	7	7	6	4	4	6	5	4	5	5	6	3	4	4	8	4	5	5	4
	Female—cornified stage	155 days	Approaches	1	0	0	1	0	2	0	0	0	0	0	1	0	0	0	0	0	0	0	0
			Contacts	1	1	1	1	2	1	0	1	0	1	1	1	0	0	1	0	0	1	1	1
			Crossings	8	8	5	6	7	5	5	6	8	7	7	7	6	3	7	6	6	5	8	3
		1 hour	Approaches	0	0	0	1	1	0	2	0	0	0	0	0	0	0	0	0	1	0	1	1
			Contacts	0	0	0	0	0	2	0	0	0	0	0	0	1	0	0	0	0	0	0	0
			Crossings	6	6	5	1	4	3	1	5	1	3	2	3	0	3	0	1	1	1	1	0

crossings to a receptive female as the result of 155 days of segregation is practically the same as that for no segregation. This change, however, will be understood when we are careful to note the differences in the life history of the animals prior to segregation. The 155-day group have had no opportunity to become conditioned to males. On the other hand, they have been surrounded continuously by females since weaning. In other words, they had a chance to become habituated to sexual stimulation by receptive females but have had no chance whatever to become adapted to stimulation by males. The general activity due to internal stimulation would naturally cause a large number of crossings and the number

TABLE 18

Showing temporal distribution of approaches, contacts, and crossings of the female groups during the twenty-minute test period in five-minute intervals

Group	Stimulus animal	Deprivation of test animal	Activity	0 to 5th Minute		6th to 10th Minute		11th to 15th Minute		16th to 20th Minute		
				Number	Per Cent	Number	Per Cent	Number	Per Cent	Number	Per Cent	
Unsegregated	Male	0 days	Approaches	15	37.5	10	25.0	2	5.0	13	32.5	40
			Contacts	4	26.7	2	13.3	2	13.3	7	46.7	15
			Crossings	49	34.5	33	23.2	33	23.2	27	19.0	142
	Female in diœstrum—recuperative stage	0 days	Approaches	2	15.4	6	46.1	3	23.0	2	15.4	13
			Contacts	4	44.4	3	33.3	2	22.2	0	0	9
			Crossings	40	30.1	34	25.5	29	21.8	30	22.5	133
	Female—cornified stage	0 days	Approaches	0	0	2	66.6	1	33.3	0	0	3
			Contacts	3	60.0	1	20.0	0	0	1	20.0	5
			Crossings	36	30.0	30	25.0	28	23.3	26	21.6	120
Segregated 35 days	Male	35 days	Approaches	49	43.0	29	25.4	23	20.2	13	11.4	114
			Contacts	8	12.1	16	24.2	16	24.2	26	39.4	66
			Crossings	59	19.2	45	14.7	81	26.4	22	39.7	307
	Female in diœstrum—recuperative stage	35 days	Approaches	3	20.0	2	13.3	4	26.7	6	40.0	15
			Contacts	0	0	4	33.3	7	58.3	1	8.3	12
			Crossings	43	41.7	23	22.3	21	20.4	16	15.5	103
	Female—cornified stage	35 days	Approaches	21	25.3	28	33.7	14	16.9	20	24.1	83
			Contacts	15	41.7	10	27.8	3	8.3	8	22.2	36
			Crossings	34	59.6	11	19.3	6	10.5	6	10.5	57
Segregated 155 days	Male	155 days	Approaches	1	25.0	1	25.0	1	25.0	1	25.0	4
			Contacts	5	50.0	4	40.0	0	0	1	10.0	10
			Crossings	37	29.8	33	26.6	30	24.2	24	19.3	124
		1 hour	Approaches	2	25.0	5	62.5	0	0	1	12.5	8
			Contacts	7	43.7	3	18.7	6	37.5	0	0	16
			Crossings	34	36.9	23	25.0	19	20.6	16	17.4	92
	Female in diœstrum—recuperative stage	155 days	Approaches	1	50.0	0	0	0	0	1	50.0	2
			Contacts	2	28.5	5	71.4	0	0	0	0	7
			Crossings	31	33.0	22	23.4	22	23.4	19	20.2	94
		1 hour	Approaches	2	50.0	2	50.0	0	0	0	0	4
			Contacts	3	33 3	4	44.4	0	0	2	22.2	9
			Crossings	32	30.8	24	23.0	22	21.1	26	25.0	104
	Female—cornified stage	155 days	Approaches	2	40.0	2	40.0	1	20.0	0	0	5
			Contacts	6	40.0	3	20.0	3	20.0	3	20.0	15
			Crossings	34	27.6	31	25.2	30	24.4	28	22.7	123
		1 hour	Approaches	2	28.5	2	28.5	0	0	3	42.8	7
			Contacts	0	0	2	66.6	1	33.3	0	0	3
			Crossings	22	46.8	13	27.6	8	17.0	4	8.5	47

of crossings is probably not lowered by negative conditioning. Perhaps it may even be increased by a slight positive conditioning to receptive females.

The lack of a strong differential response to the three stimuli in the case of the 155-day group can be accounted for by the fact that

TABLE 19

Showing results of the female groups

Groups	Stimulus animal	Deprivation of test animal	Approaches: Median	Approaches: Range	Approaches: Average	Approaches: Mean deviation	Approaches: Coefficient of variation	Contacts: Median	Contacts: Range	Contacts: Average	Contacts: Mean deviation	Contacts: Coefficient of variation	Crossings: Median	Crossings: Range	Crossings: Average	Crossings: Mean deviation	Crossings: Coefficient of variation	Crossings: Temporal median*
Unsegregated	Male	0 days	3.5	0—14	4.0	2.8	87.7	1.0	0—5	1.5	1.6	133.3	15.5	6—19	14.2	3.6	31.6	8.3
Unsegregated	Female in diœstrum—recuperative stage	0 days	.9	0—4	8	.9	141.2	7	0—4	6	8	166.6	9.7	0—15	8.9	3.1	43.8	8.6
Unsegregated	Female—cornified stage	0 days	6	0—2	3	.5	210.0	.8	0—2	.5	6	150.0	13.0	6—20	12.0	3.2	33.3	9.2
Segregated 35 days	Male	35 days	5.0	0—14	5.4	2.9	66.6	3.0	0—9	3.1	2.1	82.8	16.0	2—24	14.1	4.2	37 1	14.1
Segregated 35 days	Female in diœstrum—recuperative stage	35 days	1.6	0—3	1.2	.6	62.5	.6	0—7	1.0	1.5	188.0	10.0	0—15	8.5	3.9	56.4	6.3
Segregated 35 days	Female—cornified stage	35 days	5.2	0—14	5.5	3.7	84.3	3.1	0—6	2 4	1 8	93.7	3.5	0—16	3.8	2.6	84.2	2.7
Segregated 155 days	Male	155 days	.8	0—1	.4	.5	157.5	1.4	0—3	1.0	.6	75.0	13.0	8—16	12.4	2.4	24.1	8.8
		1 hour	9	0—4	9	1.0	138.8	1.5	0—5	1.6	1.7	133.3	11.0	2—19	10.2	5 1	62.6	7.6
		Total	.9	0—4	.6	.7	146.6	1.4	0—5	1.4	1.2	107 1	11.7	2—19	11.4	3 9	42.9	
Segregated 155 days	Female in diœstrum—recuperative stage	155 days	7	0—1	3	.4	166.6	1.4	0—3	1.0	.6	75.0	14.2	10—16	13.4	1.5	14.0	8.8
		1 hour	1.0	0—1	.5	.5	126.0	1.0	0—5	1.1	1.2	136.3	13.5	11—15	13.0	1.2	11.5	9.2
		Total	.8	0—1	.4	.5	157.5	1.2	0—5	1 1	.9	102.7	13.8	10—16	13.2	1.4	13.2	
Segregated 155 days	Female—cornified stage	155 days	.6	0—3	.5	.8	200.0	1.5	0—5	1.5	1.3	108.6	13.0	2—20	12.3	4.7	47.8	9 5
		1 hour	1.0	0—3	1.1	1.2	136.3	.7	0—2	.5	.6	150.0	9.2	3—9	7.8	1.6	25.6	5.5
		Total	.7	0—3	7	1.0	178.5	1.2	0—5	1.1	1.0	113.6	9.7	2—20	10.6	4.1	48.4	

*Temporal medians were computed by the writer. The temporal median is that time of the twenty-minute test period at which fifty per cent of the total crossings have occurred (See Table 17).

the sex life of these females has never been complicated by the stimulative effect incident to copulation. Since the copulatory response has never been elicited, they have never had an opportunity to become specifically orientated toward a given stimulus. The sex habit can never have developed beyond nosing or mounting. Although the specificity is low, nevertheless, it seem that sex activity is directed toward the homosexual stimulus, instead of the heterosexual — in fact, the incentive index for the non-receptive stimulus is slightly positive. Furthermore, the indication of homosexuality may be greater than is shown by the numerical data, since the male is a novel stimulus to the test animal. The introduction of this element of novelty into the situation may produce enough additional

TABLE 20

Reliability of the difference between the average number of crossings for females for various deprivation periods

GROUPS (deprivation periods)	STIMULUS ANIMAL	THE DIFFERENCE	STANDARD DEVIATION OF THE DIFFERENCE	DIFFERENCE / S.D. OF THE DIFFERENCE	CHANCES IN 100 OF A TRUE DIFFERENCE GREATER THAN 0
Unsegregated and 35 days	Male	.1	1.82	.05	52
	Diœstrum	.4	1.71	.23	59
	Cornified	8.2	1.50	5.46	100
35 days and 155 days	Male	1.7	1.48	1.15	87
	Diœstrum	4.9	1.55	3.16	100
	Cornified	8.5	2.03	4.18	100
155 days and 155 days — copulatory activity	Male	2.2	2.33	.94	83
	Diœstrum	.4	.88	.45	67
	Cornified	4.5	2.03	2.21	98.6
Unsegregated and 155 days	Male	1.8	1.71	1.05	85
	Diœstrum	4.5	1.22	3.68	100
	Cornified	.3	2.24	.13	55

activity on the part of the test animal to raise appreciably the average number of crossings in response to a male, in spite of the fact that the male may have very little stimulation value as a sex object.

In the 155-day group the slight advantage of the non-receptive over the receptive female as concerns their capacity to elicit a response from the female test animal is what should be expected as a

TABLE 21

Total and partial averages and average deviations for crossing of the female groups

Stimulus animal		Unsegregated		35 days		155 days		155 days—Copulatory activity	
		Total	Partial	Total	Partial	Total	Partial	Total	Partial
Male	*Av.*	14.2	9.3	14.1	11.8	12.4	8.7	10.2	6.4
	A. D.	3.6	2.3	4.2	*	2.4	2.3	5.1	4.7
Diœstrum	*Av.*	8.9	6.2	8.5	5.0	13.4	9.0	13.0	9.0
	A. D.	3.1	3.5	3.9	3.2	1.5	1.1	1.2	1.2
Cornified	*Av.*	12.0	8.4	3.8	1.5	12.3	8.9	7.8	4.1
	A. D.	3.2	2.5	2.6	1.9	4.7	3.7	1.6	1.4

*No partial values calculated for Warner's group here, because of lack of original data.

TABLE 22

Motivation indices for the female groups
(Average for unsegregated group as the standard of comparison)

Stimulus	Male			Diœstrum			Cornified		
Deprivation period	35 days	155 days	155 days—Copulatory activity	35 days	155 days	155 days—Copulatory activity	35 days	155 days	155 days—Copulatory activity
Total	—.003	—.067	—.163	—.022	.201	.187	—.518	.012	—.212
Partial	.118	—.033	—.184	—.107	.184	.184	—.699	.028	—.344

result of the conditions of segregation. The opportunity for contact with receptive females is less than that for non-receptive females. Since occasionally two receptive females would be in oestrum at the same time, the test animals may have been somewhat conditioned to receptive females. But, on account of the limited number of animals in a cage, this would be the exception and not the rule. Consequently, segregation would facilitate sexual conditioning to the non-receptive female.

TABLE 23

Reliability of the difference between the average number of crossings of females for various stimulus animals

GROUPS (STIMULUS ANIMALS)	DEPRIVATION PERIOD	THE DIFFERENCE	STANDARD DEVIATION OF THE DIFFERENCE	DIFFERENCE / S. D. OF THE DIFFERENCE	CHANCES IN 100 OF A TRUE DIFFERENCE GREATER THAN 0
Male and Diœstrum	0 days	5.3	1.73	3.06	100
	35 days	5.6	1.80	3.11	100
	155 days	1.0	1.18	.84	80
	155 days — copulatory act.	2.8	2.19	1.27	89
Diœstrum and Cornified	0 days	3.1	1.61	1.92	97
	35 days	4.7	1.61	2.91	99.8
	155 days	1.1	1.98	.55	71
	155 days — copulatory act.	5.2	.97	5.36	100
Male and Cornified	0 days	2.2	1.90	1.15	87
	35 days	10.3	1.41	7.30	100
	155 days	.1	2.08	.04	51
	155 days — copulatory act.	2.4	2.28	1 05	85

B. Individual differences

The changes in variability tend to support the conclusions drawn from an analysis of the differences in the average number of crossings for the various groups. The magnitude and characteristics of the scatter around the averages is shown in Table 25. The graphs of Figure 9 show the relation of the changes in variability to the changes in the average number of crossings for various periods of segregation. Figure 11 is designed to facilitate a comparison of the variabilities for the different incentive stimuli.

Variability of response to the normal incentive, as measured by

TABLE 24

Incentive indices for the female groups

(Average for the normal sex stimulus as the standard of comparison)

Depriva-tion period	Unsegregated		35 days		155 days		155 days—Copulatory activity	
Stimulus animal	Diœstrum	Cornified	Diœstrum	Cornified	Diœstrum	Cornified	Diœstrum	Cornified
Total	—.229	—.083	—.247	—.575	.038	—.004	.120	—.133
Partial	—.200	—.050	—.404	—.774	.016	—.011	.168	—.219

TABLE 25

Measures of variability for females

STIMULUS ANIMAL	DEPRIVATION PERIOD	AVERAGE DEVIATION	QUARTILE DEVIATION	COEFFICIENT OF VARIATION	RANGE	RELATIVE DENSITY*			
						1st quartile	2nd quartile	3rd quartile	4th quartile
Male	0 days	3.6	3.5	31.6	6—19	18	25	33	66
	35 days	4.2	2.3	37.1	2—24	8	47	40	15
	155 days	2.4	2.8	24.1	8—16	40	40	32	111
	155 days — copul. act.	5.1	6.4	62.6	2—19	38	15	15	38
Female in diœstrum — recuperative stage	0 days	3.1	2.5	43.8	0—15	14	50	47	23
	35 days	3.9	3.5	56.4	0—15	18	22	40	28
	155 days	1.5	1.2	14.0	10—16	37	66	100	55
	155 days — copul. act.	1.2	1.5	11.5	11—15	100	66	66	100
Female — cornified stage	0 days	3.2	2.9	33.3	6—20	37	23	66	15
	35 days	2.6	1.5	84.2	0—16	45	76	58	9
	155 days	4.7	5.3	47.8	2—20	15	22	16	55
	155 days — copul. act.	1.6	.5	25.6	3—9	18	142	250	250

* Relative density $= \frac{100}{\text{quartile range (continuous series)}}$

both the coefficient of variation and the average deviation increases slightly with 35 days of segregation as compared with the unsegregated group. The range also indicates increased variability, being 6—19 for the unsegregated group, and 2—24 for the 35-day group.

Although the central tendencies of the unsegregated and 35-day groups are very similar in terms of the average number of crossings to the normal incentive, individual differences are increased slightly by segregation. It may be that the drive of some females is waning, while for others the presence of a male again during the test reinforces a drive made more intense through inability to reach consummation.

As judged by all three measures — namely, average deviation, coefficient of variation, and range, for the normal sex drive, the 155-day group are less variable than either the unsegregated or 35-day females. The reason for the lack of variability of the 155-day group is better appreciated when we are reminded of the fact that this group has lived under conditions less favorable than the other groups for the development of individual differences. Let us examine these conditions for the various groups. First, all groups, other than the 155-day group, were segregated after puberty — a condition which has permitted the formation of very definite sex habits in response to a normal stimulus. The 155-day group, on the other hand, were segregated before puberty—a condition which precludes any possibility of the formation of highly integrated sex habits in response to a normal and, therefore, definitely oriented sexual stimulus. The first copulatory experiences at puberty would probably have a tendency to condition the drive of different females unequally. But the 155-day group have never had an opportunity for the stimulative effect of copulation, whereas both the unsegregated and 35-day group have had an opportunity to copulate since puberty. Consequently, differences in aggressiveness in the case of the unsegregated and 35-day groups are probably accentuated and are manifested objectively in the form of mating behavior. By aggressive mating behavior is meant responses on the part of the female which tend to stimulate the male and induce copulation.

The response of the unsegregated group to a normal stimulus varies more than that of the 155-day group for another reason — namely, the unsegregated females may have had widely differing amounts of copulatory activity before being tested, because the experimenter had no way of measuring how long the animal had been in oestrum before the test. This should tend toward greater variability in the response to the normal stimulus than would be found in the case of the 155-day group. The amount of sex activity immediately previous to the test period might be just sufficient with some animals to reinforce the drive; in others enough to lessen the drive.

Neither of these conditions obtains for the 155-day group. We may, therefore, safely conclude that the dispersion of the 155-day group for the response to a normal stimulus is less than that of the other groups because, not having had a chance to become habituated to the normal stimulus, the possibility of unequal conditioning through practice and the inequalities in the rate of disintegration of sex habits resulting from lack of exercise, are thereby eliminated.

Differences in variability of response to non-receptive females follow the same trend as those for the normal stimulus, although the differences between the various segregated groups tend to be greater for the non-receptive female than for the normal stimulus. The greatest variability, as measured by average deviations and the coefficient of variation, occurs with the 35-day group and the least variability with the 155-day group. Since a strong drive produces homogeneity in a group, an increase in the relative importance of other drives may be the cause of variability. This may help to explain the increased variability of the 35-day group.

From the large number of crossings of the 155-day group and from the lack of opportunity to become conditioned to the normal stimulus, we have concluded that they may be homosexual. The unity of the drive to cross to a non-receptive female further supports this conclusion, for we have noted repeatedly in the case of the male groups (see section on Males) that a strong drive causes a unified response so that individual differences approach a minimum.

When a variability of the drive to receptive females is examined, it is noted that the order of the differences for the average deviations is not the same as that for relative variability. To be specific, the average deviation is smallest for the 35-day group, but the coefficient of variation is by far the largest for this group. The absolute variability, i.e., average deviation, is important in itself, but, since groups are being compared, relative variability is the most suitable measure. When the groups are compared in terms of relative variability, the unsegregated animals are the least variable, the 155-day group more variable, and the 35-day group extremely variable.

As previously explained, there seems to be considerable justification for assuming that a strong drive is indicated by homogeneity of response. Likewise, when variability is great, the drive is weak. Upon this basis the increased variability of the 155-day group to receptive females can be explained. If the drive is weak, the crossings which occur are in response to various incentives and, therefore, cause increased variability for the groups. In other words, even though the fairly large number of crossings indicates a large amount of activity due to internal stimulation, the variability indicates that the specific drive to a receptive female is weak, thus enhancing the relative significance of exploratory behavior and general activity. The increased variability of the 35-day group, on the other hand, indicates considerable individual variability in negative conditioning. In other words, some females have been negatively conditioned to receptive females more strongly than others.

A brief survey of the differences in relative effectiveness among the three types of incentive stimulus, as measured by variability, lends further support to conclusions previously drawn. Figure 11 reveals the fact that the variability of the unsegregated group is fairly constant for the three incentives. The 35-day group resembles the unsegregated group in terms of absolute variability. The coefficient of variation, however, discloses genuine differences. The drive to males is most homogeneous (37), that to non-receptive females is next (56), while that to receptive females is extremely heterogeneous (84). The data of the 35-day group would, therefore, seem to suggest rather conclusively that the normal sex drive is very strong, while the drive to receptive females is very weak. Since receptive females are so ineffective, the crossings are probably due to the functioning of various other drives.

The differences in variability of the 155-day group are perhaps more significant than those of the other groups. The drive to non-receptive females is most homogeneous, that to males less homogeneous, and that to receptive females quite heterogeneous. These differences support the view, treated more fully in the previous section, that the response to the non-receptive female is due to conditioning toward the non-receptive female; that to males, is probably due largely to the novelty of the stimulus; and that to the receptive female is due to the combined effect of several drives.

It is interesting to note that, although the averages of the 155-day group in general do not differ widely from those of the unsegregated, nevertheless the indices of variability indicate that the 155-day group is in general more homogeneous than the unsegregated group. This would seem to indicate that the unsegregated group exhibits more general activity than the 155-day group. That this long period of segregation has had a tendency to make these animals sluggish is also evidenced by the relative number of approaches and contacts for the two groups — the number of approaches and contacts is uniformly low for the 155-day group.

C. Persistence of the sex drive as shown by the temporal distribution of drive behavior during the test

The third measure of the strength of the sex drive, which has been employed in this investigation, is its persistence throughout the twenty-minute test period. The constancy of the drive is indicated by the percentage changes for the five-minute intervals (cf. Table 18 and Figure 13). On the strength of the data, not much can be said definitely about persistence. The differences, as indicated by the variations in the number of crossings for the five-minute intervals, are not great enough to warrant any sweeping conclusions. The number of crossings is usually greater at first and then tends to fall off somewhat before the end of the period.

The data seem to indicate that the drive of both the unsegregated and the 155-day groups is quite persistent regardless of the nature of the incentives. For both these groups external conditions have been constant. That is, no outside agency has interfered with the development of the sex drive.

The 35-day group was originally conditioned to males and allowed to copulate freely until 150 days of age. In the last 35 days before the test period segregation has prevented the normal functioning of the sex drive. The result, as indicated by the temporal distribution, has been to cause the drive to the normal stimulus to become increasingly stronger as the test period continues, that to the non-receptive female to be less persistent, and that to the receptive female to approach zero after the first five minutes (over 60 per cent of the crossings occur in the first five minutes). An examination of the temporal distribution for evidences of persistence in response to the three types of stimuli gives no significant differences, except for the 35-day group. Persistence, then, is a poor measure for differences in the relative potency of the three types of incentives to elicit a response on the part of the female test animal.

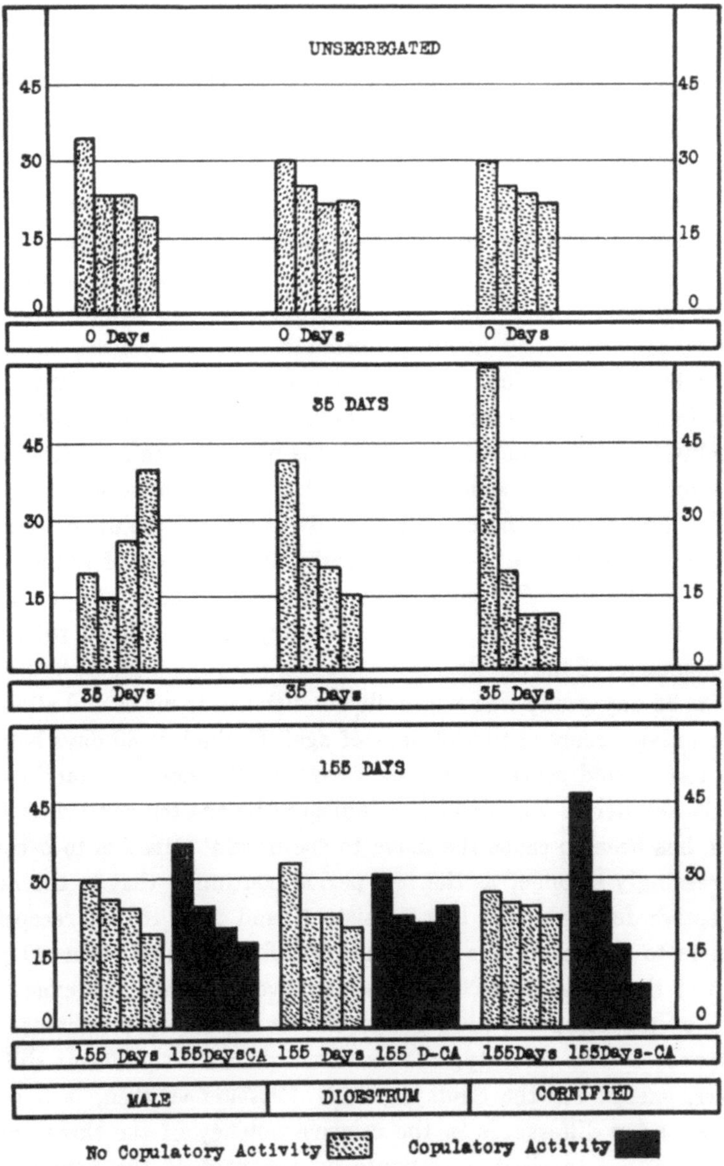

FIG. 13. Column diagram showing temporal distributions of crossings of the female groups during the twenty-minute test period in five-minute intervals.

D. The effect of copulatory activity (five responses) upon the test one hour later

1. *Variations in strength of the drive and the relative incentive values of the stimuli.* Under the section on method, the reasons for giving an interval of copulatory activity (five copulations one hour previous to the test) to the females of the 155-day group were stated. The results for this group are presented (Tables 19, 21, 22, 23, 24, and 26, and Figures 9, 10, 11, 12, and 13).

One of the significant results is that copulatory activity causes a decrease in the number of crossings. The average number of crossings to a male decreases from 12.4 to 10.2. The average number of crossings to a receptive female after copulatory activity decreases even more, dropping from 12.3 to 7.8. Copulatory activity, however, produces practically no effect upon the response to the non-receptive female, the average with copulatory activity being 13.0, and without, 13.4. Partial averages further emphasize these relationships. The incentive indices also show very clearly that the tendencies, as indicated by the direction of the small relative differences for the receptive and non-receptive stimuli, are accentuated by copulatory activity. The index for the non-receptive female becomes still larger in the negative direction, while that for receptive females becomes even larger but in the positive direction. A negative index indicates that the incentive value is higher for the non-receptive female than it is for the normal stimulus. This fact in itself suggests the development of homosexuality. The receptive female, on the other hand, loses in effectiveness after copulatory activity. When the index is calculated for the last fifteen minutes, these differences are even more strikingly exhibited.

All of the quantitative measures indicate that five copulatory acts, one hour before the test, effect a decrease in the drive to cross to males. An examination of the individual records suggests that copulation was painful. More than 50 per cent of the females which crossed less than eleven times gave evidences of some attempt to avoid the copulatory advances of the male. On the other hand, all of the animals which crossed more than eleven times were decidedly aggressive in stimulating the male during the copulatory interval. Escape behavior is well exemplified in the following example:

1st min. Male copulates. Female remains inactive.
2nd min. Male approaches. Female runs.
Male noses female genitals. Female kicks.
3rd min. Male noses female genitals. Female lies on her back and kicks.
4th min. Male licks his genitals. Female approaches.
5th min. Male copulates. Female runs and squeals.
6th min. Male licks his genitals. Female remains inactive.
7th min. Male washes his body. Female remains inactive.
8th min. Male copulates twice although female kicks and squeals.
9th min. Male noses female genitals. Female retreats.
10th min. Male noses female genitals. Female kicks.
11th min. Male noses female genitals. Female kicks.
12th min. Male attempts to mount. Female turns and fights.
13th min. Male copulates. Female kicks and squeals.

The following is a fair example of aggressive behavior of female (group crossing more than eleven times) during the copulatory interval:

1st min. Male noses female genitals. Female takes short run characteristic of mating.
Male copulates. Female walks about the cage.
Male copulates—then licks his genitals. Female approaches, noses male genitals.
2nd min. Male licks his genitals. Female again approaches and noses male genitals.
Male explores cage. Female follows and noses genitals of male
Male stands still. Female approaches and then gives characteristic run.
3rd min. Male copulates. Female approaches again and then hops a short distance.
Male copulates. Female noses and then gives short run.
4th min. Male licks his genitals. Female approaches and then gives a short run.
5th min. Male stands. Female noses male genitals and then hops a short distance.
6th min. Male copulates. Female again noses male genitals.

From these records it seems justifiable to conclude that the first copulatory experience of mature animals is very effective in conditioning the female either *positively* or *negatively* to the male.

While five copulatory acts may induce or inhibit a sex drive to the male, it does not serve to break down the homosexual drive to the non-receptive female. In other words, after copulatory activity the homosexual tendencies seem to be about as firmly fixed as ever. Not only is there a decrease in the number of crossings to a male, but there is also a quite decided decrease in the number of crossings to a receptive female, facts which indicate considerable increase in specificity of response to non-receptive females. The number of cases is too small to generalize from with assurance. We can at least state that, for the cases tested to a receptive female as the stimulus, the average fell from 12.3 to 7.8 after copulatory activity and that no female crossed over nine times, while 70 per cent of the animals which were tested without having sex activity crossed more than nine times. The effect of copulatory activity, therefore, in the 155-day group, is to increase differences in incentive values. It may be possible that if the 155-day females were given more copulatory activity in successive periods of œstrum, they might be reconditioned, so that they would give an avoiding reaction to non-receptive females and display aggressive mating behavior in the presence of a male.

2. *The effect of previous copulatory activity as measured by variability.* The measures of variability shed some light upon the conclusions previously drawn. We find that five copulations increase the variability of response to a male, supporting the hypothesis that copulatory activity may condition some females positively to a male and others negatively. An inspection of the table of relative densities will show that this conclusion is still further supported by evidence of bimodality.

Copulatory activity does not affect variability of response to a non-receptive female, a fact which supports the conclusion, already drawn, that the homosexual drive to a non-receptive female is not changed by previous copulatory activity. Copulatory activity, on the other hand, lowers variability of response to a receptive female, a fact which lends support to the conclusion that copulatory activity tends to enhance the differences in incentive value. The decrease in variability is one indication of greater homogeneity of response and shows the more predominant operation of a single drive. This single drive is probably manifested in the form of exploratory behavior, the decrease in the number of crossings being due to a diminution of sex behavior (already small) toward the receptive female. The evidence for specificity of response to the non-receptive female merely adds further support to the argument for homosexuality in the 155-day group.

3. *The effect of previous copulatory activity as measured by persistence.* Examination of the temporal distributions indicates that previous copulatory activity tends to make the behavior toward males and receptive females more transient in character. There is a considerable drop in persistence in the case of the male stimulus and a very decided drop in persistence in the case of the receptive female. It seems to the writer that this is still further evidence that the chief effect of copulatory activity, in the case of the 155-day group, is to increase the specificity of response and, since it is shown especially toward the non-receptive female, there is very strong evidence of homosexual behavior.

V. Summary of Conclusions

A. Male

Effect of segregation

1. Unsegregated males are fairly active but do not seem to be motivated strongly by the sex drive when tested at 185 days of age, all three incentives (male, non-receptive and receptive female) being about equally effective.

2. One day of segregation, at 185 days, has the effect of doubling the number of crossings to the normal incentive (receptive female), and of increasing very materially the number of crossings to the non-receptive female.

3. Thirty-five days of segregation (150-185 days) decrease the normal sex drive, due probably to a waning of the drive in some animals and to the development of homosexual tendencies in others. The non-receptive female appears to act to some extent as a substitute stimulus for the receptive female in both this and the 1-day group.

4. Segregation before puberty (at 30 days) seems to lower the activity of the group toward both the receptive and the non-receptive female. The sex drive in general appears to be subnormal but more crossings occur to the male than to either the receptive or the non-receptive female.

Influence of the mating period on segregation

1. A 2-hour mating period interpolated 24 hours before the test, seems to counteract in part the decrease in the normal drive resulting from segregation. However, the longer the animals have been segregated, the less effective is the mating period.

2. The mating period seems to cause the 35-day group to become sexually more active to the male as well as to the female incentive.

3. The mating period evidently decreases homosexuality in the 155-day group.

4. During the 2-hour mating period the number of males which actually copulate was small, regardless of segregation conditions. However, the mating period has a decided effect on the number of crossings in the test 24 hours later.

B. Female

Effect of segregation

1. The 35-day period of segregation does not appear in the 185-day animals to decrease the sex drive to the male. The number of crossings was much lower to the non-receptive female and extremely low to the receptive female. In the latter case, there was strong evidence of negative conditioning having taken place in the living cage previous to the test; more than half of the crossings occur within the first three minutes of the test and the total crossings are fewer than when no incentive whatever was employed.

2. Segregation before puberty (at 30 days) in general weakens the drive for the normal, or male, incentive. The female appears to be stimulated most by a non-receptive female and least by a receptive female (which reacts against sex advances) thus showing the operation of the homosexual factor.

Influence of the mating period on segregation

In females segregated before puberty (see 2 above) five copulatory responses one hour before the test apparently have no effect in changing the direction of the homosexual drive. Mating in this manner decreases the drive to the normal incentive and also increases variability. The detailed records suggest that copulation was painful with the result that an avoiding reaction to males was noted in the test one hour later.

C. *Comparison of the male and the female*

A detailed comparison of the sexes from group to group seems hardly necessary in view of the analysis already made. However, the following generalizations regarding sex differences seem to be warranted.

1. The values for the females are consistently higher than for the males under both segregated and unsegregated conditions, although the difference is not great in certain cases. This suggests that the female drive tends to be stronger than the male under the conditions imposed. The difference between sexes is practically negligible when the points of maximal sex activity are considered. The unsegregated female and the 1-day segregated male, representing such optimal drive conditions, differ only slightly (14.2 and 13.5 respectively).

2. The relative effect of segregation is greater for males than for females in cutting down crossings to the normal incentive. Segregation seems to bring about the appearance of the homosexual tendency earlier in the male, although when segregation is begun before puberty, the effect is approximately the same for both sexes.

Bibliography

(1) Avery, G. T. 1925. Notes on reproduction in guinea pigs. *J. Comp. Psychol.*, 5, 373-96.

(2) Craig, W. 1914. Male doves reared in isolation. *J. Animal Behav.*, 4, 121-33.

(3) Hamilton, G. V. 1914. A study of sexual tendencies in monkeys and baboons. *J. Animal Behav.*, 4, 295-319.

(4) Jenkins, T. N., L. H. Warner, and C. J. Warden. 1926. Standard apparatus for the study of animal motivation. *J. Comp. Psychol.*, 6, 361.

(5) Kempf, E. J. 1917. The social and sexual behavior of infra human primates with some comparable facts in human behavior. *Psychoanal. Rev.*, 4, 127-54.

(6) Long, J. A., and H. McC. Evans. 1922. The oestrus cycle in the rat and its associated phenomena. Berkeley, Univ. of Calif. Press, 148 pp. (See pp. 20 and 71.)

(7) LOUTITT, C. M. 1927. Reproductive behavior of the guinea pig. I. The normal mating behavior. *J. Comp. Psychol.*, 7, 247-65.
(8) MOSS, F. A. 1924. A study of animal drives. *J. Exper. Psychol.*, 7, 165.
(9) STONE, C. P. 1922. The congenital sexual behavior of the young male albino rat. *J. Comp. Psychol.*, 2, 95-153.
(10) STONE, C. P. 1923. Further study of sensory functions in the activation of sexual behavior in the young male albino rat. *J. Comp. Psychol.*, 3, 469-75.
(11) STONE, C. P. 1924. A note on "feminine" behavior in adult male rats. *Amer. J. Physiol.*, 68, 39-42.
(12) STONE, C. P. 1924. The awakening of copulatory ability in the male albino rat. *Amer. J. Physiol.*, 68, 407-25.
(13) STONE, C. P. 1926. The initial copulatory response of female rats reared in isolation from the age of twenty days to the age of puberty. *J. Comp. Psychol.*, 6, 73-83.
(14) WARNER, L. H. 1927. A study of sex behavior in the white rat by means of the obstruction method. *Comp. Psychol. Monog.*, 4, No. 22, 1-68.
(15) WATSON, J. B. 1903. Animal education. *Univ. Chicago Contrib. Philos.*, 4, No. 2, 53.
(16) WHITMAN, C. O. 1919. Orthogenic evolution in pigeons, 3. The behavior of pigeons. *Publ. Carnegie Instit. Wash.*, No. 257.

3. THE EFFECTS OF GONADECTOMY, VASOTOMY, AND INJECTIONS OF PLACENTAL AND ORCHIC EXTRACTS ON THE SEX BEHAVIOR OF THE WHITE RAT [1]

H. W. NISSEN

I. INTRODUCTION

A. Nature and scope of the problem

The general aim of this investigation was to measure the effects of the internal secretions of the gonads on sex behavior of male and female albino rats. More specifically, the attempt was made to determine the influences of gonadectomy, of injections of the female hormone into males and females, of injections of several types of testicular extract into males, and of vasotomy on such behavior. Among the more important problems of endocrinology, at which these experimental procedures were especially directed, may be mentioned the following: (1) The function of the gonads in controlling sex behavior and the probable manner in which this control is exercised in the male and female, respectively. (2) The relatively immediate effect of the female hormone on the male; sex-specificity of the gonadal substance and the relation between the heterologous sex gland hormones. (3) The analysis of the testicular substance into its efficient and incidental components. (4) The question of the efficacy of vasotomy in restoring sexual vigor in old males.

This study differs from two others (59, 146) included in the same general project on animal drives and also dealing with sex behavior in that the latter concerned themselves with what may be termed "normal" variations in intra-organic and environmental conditions, whereas in the present series of tests arbitrary and "artificial" physiological conditions were imposed.

[1] Reprinted with modifications from *Genetic Psychology Monographs*, 1929, 5: 451-547.

The present investigation differs in an important respect, also, from most or all previous physiological studies in the field. Earlier workers have contented themselves, for the most part, with one or more of the following types of criteria for judging the effects of their experimental conditions: (1) Morphological and histological changes, including general growth, distribution of fat, distribution and nature of hair or feathering, histology of the vaginal and uterine mucosae, size and histological character of other organs. (2) Mating activity and non-quantitative observations of cage behavior; measurements of spontaneous activity. (3) Miscellaneous, including clinical findings and reports of patients, metabolism determinations, physiological effects on various organs and on contractibility and fatigue of isolated muscles or muscle groups. In the work here reported, on the other hand, the effects of the various physiological conditions induced were measured in terms of the intensity of the drive of the test animal towards a normally adequate sex object. The indices obtained, therefore, are neither of spontaneous activity nor of ability to copulate, but are measures of the strength of the internal stimulation (motivation) causing the test animal to cross an obstruction (constant electric shock) in order to reach a sexually active animal of the opposite sex. The scores yielded by the Columbia Obstruction Method are, of course, the resultants of two kinds of stimulation: intra-organic and environmental; since the latter, including as its most important parts the shock and the incentive, was kept constant, for each sex, throughout the study, the differences in scores obtained can be attributed only to the changes produced in the intra-organic factor.

It has been long recognized (88) that the sex drive of most of the vertebrates can be divided into two distinct behavior sequences, commonly termed the contrectation and detumescence drives (39). The former brings the animal into the proximity of a sex-object and, in rats, leads to nosing of the genitals, gentle biting, and other "sex play," whereas the latter culminates in coitus. The term "detumescence drive" is, obviously, more descriptive of the activity of the male, who actually ejaculates the accumulation of seminal fluid from the genital tract in copulation, than it is of the female who, in certain species, ovulates some time after mating (86).

Nevertheless, the descriptive division of the sex drive into two constituent parts is applicable to both sexes, and, although "copulatory drive" probably is a more exact designation for the second component, the writer has retained here the more conventional terms already mentioned.

It is possible, of course, to consider the contrectation drive as forming merely the "preparatory reactions" of the true sex drive (150) or to think of the two components as constituting successive links in a chain drive or instinct (82, p. 481). Some of the observations described later in this report (page 283) lead the writer to conclude, however, that these "preliminary" activities have an independent end or consummatory reaction of their own. They are brought to a conclusion — the contrectation drive is "satisfied"— apparently, by the olfactory, tactual, visual, and perhaps even auditory and kinaesthetic stimulation which the sex-object affords. This viewpoint receives support from the observation made by two other writers (59, 146), as well as by the present author, that there is no correlation between copulatory strength or ability and the indices of the contrectation drive obtained by the Obstruction Method. The contrectation drive, at least in rudimentary form, is probably a necessary antecedent to the detumescence drive, but the latter certainly does not always follow the manifestations of the former. It seems probable that the contrectation drive has its principal biological significance in its relation to the copulatory or detumescence drive. It may well be that phylogenetically or even ontogenetically the detumescence drive comes first; the preparatory reactions of this drive, then, may develop their own self-sufficient end-reaction in accordance with some such principle as Wundt's "heterogeny of ends." As far as the writer is aware, the present study is the first one to attempt the quantitative determination of gonadal influences on the contrectation drive.

B. Physiology of the sex glands

The present section aims to give only a brief account of those facts known about the sex glands which have a direct bearing on the results of this investigation and their interpretation. In view of the great volume of the literature on the gonads — at least six hundred, and probably over a thousand articles and books on the topic have appeared during the last decade alone — a comprehensive survey of the field obviously cannot be undertaken here. More or less complete compilations have been made by Lipschütz (72), Biedl (12), Marshall (86), and Harms (47).

1) *The female gonads.* It is well known that among mammalia other than primates, the female will mate only during the rhythmically recurrent periods of rut or heat; in the intervals between such periods the animal is sexually quite inactive. These alternations in sex behavior are only one item in a syndrome of changes going on in the female organism during the oestrous cycle, among which modifications are the regular growth and rupturing of the Graafian follicles and formation of corpora lutea in the ovary, and the cyclic distension and degeneration of the uterine epithelium. The correlated changes in the histological nature of the vaginal mucosa form the basis of a convenient and accurate method, developed by Stockard and Papanicolaou (130, 131), for determining the various stages of the oestrous cycle. An excellent description of the cycle in the rat is given by Long and Evans (80), who have shown that the periods of heat are characterized by a vaginal mucosa composed exclusively of cornified cells.

After extirpation of the ovaries the phenomena of the female cycle come to an end (80, p. 94). The effect of spaying on behavior has long been recognized, as is apparent from the following statement by Aristotle (6): "The ovaries of sows are excised with the view of quenching in them sexual appetites." It is clear, therefore, that mating behavior, in common with other oestrual manifestations, is dependent on the ovaries. The proof that the ovarian influence is chemical rather than nervous is found in experimental and therapeutic work in which the effects of ovariectomy were overcome by transplantations of the female sex glands or by injections of ovarian extracts. Knauer (62), using rabbits, was apparently the first (in 1895) to perform successful auto-transplantations of ovaries. Subsequent work with homoio- and hetero-transplants has quite conclusively demonstrated the chemical nature of the ovarian control of the oestrous cycle. The grafted ovary can maintain its normal structure and the rhythmical character of its functioning; some cases are reported (42) in which the ova of the transplant have been fertilized and have developed normally. Feeding and injection of ovarian substance have been tried, with varying degrees of success, for the past 30 to 35 years. Definite proof that the oestrous cycle can be re-established in spayed mammals by injections of an extract of the follicular fluid has been given by Allen and Doisy (2, 3) and others. Steinach (128) has shown that extracts of ovary and of placenta will induce oestrum in infantile spayed rats who have never had a normal cycle, and that such injections will cause the reappearance of the sexual cycle in senile females whose ovaries had ceased their rhythmical functioning two to five months earlier. Slonaker (116) was able to produce the typical vaginal changes of oestrum by injections of the fresh follicular hormone into female albino rats 538 to 971 days old, some of whom had been previously spayed. Doisy (28) points out that an overdose of the hormone is not harmful (one rat was injected with 20 rat units), and Courrier (26) found that varying the amount of the injection did not influence the nature of the changes in the vaginal epithelium.

The exact part of the ovary responsible for the production of the oestrous-inducing hormone and the mechanism of its elaboration have not been conclusively determined. The fact that this substance is regularly found in the follicular cavity and walls has led many writers, including Allen and Doisy (2), to the conclusion that the hormone is produced by the follicles, perhaps under stimulation from the maturing ova. Pugh believes that the cystic condition of the larger follicles, which he found in nymphomaniac cows and heifers, may have stimulated the growth of the follicular epithelial cells. Recent work by Parkes and his associates (17, 100), indicating that the hormone is produced after the germinal epithelium has been destroyed by irradiation, leads this investigator to conclude that the cortex of the ovary is responsible for its elaboration. Steinach (122), Lipschütz (72), and others believe that the *Pubertätsdrüse* or interstitial cells of the ovary secrete the female hormone. The oestrous-inducing hormone has been found in the placentae of woman, of the cow, mare, and other animals. The source of the placental hormone is not definitely known; Parkes and Bellerby (101) suggest that it is derived partly by withdrawal from the maternal circulation and partly by elaboration of the placenta itself. The female hormone manifests complete lack of species-specificity. Thus extracts of human placenta have been injected with positive results into mice and rats (27, 151). Allen and his associates (4) were able to produce oestrous growth in the genital tract of spayed rats by injections of an extract of the ovaries of laying hens. Fellner (32) had positive results with extracts of bird eggs and fish ovaries injected into rabbits. Loewe (77) has demonstrated the presence of the oestrous-inducing substance in the blood of women, rabbits, and cows during the pro-oestrous period. Concerning the little-known chemical composition of the female hormone, it may suffice to note that Allen and Doisy (2) found that the active principle followed the lipoid fraction of the follicular liquid.

Several authors (35) have reported the presence of the oestrous-inducing substance in the corpora lutea. There is good evidence, also, that this part of the ovary produces a substance which has, among other possible functions, an inhibitory influence on the follicular hormone. Loeb (76) and others have been able to hasten the occurrence of oestrum by destruction or removal of the corpora lutea, and Papanicolaou (99) has suspended or suppressed the female cycle for months at a time by weekly injections of a purified corpus luteum extract. It seems, therefore, that the corpus luteum is either a "mixed gland," producing two distinct hormones, or that, while actively secreting only a non-follicular agent, it also contains some of the hormone produced by other parts of the ovary.

Wang (143) has shown that correlated with the other manifestations of the oestrous cycle there are rhythmical variations in the spontaneous activity of female rats. During dioestrum the animals are relatively quiet whereas at oestrum they show great activity. After ovariectomy these fluctuations cease and the activity level remains constantly low.

The influence of the ovarian hormone or hormones extends beyond those conditions primarily associated with the oestrous cycle. The morphological secondary sex characters are gradually lost after spaying, the animal tending towards an asexual or even masculine type. The uterus and other parts of the genital tract not only no longer show the rhythmical variations associated with the female cycle but undergo a progressive atrophy. There is often an increase in weight and decrease in basal metabolism after ovariectomy. Lee (66) found a decrease of about 10% in heat production, following extirpation of the ovaries, and interprets his result as demonstrating the stimulating effect of the female gonads on the thyroid gland. Loewy and Richter (79) report a decrease of 14-20% in metabolism following ablation of the ovaries in dogs; the same writers state that injection of ovarian substance into spayed animals results in raising the metabolism 30-50% above normal. Livingston (75) found that the pituitary gland often enlarges after ovariectomy of rabbits.

2) *The male gonads.* Animals castrated before puberty usually show no signs of a sex drive at any time of life. Steinach (119) states, however, that male rats, prepuberally castrated, may show slight and weak signs of heterosexual interest at the normal age of maturity. Steinach (119), Lichtenstern (67), and Busquet (20) have found that male rodents and horses castrated after the time of puberty continue to manifest sex activity for some time after the operation. This was not an altogether new finding, for we find in Aristotle (6) the following statement: "If a full-grown bull be mutilated, he can still to all appearance unite sexually with the cow." Rieger (108) quotes a number of authors, as far back as Juvenal and Lucian, to the effect that the sex drive (in man) is not extinguished after castration. Jager (58) differentiates between the eunuch and the castrate, the former being deprived of both testes and penis, the latter lacking only testes; he remarks that the eunuch seems incapable of any sexual feeling, whereas the castrate does not lose the ability or desire to have intercourse. Stone (134) has recently made an exact study of the length of time during which copulatory ability continues in rats castrated when 90 days old. (According to Stone's figures (133), the upper age limit for the first copulatory act in male rats is 72 days.) He found that three months after the operation about 43% of his animals were still copulating. Unfortunately, Stone does not tell us what kind of rats he used, so that his results are not directly comparable with our own.

Probably the most generally accepted explanation for the gradual extinction of the sex drive following castration is the one summarized in the following quotations from Lipschütz (72): "Persistence of some signs of the sexual activity which diminish very slowly after postpuberal castration is by no means incompatible with the contention that the *erotization* of the nervous system (Steinach, 1910) depends on the sexual glands." In speaking of the slower extinction of sexual libido in man as compared to animals, Lipschütz says, "I think that this is to be explained by the reflex actions responsible for erection and coitus being fixed and brought into play in the normal man by more manifold external factors than in the lower mammal." If I interpret the implications of these statements correctly, this means that the sexual responses, originally stimulated or facilitated directly by the testicular hormone, become "attached," in accordance with the laws of learning, to certain external stimuli which thereafter can evoke sex behavior independently.

The most obvious effect of gonadectomy is, probably, the loss of secondary sex characteristics. Pelikan (102), Tandler and Gross (135), Koch (63), and others have described the results of castration among the Skopecs, a religious sect of Russia and Roumania, which includes this operation as part of its ritual. Supplementing these accounts by the great mass of clinical literature on eunuchoidism and on operative or accidental castration, we find among the more salient and common consequences of human male gonadectomy the following: abnormal bodily proportions, often but not always accompanied by obesity, feminine distribution of fat, scanty or absent facial, axillary, and pubic hair, but usually luxuriant head hair, very small penis, sallow complexion, and, if the testes were extirpated before puberty, a childish treble voice. Birds, having pronounced secondary sex characters, show the effects of castration in a striking manner; the caponized rooster loses his male feathering, often ceases to crow, and, if the operation was carried out early in life, does not develop comb and wattles.

In 1849 Berthold (10) transplanted the testes of cocks from their normal site to other parts of the body and found that the usual consequences of castration did not appear. Although John Hunter (57) had done similar work in the preceding century, Berthold's results are generally accepted as giving the first definite proof that the testicular influence is chemical rather than nervous. In 1889 Brown-Séquard (19) reported remarkable physical and mental improvements, following injections, into himself, of emulsions of testicular substance. Since this time a large amount of work has been done with auto-, homoio-, and hetero-transplants and with implantations and injections of testicular substance and extracts. Many investigators report excellent results in overcoming symptoms of senescence and eunuchoidism with testicle transplants and with implantations of orchic substance. Among the beneficial results to human patients, mentioned by such workers as Voronoff (140, 141), Stanley (118), Thorek (138), Hunt (55), Lydston (81), Retterer (105), Falcone (31), Muhsam (94), and others, are: loss of obesity, growth of facial and body hair, increase in size of penis, return of potency and erection, lower pitched voice, improvement in memory and general intelligence, increased strength, libido and feeling of well-being, and improvement in asthenia, acne vulgaris, functional disorders, and psychoses. It should be noted, as some of the above authors themselves point out, that suggestion may have been responsible for some of the improvements reported; in some cases, on the other hand, it is claimed that the patients did not know what results to expect. The effects of transplants often do not survive very long, although Voronoff (141) reports the indefinite survival of anthropoid grafts in man, and recently Moore (93) has been able to preserve the normal structure of testes, transplanted into animals, for long periods of time. Implanted testicular tissue soon becomes necrotic but exerts its influence during absorption. The effect of a single injection of testicular extract is at best transitory, and for practical reasons this method has not been employed extensively in human work.

In 1911 Pézard (103) demonstrated the growth of comb and wattles and the retention of normal sex behavior in the castrated cockerel following injections of an extract of pig testicles. Since that time many workers, including especially Steinach (120), Sand (109), Lipschütz (72), and Voronoff (140) have performed experiments on animals in which the effects of castration, senescence, and eunuchoidism were overcome by testicular grafts or injections. A few examples from the more recent work may be cited: Wilhelm (149) reports that homoio-transplants into three senile dogs caused increase of weight, return of sexual potency, and partial rejuvenation of skin. Busquet (21) was able to produce male characteristics in capons and senile cocks by injections of a serum from young bulls, stallions, and rams. Harms (49) found rejuvenescence in old dogs following testicular transplantations. Thorek (138) castrated six apes and after all of them had lost sexual potency (7 to 11 months after the operation) made homoio- and hetero-transplantations. Four of the apes regained completely the ability to copulate, one recovered partially, and the other gave negative results. Bouin and Ancel (14, 15) were able to minimize the effects of castration in guinea pigs by subcutaneous injections of an extract of retained testicles prepared with glycerine and water. Steinach (120) brought back the clasping reflex in castrated frogs by injections of a pulp of testicular substance; the reflex appeared 12-24 hours after the injections and disappeared again in 3-4 days. Loewy (78) was able to prevent some of the effects of gonadectomy in capons by feeding testes, but Steinach (120) found oral administration of testicular substance to be without effect in castrated rats.

There are three different kinds of tissue in the testicle which might produce the male hormone: the germinal epithelium, the interstitial cells of Leydig, and the Sertoli cells. Of these, the first two have usually been considered the most probable source of the efficient substance. Reinke (104) was apparently the first, in 1896, to attribute to the interstitial cells the function of producing the internal secretion controlling sex behavior. Without going into the details of the controversy regarding the "interstitial gland" (Bouin and Ancel, 13), the principal lines of evidence favoring the view that the interstitial cells elaborate the male hormone may be briefly summarized as follows: All of the secondary sex characters may remain after the process of spermatogenesis has been stopped by (*a*) spontaneous or artificial cryptorchidism (72), (*b*) by ligature of the vas deferens (127), (*c*) by irradiation (1, 17), and (*d*) by disease (72). Lillie and Bascom (69) point out that the free-martin, whose occurrence is supposed to depend on the influence of the male hormone, comes into being at a time of foetal life when the male germinal epithelium is not yet functioning but when the male interstitial cells are active. Grafted testes exerting a masculine effect often show an absence of germinal epithelium which is accompanied by a hypertrophy of the interstitial tissue (114). Histologically, interstitial cells look more like typical gland cells than does the epithelial tissue (47). Certain animals showing seasonal periods of sex activity with elaboration of secondary sex characters at this time manifest a synchronous proliferation of the interstitial cells.

Although the majority of workers today probably accept the interstitial cells as the source of the male hormone, this view is by no means universal. Champy (23) points out that potent testicular transplants occur which show only a very few interstitial cells and that some fish with definite sex characters have none of this tissue, whereas the mole, with many interstitial cells, has no marked secondary sex characteristics. Harms (48) points out that testicular grafts do contain unripe spermatogenes and Sertoli cells. Oslund (96) found germinal tissue in the indifferent stage present in cryptorchic testicles. Moore (92) and Oslund (97) were unable to constitute hypertrophy of the interstitial tissue after vasectomy. Ceni (22) reports a case of hyperplasia of interstitial cells with loss of secondary sex characters in a rooster. The negative viewpoint is summarized in the following statement by Kingsbury (61): "Aside from the fact that the interstitial cells look like gland cells, failure to appreciate that the 'indifferent' cells composing the tubules of the testes in the absence of any spermatogenic process are nevertheless parenchyma and a possible source of substances of endocrine function is responsible for the acceptance of the interstitial gland interpretation. As far as I can ascertain, in no instance has a gonadal effect been demonstrated in the absence of parenchyma."

In addition to the changes in copulatory ability, noted above, castration is followed by a significant diminution in spontaneous activity. Hoskins (53) found that when male rats were gonadectomized at the age of 70 days, the activity was reduced to about one-fifth that of normal animals. Gans (37) showed, however, that if the castration is performed during the first 12 days of life this diminution does not follow. Hoskins (54) was unable to raise the activity level of his castrated animals by grafting testes into them, although he did succeed in increasing the activity of two senile males by such transplants. Smith and Boukalik have demonstrated a slight increase following implantations, by the Stanley syringe method, of testicular substance from young males into castrated and senile animals. Feeding of dessicated testis and subcutaneous injections of testis extracts — saline, glycerine, and lipoid solvent extracts and dialysates of several kinds — were tried by Brown and his collaborators with no significant results. Gans and Hoskins (38) showed that the gastrocnemius muscle of castrated rats is heavier than that of normals, although the proportion to total body weight was the same in both cases. The original strength of the muscle was the same, but those of castrated animals fatigued more quickly and, therefore, did less work. These authors state that their results need not be ascribed to lessened muscular efficiency per se, but are more probably due to inadequacy of supporting functions, such as circulation or respiration. Athanasiu and Pézard (7) found that the action current of the gastrocnemius muscle during voluntary walking in capons is only 20 per cent of that in normal cocks. Bukalik and Hoskins (16) gave repeated implantations of fresh rat testicle substance to castrated and senile rats, but found an increase in spontaneous activity in the case of only one senile.

Loewy and Richter (79) found the same decrease in basal metabolism in gonadectomized male dogs as has been reported above for females. Tandler and Gross (135) report that the thyroid glands of the Skopecs are extremely small. Bertschi (11) found no change in respiratory exchange after castration or following subsequent injections of testicular extracts. Korenchevsky (64) found the gaseous metabolism unchanged after injections of testicular emulsions. Tsubura (139), on the other hand, reports a diminution of respiratory exchange following gonadectomy; this effect could be overcome by implantation of testes but not by feeding testicular substance. Wheelon (147) reports a fall of blood-pressure after castration and a decrease of blood-pressure reactions to nicotine; he interprets this result as showing that extirpation of the testicles affects the sympathetic nervous system itself. In a later article (148) the same author reports that these consequences of castration are partially overcome by implants of slices of testicles. Fichera (33) found that after gonadectomy the pituitary enlarges to about twice its normal size in many animals. Hatai (52) also found this increase, but Livingston (75) reports that it is not an invariable accompaniment of castration. Steinach and Kun (129) report that sexual development could be hastened by injections of pituitary extract into infantile rats; the characteristics of eunuchoidism in rats were overcome by the same treatment. Smith (117) obtained similar results with transplantation of the pituitary gland but found that the results were negative when the host had been previously castrated. Donaldson and Hatai (29) and Hatai (51) found a diminution in the weight of the nervous system following extirpation of the male gonads.

Atrophy of the penis, the prostate gland, the seminal vesicles and Cowper's gland,[2] following castration, has been found by many investigators (40, 72, 86, 112, 119). The seminal vesicles probably have an external secretion (34), but the writer has found no evidence of an internal secretory function of this organ. There is some evidence indicating that the prostate gland elaborates a hormone which passes into the blood stream. Serrelach and Parès (113) found that ejaculations ceased after prostatectomy in dogs. Ott and Scott (98) report that injections of prostatic extract are followed by erections. Various psychical disturbances, including neurasthenia, following extirpation of the prostate in man have been interpreted as indicating an internal secretion of this gland (41). Macht and Ulrich (84) found that prostatectomized white rats learned a problem involving muscular coordination and efficiency more slowly and less well than did normal animals. Hunt (56) believes that testicular transplantation is more successful when accompanied by injections of prostatic extract. Wallace (142) and Griffiths (40) point out that the prostate exhibits seasonal variations in those animals which pair only at certain times of the year. Steinach (127) reports hypertrophy of the prostate after vasectomy, the latter operation, according to this author, increasing the internal secretion of the testes. Macht (83) found

[2] Nagel (95), however, reports that Cowper's gland does not undergo hypoplasia after gonadectomy.

that feeding of prostate accelerates the metamorphosis of frog and salamander larvae. Korenchevsky and Carr (65) state that injections of testicular emulsions caused a slight decrease in nitrogen metabolism, whereas if prostate emulsion was injected at the same time, there was an increase of as much as 11 per cent.

3) *Hermaphroditism and sex reversal.* In true hermaphroditism, rare occurrence among mammals, the gonads of both sexes are present in the same organism. One or the other sex may predominate in regard to behavior, although various degrees of bisexuality are often present. The morphological sex characters are usually mixed, in varying proportions. Sand (109, 111) has succeeded in producing hermaphroditism experimentally, demonstrating that testes and ovaries may survive in the same animal without inhibiting each other in development or function. These hermaphrodites, Sand reports (111), often showed rapid alternations in behavior, changing from the character of a placid female to that of an aggressive male within an hour. Lipschütz and Vinals (74) transplanted ovaries into two non-castrated male guinea pigs; at a time when their mammary glands were actively secreting, these animals impregnated three normal females. Steinach (125) reports that if ovaries are transplanted into normal males the behavior for the first few weeks is definitely male but then passes to a periodical alternation between that characteristic of male and that characteristic of female.

In pseudo-hermaphroditism the gonads of only one sex are apparently present, but the organism has also some of the secondary characters of the other sex. Whether these cases arise as the result of a perversion of some of the hormone-producing gonadal tissue or whether there are present, in addition to the gonads of one sex, some of the heterologous secreting cells, has not been definitely determined (45). Steinach (126) has reported that in a number of human cases, showing no marked abnormalities except homosexual behavior, he found unusual elements in the testes, resembling, in several respects, cells of the corpus luteum.

Steinach (121, 122, 123, 125) gonadectomized male and female guinea pigs and rats and grafted into them the heterologous sex glands. After a period of time these animals developed the morphological and behavior characteristics belonging to the sex of the grafts. Moore (90, 91) has confirmed the possibility of feminizing male and masculinizing female guinea pigs and rats. By the same method Brandes (18) has produced sex reversals in deer, and Guthrie (42), among others, in fowl. Lipschütz (71) and others have found that when a spayed female is masculinized, the clitoris develops into a penis-like organ. Wang, Richter, and Guttmacher (144) were able to raise markedly the spontaneous activity level of castrated rats by transplanting ovaries into these animals; they report that a number of these feminized males showed the rhythmical variations in activity characteristic of normal females.

A few cases are reported in which the characteristic sex activity of male animals has been facilitated by the female hormone. Steinach (120) found that the clasping reflex of castrated male frogs could be elicited by injections of triturated ovaries of females in heat. Harms (46) and Meisenheimer (87) have confirmed this finding.

II. Method and Procedure

A. Age and care of animals

This study is one of a series of investigations which together constitute a comprehensvie survey of drive behavior in the white rat. In all of these studies laboratory conditions and such factors as the strain and age of animals used have been kept as much alike as possible. A complete description of such conditions is given in the first monograph of the series (146, pp. 24-27) and will not be repeated at length here.

The animals used were albino rats from the "Experimental Colony Strain" of the Wistar Institute of Anatomy, Philadelphia. All of the females, and all but 30 of the 168 males tested, were born at the Wistar Institute and were kept there until at least two months old. The other 30 males were of the first generation born of Wistar animal parents in this laboratory. No animal was used which had not been living in the laboratory for at least 35 days before the test. The animals included in the two senile groups ranged in age from 16 months 0 days to 16 months 22 days, averaging 16 months 9 days, at the time of the test. All other animals were approximately 185 days old when tested, the range being from 175 to 196 days.

An over-sufficient supply of food (McCollum's mixture) was kept in the cages at all times; lettuce was given weekly. This diet was never supplemented during the 35 days immediately preceding the test, but prior to this time, especially in the case of the very young animals and the seniles, a meat and bone powder, made into a paste by adding boiling water, was given in small quantities every 7-10 days. An ample supply of fresh water was available at all times.

B. Aparatus and procedure

(1) *Apparatus.* A complete description of the apparatus used in this investigation is given by Jenkins, Warner, and Warden (60). The modified grid (146, pp. 28-31) and the improved control mechanism, used on the fifth preliminary crossing and described by Jenkins (59, pp. 471-474), were used. An experimental analysis of the most important features of the apparatus is given by Warden and Nissen (145). For Figure 1 of this monograph see Part I of this volume.

(2) *Social factors.* The animals were kept in the laboratory for at least 35 days before being tested. There were never less than 5 and never more than 12 rats in one living cage; the usual number was 6 to 8. The cages were wire mesh, of 5,900 cubic inches capacity. We hoped to reduce to a minimum, by this precedure, the purely social or gregarious drive to other animals, regardless of sex. That we were successful in this attempt is shown by the work of Jenkins (59), who found that the number of crossings of males to a male incentive, and of females to a receptive female incentive, is not significantly different from the number of crossings to an empty incentive compartment.

Thirty-five days before the test the sexes were segregated; that is, for about a month preceding the test, males and females were kept in separate cages. Warner (146, pp. 25-26) has given the practical reasons for adopting this procedure, and Jenkins (59) has shown that this period of segregation does not materially affect the results. Before the time of segregation the two sexes were present in about equal numbers in each cage. This means that for about three months following the average age of puberty (see Stone, 133) all of our animals had what is probably normal opportunity for sex activity. In the case of the two senile groups this period of normal sex life was, of course, much longer.

(3) *Incentive animals.* The incentive for the female test animals was always a sexually aggressive male. In order to keep the incentive for the females as constant as possible, only four different males were ever used for this purpose. These males were chosen for uniformity of behavior and for sexual vigor. The writer has found that a receptive female[3] is usually considerably more active than the average male, so that in order reasonably to equate the incentives for the two sexes it was necessary to select especially energetic incentive males. When a number of females were being tested in succession it was often found advisable to change incentive animals occasionally, since these rats would sometimes become sluggish after being in the apparatus for several hours. It was found that allowing the incentive male to copulate a few times, outside the apparatus, would often revive his activity. Before beginning a test, the incentive animal was always given about 20 minutes to adapt itself to the apparatus; it would then direct its entire activity to the test animal when the latter was introduced.

The incentive for all male test animals was a receptive female. Since 60 or more animals were usually available for this purpose, there was, with occasional exceptions, not much difficulty in securing two or more females who were in heat. A female was used as incentive only after it had copulated several times with at least two males. When two or more tests were made in succession, the female incentive animal was taken out of the apparatus between tests and given opportunity for copulation. If this activity did not occur in typical and characteristic fashion (indicating that the animal was passing out of oestrum), it was discarded and another incentive female substituted. The individual differences in mating behavior on the part of receptive females are greater than would be supposed

[3] Throughout the paper the terms "receptive female," "female in heat," "female in cornified stage," and "female in oestrum" are used interchangeably.

from cursory observation (see page 281). Care was taken to select always those females which showed the most typical and what seemed to be the most provocative pattern of behavior. In this connection it should be noted that the incentive animals used for the first 61 males tested (see page 281) were about 1 year old whereas the incentive females used with the other males were from 3½ to 9 months old. Although all incentive females used in the study were active and vigorous, showing the typical pattern of copulatory activity, it is not impossible that the difference in age mentioned had some slight effect on the results, probably in the direction of reducing the scores of the first 61 test males.

(4) *Control of normal physiological variations in the female.* Warner (146) has found that the contrectation drive is at its maximum at the time when the vaginal mucosa is composed exclusively of cornified cells. It is obvious that the effects of our control operation (laparotomy without extirpation of glands) would be most apparent if the animals were tested during oestrum. In order to determine when these females were in the cornified stage of the oestrous cycle, it was necessary to rely on behavior criteria not involving the use of a male animal. It would have been easier to select the animals on the basis of their reactions to a male, but such a procedure would have been, in effect, the interpolation of a period of sex stimulation before the test, which would have made our results non-comparable with those of Warner and Jenkins. Another possible and very definite basis of selection would have been the taking of a vaginal smear before the test, but this would have involved the danger of interfering with the normal expression of the sex drive during the test period immediately following. Careful observation of cage behavior, therefore, formed the sole criterion for selecting these females, but immediately after the test period a vaginal smear was taken from each animal. These smears were stained with haemotoxylin and eosin and permanently mounted in balsam; they are on file and open to inspection at the Columbia Laboratory. The results of any animal whose smear did not show the typical picture of cornification, with absence of all other types of cells, were discarded.

The test females which were spayed and injected with placenta extract were uniformly tested 48 hours after the second of two injections, preliminary work having shown that with this procedure the animals were quite regularly in oestrum at the time of the test. Vaginal smears as above were taken immediately following the test period, but the requirement of complete absence of all elements other than cornified cells was not as rigidly adhered to here as in the control group. Three of the smears show a few leucocytes and epithelial cells, although in all cases cornified cells predominate overwhelmingly. The writer would have eliminated the records of the animals giving these slightly atypical smears had it not been found that such elimination would not have changed the results in any significant way. Each animal of this group was placed in a cage with a sexually aggressive male a short time (10 to 60 minutes) after the test; in every case copulations took place within 8 minutes.

The spayed, non-injected females show no cyclic variations in the histological character of the vaginal mucosa. It was found that the smears taken from these animals conform closely to the following description given in a private communication to the writer by Professor J. A. Long: "After ovariotomy and the subsequent cessation of the cycle the vaginal smear picture is that of the interval, that is, the cells are epithelial and leucocytes. As in the ordinary interval there will be considerable variation in the proportion of these two kinds of cells."

(5) *Control of normal physiological variations in the male.* Spontaneous, rhythmical variations in the physiological conditions underlying sex behavior, such as those found in the oestrous cycle of the female, are apparently absent in the male. The "normal" factor determining the intensity of the drive in the latter seems to be the length of time elapsed since the last previous period of sex activity (146), provided, of course, that contributing factors, such as age and diet, are kept constant. It has been found (146) that the sex drive of the male albino rat, as measured by the Columbia Obstruction Method, is at its maximum 24 hours after a two-hour period of copulatory activity. Therefore, in order to keep the factor of normal variation in the drive at a constant and optimum level, every animal included in the male groups was given a two-

hour period of opportunity for copulatory activity 24 hours before the test. The procedure was as follows: A receptive female which had copulated several times with at least two indicator[4] males was placed by itself in a small cage (15x8x6 inches) and given 10 to 15 minutes in which to adapt itself to the new situation. At the end of that time the test male (one to be tested the following night) was introduced into the small cage. Careful observation was made of the behavior occurring during the following two hours, all copulations taking place being recorded. The latter are summarized in Table 1. These mating periods always took place between 7:30 P.M. and 4:30 A.M.; during the summer months they were never begun before 10 P.M.

Interesting individual differences came to light in the course of these mating periods. Certain males were relatively passive and could be "teased" into copulatory activity only after prolonged activity of the female. A short little run with a sudden stop, often followed by depression of the lumbar, and elevation of the sacral regions of the back, was the most frequent and characteristic behavior pattern on the part of the female. Often she would nose the male genitals repeatedly, and, if this failed to evoke a response, she might straddle her hind legs over the back of the male and then move forward, thus bringing her vulva into direct contact with the nose of the male. On the whole, the females were far more active than the males and would only rarely resist copulation. Such resistance was sometimes induced when an especially virile male had mated a large number of times in quick succession. Occasionally, also, a male, after mounting, would maintain his hold on the flanks of the female for an abnormally long time instead of releasing her, as usual, a second or so after gaining intromission; this often elicited squeaking and a temporarily aggressive attitude on the part of the female. Whether or not intromission continued during the entire duration of these long embraces could not be ascertained

[4] Indicator males were animals selected for exceptionally strong sex behavior; their function was to pick out receptive females with a minimum loss of time. When placed in a cage of females, these males would manifest very little exploratory behavior, but would begin almost immediately to nose the genital region of the females and would copulate as soon as a receptive animal had been found.

TABLE 1

Distribution table covering number of matings of the male groups during the two-hour period twenty-four hours before the test

No. of copulations	Control	Castr. at 5 mos.	Castr. at 3 mos.	Castr. at 3 mos., inj. placenta	Castr. at 3 mos., inj. orchic 1	Castr. at 3 mos., inj. orchic 2	Castr. at 3 mos., inj. orchic 3	Normal, inj. placenta 6 hrs.	Normal, inj. placenta 2-3 days	Senile, normal	Senile, vasotomised
0	15	19	16	12	9	8	9	8	4	8	8
1-5			1						1		
6-10		1		1							
11-15				1							
16-20		1									1
21-25	1					1					1
26-30		2	1	1							1
31-35		2	1			1					2
36-40							1				
41-45	1									4	1
46-50	2		1		1			1		1	
51-55	1							1		3	1
56-60	1								1		
61-65	1										
66-70	2								2		
71-75	2								1		
76-80											
81-85	1										
86-90									1		
91-95											
96-100											
101-105											
106-110									1		
Frequency	27	25	20	15	10	10	10	10	11	16	15
Percentage copulating	44	24	20	20	10	20	10	20	64	50	47
Average of those which copulated	59	25	27	17	48	28	38	51	67	48	33

with certainty. Such behavior on the part of the male usually appeared at the end of a long-continued series of copulations, although with certain individuals it was a regular occurrence. The behavior of the males who copulated most often was usually quite independent of the provocative activity of the female. These animals often mated 5 to 20 times in quick succession and then remained quiescent for a period of 2 to 10 minutes. It was found that some of the more passive males would not mate with one female although they would copulate with another; this in spite of the fact that both females were receptive and both mated freely and typically with other males. The writer, therefore, made it a practice, when a given male had not copulated at all during the first 30 minutes of the period, to put a different female in the cage. Three cases are recorded in which copulation followed such a substitution, although in all instances the number of matings occurring was small.

Very striking was the behavior of many animals, both castrated and non-gonadectomized, who seemed very eager to sniff at the genital region of the female and to engage in all kinds of "sex play" but who showed no attempts at copulation. This sort of behavior often continued, with intermissions, throughout the mating interval. Similar observations are reported by Jenkins (59, pp. 479-480). Other males would partly mount the female without actually copulating, this pattern of response being observed in animals which did not mate at all, as well as in those who, at other times during the period, did copulate. At times it was difficult to determine with certainty whether a real mating had occurred; for the sake of uniformity in recording results, all of the following criteria were required before an act was considered a true copulation:

(1) The typical depression of the lumbar, and elevation of the sacral parts of the back by the female.

(2) Complete mounting by the male so that intromission seemed physically possible and probable.

(3) Licking of the penis by the male before another mounting was attempted.

Any mounting which did not fulfil all three of these requirements was recorded as an incomplete attempt and was not included in the figures of Table 1, which shows only true copulations. It is possible that licking of the penis does not follow every complete mating, but it is such a frequent accompaniment of the other criteria mentioned that the figures would not have been changed significantly had this sign been disregarded.

A comparison of the scores made during the mating interval and those obtained in the test period shows that there is no correlation between these two sets of indices. This finding corresponds with the results of other studies in which measures of both drives were obtained from the same animals (59, 146). The only reasonable interpretation of this fact seems to be that there is no correlation between the contrectation and detumescence drives; that a male may be strongly motivated by the stimulation afforded by an animal of the opposite sex without necessarily having a drive to copulate, and vicé versa. In females the correlation between contrectation and detumescence drives is apparently very high (146).

(6) *Test procedure.* The precise technique of testing animals by the Columbia Obstruction Method has been described in detail in previous publications (146, 59); this technique was followed as closely as possible in the present study. The test animal was allowed to make four crossings from the entrance compartment to the incentive chamber without receiving any shock in the interposed obstruction compartment. By this procedure the association "incentive — other side" was presumably formed. On the fifth or last preliminary crossing, the circuit was closed while the animal was still on the grid but after the door leading back to the entrance compartment had been noiselessly closed. The animal was thus forced to pass from the obstruction (shock) to the third compartment where it came into contact with the incentive. For an extended discussion of the reasons for adopting this procedure as well as for a complete description of the various precautions taken to remove the experimenter from the situation and to avoid disturbing the animal in any way, the reader is referred to the abovementioned monographs. Immediately after the last preliminary crossing the test proper began. The test period was always 20 minutes in duration, and the behavior of the animal during this time

was recorded in terms of approaches (the animal brings its head into compartment *B*), contacts (the animal steps on the grid or touches it with its nose but then retreats into the entrance compartment), and crossings (the animal crosses the grid and reaches the incentive).

The test animal was returned to the entrance compartment as soon as there was any specific contact between it and the incentive animal. This "specific contact" took several forms, the most common being nosing of the genital region, biting the neck of the other animal, the short jerky run and sudden stopping on the part of the female test animal, and the straddling movement which forcibly brought the genitals of the one animal (male or female) into contact with the nose of the other. If none of these criteria was realized within 30 seconds after the test animal had reached the incentive compartment, it was nevertheless returned at the end of that time. (For reasons and justifications see 59, pp. 483-485.) It is clear, at any rate, that the incentive used in this study is directed to appeal primarily and immediately to the contrectation drive and only secondarily and indirectly to the detumescence drive.

All animals were tested between the hours of 8:30 P.M. and 4 A.M. At no time was a test begun earlier than two hours after sunset. The tests were made at the following times of year:

Females	
Control	December, 1927
Spayed at 5 months	December, 1927
Spayed at 5 months, inj. placenta extract	December and January, 1927-1928
Males	
Control (20 animals)	December, 1927
Control (7 animals)	July, August, September, 1928
Castrated at 5 months (20 animals)	December, 1927
Castrated at 5 months (5 animals)	July, August, September, 1928
Castrated at 3 months	September, October, November, December, 1928
Castrated at 3 months, inj. placenta extract	November, December, 1928
Castrated at 3 months, inj. orchic 1, 2, and 3	December, January, 1928-1929
Normal inj. placenta 6 hours and 2-3 days	December, January, 1927-1928
Seniles, normal and vasotomized	August, September, 1928

C. Operations and injections

(1) *General conditions.* All operations were performed with general ether anaesthesia. Air was forced through liquid ether in a jar, the vaporized ether passing through a tube to a rubber cone which was held over the face of the animal. The animal was held firmly in one hand until it became completely quiet. During the operation just enough anaesthetic was used to keep the animal motionless and relaxed. Injections were made without local or general anesthesia.

Aseptic conditions were maintained as far as possible. All instruments were boiled in a sterilizer for at least 20 minutes. The site of the incision was cleared of hair by the use of sodium sulphide, after which the part was thoroughly washed with alcohol. After completion of the operation the wound was painted with iodine. In the first few operations sterilized surgical silk was used for ligating and sewing, but in the great majority of the operations fine catgut, sterilized in a solution of corrosive sublimate, was employed.

(2) *Ovariectomy.* There are two general methods for performing this operation, both of which are frequently used. In the first, a single incision is made along the ventral median line, the intestines being pushed first to one side and then to the other, successively exposing the right- and left-uterine horns and ovaries. Preliminary work convinced the writer that this method involved unnecessary manipulation of the intestines with its attendant danger of visceral adhesions, and all animals whose results are included in this report were spayed by the dorsal route. Two skin incisions, each about one-half inch long, were made on either side of the dorsal median line just above the location of the ovaries. Then a somewhat shorter incision was made through the underlying musculature into the peritoneal cavity, exposing a mass of fat in which the ovary was imbedded. This fat was pulled out of the cavity with forceps, and the ovary freed from the surrounding tissue. A ligature was placed over the uterus, just below the end of the oviduct, so that when the ovary was ablated the oviduct and about one-sixteenth inch of the uterus were removed also. The fatty tissue was then replaced into the cavity and both incisions were

sewed up. In all cases both ovaries were removed. The animals recovered quickly from the operation; cage behavior seemed normal within a day or two and at the time of the test, about 35 days later, the wounds were hardly visible. At postmortem the only adhesions found were a few between skin and underlying musculature; infections or regenerations were never found. Considerable atrophy of the uterus was noted 35 days after the operation.

In the control operations the technique was exactly the same as that outlined above, except that after the ovary and its surrounding fat had been drawn out of the cavity no ligature was applied and the glands were left uninjured.

(3) *Castration.* Two incisions were made parallel and lateral to the median ventral line of the scrotum. The incision through the integument was about three-eighths inch long and a second incision through the tunica vaginalis or muscular sac containing the testicles was of about the same length. Pressure of the thumb from the abdomen caudally served to bring the testis out of its scrotal cavity. The relatively frail mesentery by which the testis and its appendages are suspended from the tunica was torn to a point just cranial from the caput epididymis. Here a ligature was placed round the entire spermatic cord (tying off the vas deferens, spermatic arteries and veins), the latter being sectioned caudally from the ligature. In this way the entire testis together with all parts of the epididymis were ablated. The incision in the tunica was not closed, but the skin incision was sewed in such a way as to reduce considerably the size of the scrotal cavity; this was done in order to reduce the chances of a subsequent hernia. In all cases both testes were extirpated. The animals were apparently not seriously affected by the operation, copulations being observed in several cases 12 hours later.

In the control operation the same technique was followed except that the testes were not removed from the scrotal cavity nor otherwise mutilated. Care was taken in sewing up the incisions not to draw the skin so tightly as to cause cryptorchidism.

(4) *Vasotomy. Two short incisions were made through the* skin just caudal from the inguinal canal. A slit was then made through the tunica vaginalis, and the vas deferens was pulled out

of the opening so produced. The vas was ligated in two places, about one-half inch apart, and a piece one-fourth inch long was excised from between the ligatures. Care was taken not to injure the blood vessels. The incisions were closed with a few stitches. In all cases of vasotomy both canals were treated as described.

(5) *Injections.* The materials for injection were prepared by Eli Lilly and Company and were given to the writer by Dr. G. N. Papanicolaou, of the Cornell Medical College. All injections were made subcutaneously, through the skin of the back. Care was taken not to get the substance near the mammary glands or the axillary lymphatics. When two or more injections were made into the same animal, the needle was applied to different parts of the skin each time.

The placental substance used was an oily preparation of ether-soluble, lipoid extract of cow placentae. It had been previously tested and found potent in producing oestrous phenomena in spayed guinea pigs by Dr. Papanicolaou. The timing of these injections into spayed females was determined by preliminary experiments: one-half cc., 72 hours before the test, one cc., 48 hours before the test. Under these conditions all but two animals of the group were in oestrum when tested; the results of the two exceptions, of course, were excluded from the tabulated data. Determination of the oestrous stage was made by the vaginal-smear method as described above.

A group of males castrated when three months old and a group of normal, non-operated males were given injections of placental extract under the same time relations as those described for females. A third group of males, non-castrated, was given one and one-half cc. of the extract in one injection, six hours before the test.

Three kinds of orchic extract were used, all of them being prepared from the entire testis and epididymis of immature cattle. The three types of extract are designated Orchic 1, Orchic 2, and Orchic 3; below is given, in reversed order, a short description of each:

Orchic 3—The ether soluble, lipoid fraction of the testis-epididymis substance; oily preparation.

Orchic 2—The first water soluble fraction made from residue remaining after removal of Orchic 3. Light in color.

Orchic 1—The water soluble residue remaining after removal of Orchics 2 and 3. Dark in color.

Three groups of castrated males were given orchic extracts, one group for each type of extract. One-half of each group was given the injections on the following schedule: one-half cc., 11 A.M.; three-fourths cc., 3 P.M.; three-fourths cc., 6 P. M.; all on the day of the test. This arrangement was adopted because it was thought that the testicular hormone exerted its effect rapidly and for a short time only. However, after observing that there were no positive results from this temporal arrangement of the injections, the following schedule was adopted for the other half of each group: one-half cc., 48 hours before the test; three-fourths cc., 24 hours before the test; and three-fourths cc., at 11 A.M., on the day of the test.

III. Results and Discussion

Of the three types of indices yielded by the Columbia Obstruction Method, approaches, contacts, and crossings, the latter is clearly the most specifically indicative of the strength or intensity of the drive. While it is probably true that an approach or a contact shows a tendency in the animal to cross over to the incentive, such behavior *may* be merely the manifestation of general activity. A crossing, on the other hand, quite definitely manifests the operation of a specific drive. Most of the following discussion, therefore, will be based on scores in crossings.

A comparison of the figures of Tables 4 and 6 shows that there is a high correlation between the average number of crossings and persistence of the drive as indicated by the percentage of scores occurring in each of the successive five-minute intervals. That is, when the number of crossings is relatively high it will be found that toward the end of the test period these scores are as frequent as, or more frequent than they were at the beginning. On the other hand, when the number of crossings of a group is low, only a small proportion of the total number usually occurs during the last 5 to 10 minutes of the period. It will be seen that this general corre-

TABLE 2

Female groups

Number of animals	Operation	Injection	Oestrous condition
21	None*	None	Oestrum
32	None*	None	Composite dioestrum
20	Control operation simulating that of ovariectomy but not involving any extirpation	None	Oestrum
20	Ovariectomy	None	Anoestrum
18	Ovariectomy	Placenta extract	Induced oestrum

*These groups were tested by L. H. Warner (146) and are included here for purposes of comparison.

TABLE 3

Frequency distribution covering approaches, contacts, and crossings of the female groups

Group	Normal cornified			Control			Spayed at 5 mos.			Spayed at 5 mos., inj. placenta ext.		
No. of responses	Approaches	Contacts	Crossings	Approaches	Contacts	Crossings	Approaches	Contacts	Crossings	Approaches	Contacts	Crossings
0	2	3		2	3		1	2	6	2	5	
1	1	5			5		2	3	4	1	4	
2	2	1	1	4	4		3	7	2	2	2	1
3	3	5		1	5		2	6	3	1	2	
4		3	1	1	1		1		1	1	1	1
5	3		1	3		1	1		1	2	1	1
6	2	1		4		1			1	1	1	
7	3			3	2		2	1	1	2	1	
8	1	1		1		2	3	1	1			2
9	2	2	1							2		1
10				1		2	1			2		1
11			1			1	1			1	1	1
12	1					1						1
13						1	1			1		2
14	1		3			2						3
15			2			2						
16			3			3	2					
17			2			1						1
18			1									
19			2			1						
20												3
21			2			1						
22												
23						1						
24			1									

spondence holds also for the male groups reported below, as well as for the findings of other investigators using the same method (146, 59). In the persistence of the drive, therefore, a check on small differences in averages is available; if two groups differ in average number of crossings by only a few points, and if, at the same time, the group with the lower average also shows less persistence, the disparity of the groups in the indicated direction is emphasized.

A. *Results for females*

In Table 2 are shown the special conditions pertaining to the female groups. All operations were carried out when the animals

TABLE 4

Results of the female groups

Group	Approaches: Median	Approaches: Range	Approaches: Average	Approaches: Average deviation	Approaches: Standard deviation	Approaches: Coefficient of variation	Contacts: Median	Contacts: Range	Contacts: Average	Contacts: Average deviation	Contacts: Standard deviation	Contacts: Coefficient of variation	Crossings: Median	Crossings: Range	Crossings: Average	Crossings: Average deviation	Crossings: Standard deviation	Crossings: Coefficient of variation
Normal; cornified	5.0	0-14	5.43	2.93	3.63	67	3.0	0-9	3.14	2.10	2.73	87	16.0-	2-24	14.14	4.21	5.14	36
Control	5.0	0-10	4.65	2.22	2.63	57	2.0	0-7	2.30	1.46	1.93	83	14.0	5-23	13.45	3.81	4.66	35
Spayed at 5 mos.	6.0	0-16	6.35	4.05	4.79	75	2.0	0-8	2.50	1.30	1.91	76	1.5	0-8	2.35	2.02	2.43	104
Spayed at 5 mos., inj. placenta	6.5	0-13	5.78	3.33	3.87	67	1.5	0-11	2.61	2.30	2.93	112	12.5	2-20	11.89	4.79	5.22	44
Normal, composite dioestrum	1.0	0-4	1.16		1.42	117	1.0	0-5	1.47		1.27	86	1.0	0-5	1.34		1.57	117

TABLE 5

Temporal distribution of approaches, contacts, and crossings of the female groups during the twenty-minute test period in one-minute intervals

Group	Activity	1st Minute	2nd Minute	3rd Minute	4th Minute	5th Minute	6th Minute	7th Minute	8th Minute	9th Minute	10th Minute	11th Minute	12th Minute	13th Minute	14th Minute	15th Minute	16th Minute	17th Minute	18th Minute	19th Minute	20th Minute
Normal, cornified	Approaches	12	4	11	12	10	7	8	7	2	5	6	7	6	2	2	4	4	2	2	5
	Contacts	3	0	1	2	2	3	3	3	3	4	3	0	3	4	6	9	5	8	2	1
	Crossings	14	11	12	12	10	7	8	12	9	9	9	10	11	16	35	26	24	30	28	2
Control	Approaches	12	6	7	10	3	5	1	5	7	8	4	7	4	4	2	2	1	3	2	0
	Contacts	4	8	5	3	5	3	1	2	2	2	1	3	1	1	1	2	0	1	0	1
	Crossings	13	15	9	10	12	10	14	15	11	14	12	13	12	18	13	18	12	15	17	16
Spayed at 5 mos.	Approaches	19	10	10	4	4	6	7	8	3	8	3	2	6	6	6	6	3	6	5	5
	Contacts	6	9	0	1	3	1	4	3	5	4	0	2	1	1	3	2	2	0	2	1
	Crossings	10	7	6	3	2	3	4	1	1	3	1	1	1	0	1	0	1	2	0	0
Spayed at 5 mos., inj placenta	Approaches	7	1	6	6	5	8	10	7	7	7	4	5	4	4	5	4	5	2	4	3
	Contacts	2	2	0	4	3	2	6	8	4	6	2	1	3	1	0	0	0	0	0	3
	Crossings	14	18	15	11	10	12	6	8	7	8	7	13	10	12	11	11	12	9	8	12

TABLE 6

Temporal distribution of approaches, contacts, and crossings of the female groups during the twenty-minute test period in five-minute intervals

Group	Activity	0 to 5th Minute		6th to 10th Minute		11th to 15th Minute		16th to 20th Minute		
		Number	Percentage	Number	Percentage	Number	Percentage	Number	Percentage	Total number
Normal, cornified	Approaches	49	43.0	29	25.4	23	20.2	13	11.4	114
	Contacts	8	12.1	16	24.2	16	24.2	26	39.4	66
	Crossings	59	19.2	45	14.7	81	26.4	22	39.7	307
Control	Approaches	38	40.9	26	27.9	21	22.6	8	08.6	93
	Contacts	25	54.4	10	21:7	7	15.2	4	08.7	46
	Crossings	59	21.9	64	23.8	68	25.3	73	29.0	269
Spayed at 5 mos.	Approaches	47	37.0	32	25.2	23	18.2	25	19.6	127
	Contacts	19	38.0	17	34.0	7	14.0	7	14.0	50
	Crossings	28	59.6	12	25.5	4	08.5	3	06.4	47
Spayed at 5 mos., inj. placenta extract	Approaches	25	24.0	39	37.5	22	21.2	18	17.3	104
	Contacts	11	23.4	26	55.3	7	14.9	3	06.4	47
	Crossings	68	31.8	41	19.1	53	24.8	52	24.3	214

TABLE 7

Reliability of the differences between the average number of crossings of the female groups

Groups	Difference	Standard deviation of the difference	Difference / S.D. of Diff.	Chances in 100 of a true difference greater than 0
Normal, cornified—control	.69	1.54	.45	67
Control—spayed at 5 mos.	11.10	1.17	9.49	100
Spayed at 5 mos.—spayed, inj. placenta	9.54	1.34	7.12	100
Control—spayed, inj. placenta	1.56	1.61	.97	83

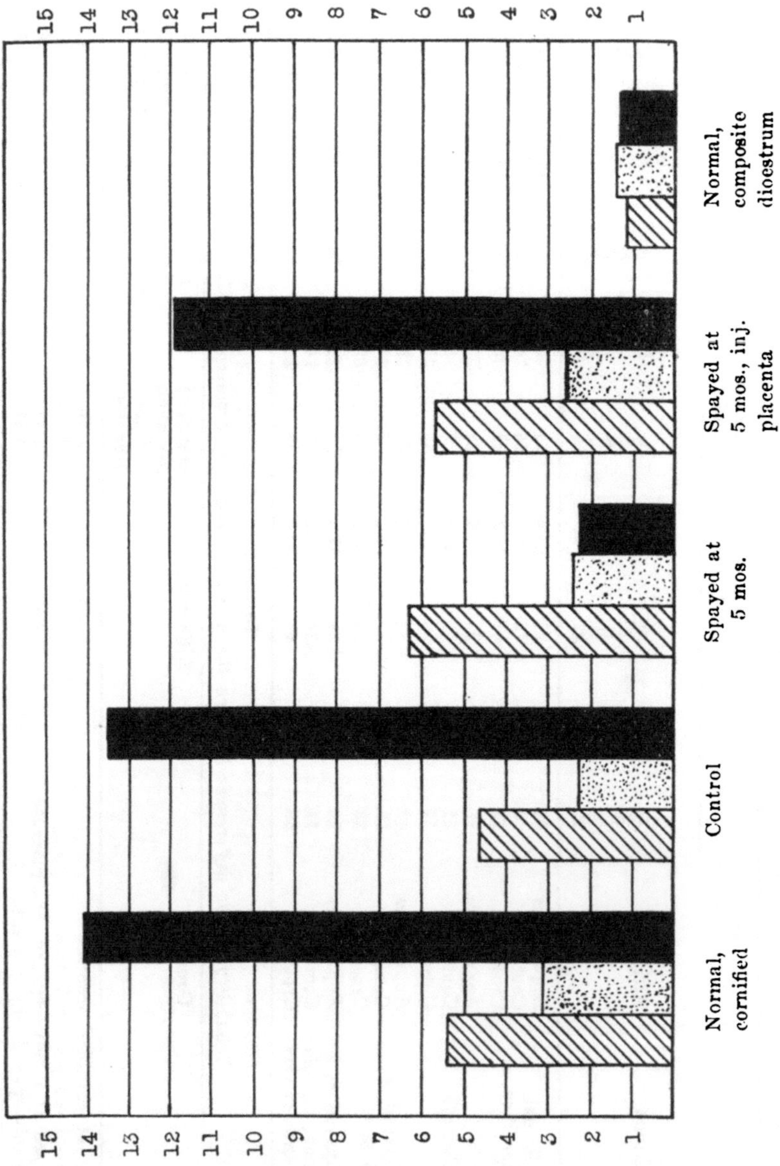

FIG. 2. Average number of approaches (hatched), contacts (stippled), and crossings (solid) for the female groups.

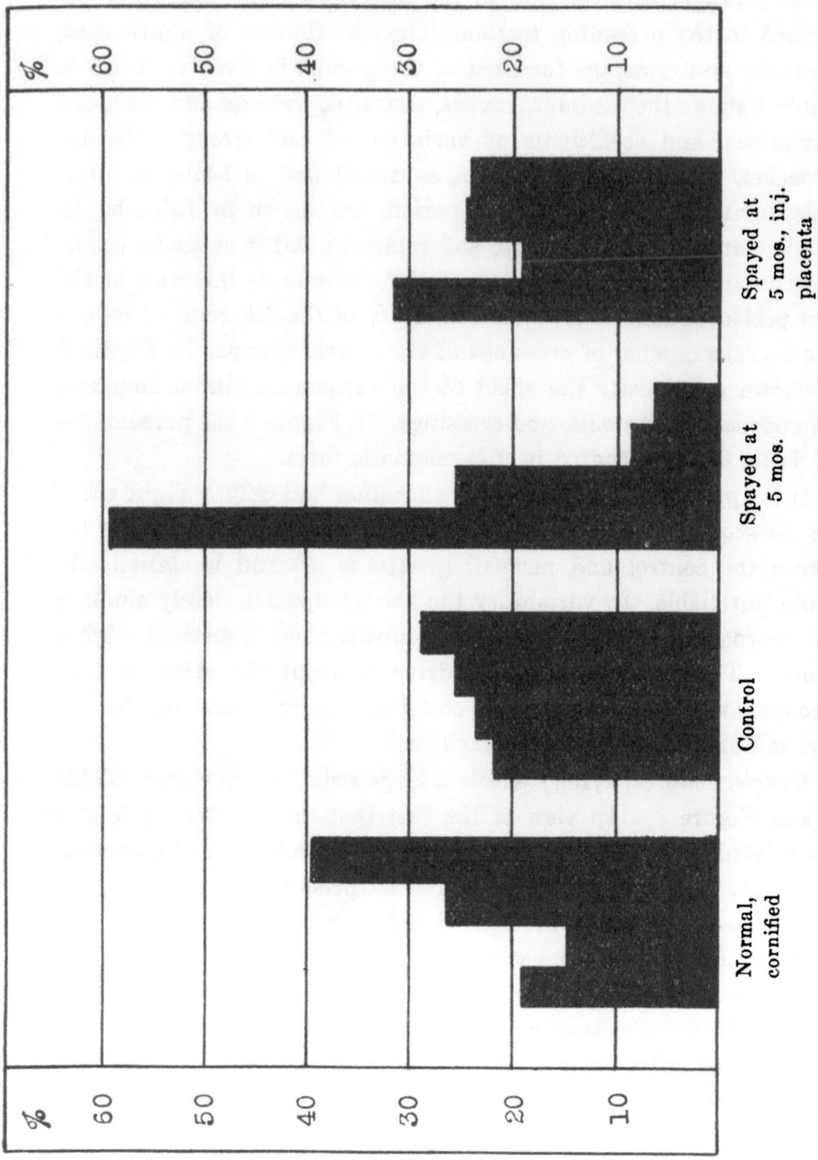

FIG. 3. Column diagram showing the percentage of crossings occurring during each 5-minute interval in the 20-minute test period for the female groups.

were approximately 150 days old (range 143-157 days); the temporal arrangements relating to the injections have been fully described in the preceding section. The distribution of approaches, contacts, and crossings for each of the groups is given in Table 3. Table 4 shows the medians, ranges, averages, average and standard deviations, and coefficients of variation of each group. The approaches, contacts, and crossings, as distributed in 1-minute intervals during the 20-minute test period, are shown in Table 5. In Table 6 are given the absolute and relative number of scores occurring during each of the four successive five-minute intervals of the test period. Table 7 shows the reliability of the differences between the average number of crossings of the several groups. In Figure 2 is shown graphically the effect of the various conditions imposed, on approaches, contacts, and crossings. In Figure 3 the percentages of Table 6 are presented in diagrammatic form.

It is apparent that the *control operation* had only a slight effect on the scores of the females. The actual difference in crossings between the control and normal[5] groups is .69 and is statistically quite unreliable. In variability the two groups are closely similar; in approaches and crossings the controls show somewhat higher scores. The persistence of the drive is about the same in both groups, over 25 per cent of the crossings occurring during the last five minutes of the period in each case.

Ovariectomy (spaying) effects a large reduction in scores (Table 4 and Figure 2). In view of the fact that the oestrous cycle and its related phenomena seem to be quite dependent on the ovarian hormone, this result is, perhaps, not surprising. Nevertheless, the outcome could not have been predicted with certainty, for although we know that the female rat will copulate only during oestrum and that oestrum does not occur in the absence of the ovarian hormone, this would not necessarily imply that the contrectation drive also disappeared after ovariectomy. The finding (146) that the sex drive, as measured by the Obstruction Method, is practically nonexistent in normal females during dioestrum would not have made prediction of the result any more possible, for there is reason to

[5] The term "normal" is used throughout this report to designate non-operated animals and is not meant to imply anything beyond that fact.

believe (see Section I) that the ovaries produce two quite different hormones; the possibility was, therefore, given for a non-follicular ovarian hormone which might have actively inhibited expression of the contrectation drive during dioestrum. The present finding rather definitely establishes the fact that the depressed sex activity of the female during dioestrum can be attributed only to a slight extent, if at all, to the inhibitory action of an ovarian hormone.

A comparison of the ovariectomized group with normal animals in dioestrum shows a slightly greater number of crossings and a considerably higher score in approaches and contacts for the former. Two interpretations for this finding present themselves. We may, in the first place, consider the differences as expressive of a higher level of *general* vitality and activity on the part of the spayed group, and this, in turn, might be explained on the basis of either one of two assumptions: (*a*) The recurrent phases of tremendous activity accompanying oestrum may leave, in their wake, periods of generally lowered physiological vigor, but these alternations of great activity and subsequent relapse into inactivity would not be found in spayed animals; (*b*) following elimination of the ovarian influence there may be a hyperplasia of some other gland, such as the thyroid, which raises the level of general activity. Although this hypothesis is very attractive it is inconsistent with the work of Wang (143) and others showing that, after spaying, spontaneous activity remains on the low level of dioestrum. There is, furthermore, good evidence of lowered metabolism following ovariectomy (79).

The second interpretation is based on the consideration that approaches and contacts are really expressions of a specific drive rather than merely of general vigor. Apparently the animals were motivated quite strongly and were actively deterred from crossing over to the incentive by the shock; it is not improbable that the operation made these animals more sensitive to the obstruction. These considerations imply, of course, that the sex drive of the spayed rats was greater than that of normal animals in dioestrum. As was pointed out above, it is not impossible that the ovary does secrete a hormone during dioestrum, which has a slightly inhibitory effect on the expression of the contrectation drive. A possible and more probable interpretation of such a difference would be that normal animals exercise their sexual mechanisms to such an extent during oestrum, stimulated by the follicular hormone, that these mechanisms, especially the neutral, go through a subsequent phase of fatigue or reduced activity. Such rhythmical recurrences of great stimulation not occurring in ovariectomized rats, the mechanisms would remain on a medium level of activity, higher than that of dioestrum and lower than that of oestrum in normal animals.

The injections of placental extract several days before the test served to raise the number of the crossings of spayed females to a point only 1.34 below that of the control group. In other words, the injection of this extract overcame almost entirely the effect of ovarian extirpation. The slight difference between the groups is probably explained by the fact that all animals were given the same amount of extract and were tested the same length of time after the injections; possibly there are individual differences (based on weight, for instance) which would vary the optimum conditions in respect to dosage and time relations. The persistence of the drive in the injected group is slightly lower than that in the control group, 24.3 per cent of the total number of crossings occurring during the last five minutes of the period. The extract used was obtained from cows, illustrating again (see Section I) the species non-specificity of the hormone.

B. Results for males

In Table 8 is given a summary of the male groups, comprising 168 animals; all of these rats with the exception of the two senile groups were tested when 185 days old. Table 9 gives the distribution of approaches, contacts, and crossings for each group. Medians, ranges, averages, average and standard deviations, and coefficients of variation are given in Table 10. The approaches, contacts, and crossings, as distributed in 1-minute intervals during the 20-minute test period, are shown in Table 11. In Table 12 are shown the absolute and relative number of scores occurring during each of the four successive 5-minute intervals of the test period. Table 13 shows the reliability of the differences between the averages of the

TABLE 8

Male groups

Number of animals	Operation and age at operation	Injection
20	None*	None
27	Control operation simulating that of castration but not involving any extirpation—5 months	None
25	Castration at 5 months	None
20	Castration at 3 months	None
15	Castration at 3 months	Placental extract, 2-3 days before test
10	Castration at 3 months	Orchic extract 1
10	Castration at 3 months	Orchic extract 2
10	Castration at 3 months	Orchic extract 3
10	None	Placental extract 6 hours before test
10	None	Placental extract 2-3 days before test
16	None (senile, tested when 16 mos. old)	None
15	Vasotomy at 14 months (senile)	None

*This group was tested by L. H. Warner (146) and is included here for purposes of comparison.

TABLE 9

Frequency distribution covering the approaches, contacts, and crossings of the male groups

Group	Normal 1 day			Control			Castr. at 5 mos.			Castr. at 3 mos.			Castr. at 3 mos., inj. placenta			Castr. at 3 mos., inj. orchic 1			Castr. at 3 mos., inj. orchic 2			Castr. at 3 mos., inj. orchis 3			Normal, inj. placenta 6 hours			Normal, inj. placenta 2-3 days			Senile, normal			Senile, vasotomised		
No. of responses	Approaches	Contacts	Crossings	Approaches	Contacts	Crossings	Approaches	Contacts	Crossings	Approaches	Contacts	Crossings	Approaches	Contacts	Crossings	Approaches	Contacts	Crossings	Approaches	Contacts	Crossings	Approaches	Contacts	Crossings	Approaches	Contacts	Crossings	Approaches	Contacts	Crossings	Approaches	Contacts	Crossings	Approaches	Contacts	Crossings
0	4	6		3	7	1	2	8	1	2	3	2	2	7		2	3		2	2	2		3	1		3		2	3		3	6	1	2	7	1
1	6	6		1	8			7		2	4		3	4		2	3	1		2		1	2	1			1		3		3	6	1	2	3	2
2	3	4		4	6	1	4	3	2	3	5	3	1	1	1		1	3		1		1	3		2	1		1			2	2	1	4	2	2
3	1	2		6	2		3	4	2	1	3	2	2	3		1			1	2	1	1	1	1		1			3	1	2	1	4	2	2	1
4	4	1	1	1		1	4	1	1	1	2	1	1		1	1			1	1	1	2	1		1	2		1			2		1	2		2
5	1			4		1	4	1	1			1			1	1	2	1	3	1	4	2					1	5	2		1			2		
6		1	1	2	1	1	1		2	3	2	4	1		2	1	1					2		1	2	2					1	1	1			
7				1	3	1	2		2	2		1	2		1			1	1	1		1		2	1		1						1			2
8						4	1		2	2		1	2		2				1		1				1	1	1			1	1			1	1	
9			2	3		1	1		1	1		1												2				1								2
10	1					2	1		3						1	1		2			1				1		1				1		2			1
11			2	1		2		1	3	1	1				1										1					2						1
12			1			2						2			1									1			2	1		1						
13			2			3			4	1					2			1						1	1		1			1			1			
14						2				1		1															1									
15			4			2						1	1																	1			1			1
16			3			1																								1						
17			1	1			1								1	1											1									
18			2				1								1															1						
19																			1														2			
20			1			1			1									1																		
21						1																								1						
22																														1						

TABLE 10

Results of the male groups

Group	Approaches						Contacts						Crossings					
	Median	Range	Average	Average deviation	Standard deviation	Coefficient of variation	Median	Range	Average	Average deviation	Standard deviation	Coefficient of variation	Median	Range	Average	Average deviation	Standard deviation	Coefficient of variation
Normal	1.5	0-10	2.30		2.35	102	1.0	0-6	1.50		1.50	100	15.0	4-20	13.45		4.03	30
Control	3.0	0-17	4.67	2.81	3.76	81	1.0	0-7	1.96	1.59	2.19	112	11.0	0-21	10.56	3.87	4.85	46
Castr. at 5 mos.	4.0	0-18	5.40	3.12	4.34	80	1.0	0-11	1.80	1.60	2.33	135	8.0	0-20	8.20	3.65	4.49	55
Castr. at 3 mos.	6.0	0-14	5.50	3.45	4.09	74	2.0	0-11	2.70	1.84	2.55	94	6.0	0-15	6.10	3.43	4.32	71
Castr. at 3 mos., inj. placenta	3.0	0-15	4.40	3.28	3.98	90	1.0	0-3	1.00	.93	1.15	115	8.0	2-18	9.33	3.82	4.20	48
Castr. at 3 mos., inj. orchic 1	3.5	0-17	4.70	3.84	5.06	108	1.0	0-6	2.10	1.94	2.21	105	6.0	1-20	7.20	4.84	5.81	81
Castr. at 3 mos., inj. orchic 2	5.0	0-19	5.60	3.44	5.10	91	2.5	0-7	2.60	1.80	2.15	83	5.0	0-10	4.50	2.20	2.94	43
Castr. at 3 mos., inj. orchic 3	4.5	1-7	4.30	1.50	1.79	41	4.5	0-4	1.50	1.10	1.28	85	7.0	0-13	6.70	3.36	4.12	61
Normal, inj. placenta 6 hours	6.5	2-13	6.90	2.90	3.51	51	3.5	0-8	3.30	2.30	2.69	82	11.0	1-17	9.90	3.72	4.48	45
Normal, inj. placenta 2-3 days	5.0	0-12	4.73	2.35	3.39	72	1.0	0-5	2.00	1.64	1.81	91	13.0	3-22	13.64	4.33	5.59	41
Senile, normal	2.5	0-10	3.13	2.28	2.85	91	1.0	0-6	1.19	1.03	1.51	126	5.0	0-19	7.38	5.22	6.06	82
Senile, vasotomized	2.0	0-8	2.80	1.65	2.07	74	1.0	0-8	1.40	1.47	2.06	147	4.0	0-15	5.67	3.78	4.28	75

TABLE 11

Temporal distribution of approaches, contacts, and crossings of the male groups during the twenty-minute test period in one-minute intervals

Group	Activity	1st Minute	2nd Minute	3rd Minute	4th Minute	5th Minute	6th Minute	7th Minute	8th Minute	9th Minute	10th Minute	11th Minute	12th Minute	13th Minute	14th Minute	15th Minute	16th Minute	17th Minute	18th Minute	19th Minute	20th Minute
Normal	Approaches	7	2	1	3	6	6	5	1	2	2	1	1	2	0	2	1	0	1	2	1
	Contacts	3	1	1	2	2	3	3	1	1	1	2	1	0	3	2	2	0	1	1	0
	Crossings	12	17	13	16	19	17	15	9	11	10	10	13	16	13	10	7	12	17	15	17
Control	Approaches	17	10	12	7	11	7	4	3	9	4	6	5	5	3	6	2	1	6	6	2
	Contacts	8	4	3	7	3	3	3	1	4		5	3	2	1		2	2		2	
	Crossings	25	17	14	9	14	11	12	10	12	16	17	17	12	12	15	16	11	15	15	15
Castrated at 5 mos.	Approaches	11	7	9	15	10	9	7	3	8	4	6	2	7	10	3	3	10	2	6	3
	Contacts	2	2	6	5	2	3	3		3	2	5	2		3			1		2	4
	Crossings	18	17	18	7	11	11	11	8	8	9	8	8	7	9	10	6	10	11	9	9
Castrated at 3 mos.	Approaches	8	9	10	9	8	3	9	4	8	2	2	5	6	4	5	4	3	5	2	4
	Contacts	6	6	7	4	3	2	5	5	2	3	1		2	1		2	2	2	1	
	Crossings	16	9	8	9	6	7	8	7	8	9	5	4	3	6	4	4	3	2	1	3
Castrated at 3 mos., inj. placenta	Approaches	4	4	6	5	4	3	8	4	1	1	1	5	3	3	3	1	3	2	3	2
	Contacts	1	1	5		1		3	1				1						2		
	Crossings	11	9	4	9	8	6	7	6	5	7	7	4	8	9	5	6	9	7	7	6
Castrated at 3 mos., inj. orchic 1	Approaches	5	2	6	6	7	4	3	1			1		3	1	2	1	1	1	2	1
	Contacts	5	5	3	2	1	1				2						1	1			
	Crossings	10	8	4	7	4	4	3	5	3	2	2	4	1	2	3	2	2	3	1	2

TABLE 11 (*Continued*)

Group	Activity	1st Minute	2nd Minute	3rd Minute	4th Minute	5th Minute	6th Minute	7th Minute	8th Minute	9th Minute	10th Minute	11th Minute	12th Minute	13th Minute	14th Minute	15th Minute	16th Minute	17th Minute	18th Minute	19th Minute	20th Minute
Castrated	Approaches	5	5	9	6	2	1	6	5	3	3	1		2	2	1	1	1	1	1	1
at 3 mos.,	Contacts	3	4	3	4	1	1	5		3					1					1	
inj. orchic 2	Crossings	7	5	4	5	5	3	3	1	2	1	1	1	1	1	1	1		2	1	
Castrated	Approaches	5		3	5	1	3	1	1	4	3	2	3	3		2		2	1	2	2
at 3 mos.,	Contacts	1	1	2	3	1	2	1			1				2	1					
inj. orchic 3	Crossings	7	7	6	4	7	7	6	4	3	3	2		1	3	1		2	1	1	2
Normal, inj.	Approaches	5	5	3	3	6	1	4	5	4	2	4	7	2	3	5	1	2	2	2	3
placenta	Contacts	2		3		5	6	2		3	3		2	1	1	1	1	1	2		
6 hours	Crossings	7	7	6	8	4	4	4	4	5	6	2	7	5	7	3	5	5	4	4	2
Normal, inj.	Approaches	3	3	3	2		6	4	5	3	3		4	1	5	1			4	3	2
placenta	Contacts		1			2	7	4	2	4						1	1				
2-3 days	Crossings	10	9	9	9	6	7	6	8	3	5	7	10	9	7	9	10	4	7	8	7
Senile,	Approaches	2	2	7	1	6	4	3	4	2	4	2	2		5	1	1		2	2	
normal	Contacts	4		2	2	2	2		2				1				2		2		
	Crossings	13	13	7	9	8	6	7	5	3	5	5	4	2	5	5	3	5	4	3	6
Senile,	Approaches	3	2	3	3	1	3	3	2	1	2	1	2	2	1	1	1	2	2	5	2
vasotomized	Contacts	6	2	1	1	4	2				1						1	2			1
	Crossings	10	7	8	8	2	5	6	8	4	2	5	1	3	3	3		1	3	1	5

TABLE 12

Temporal distribution of approaches, contacts, and crossings of the male groups during the twenty-minute test period in five-minute intervals

Group	Activity	0 to 5th Minute		6th to 10th Minute		11th to 15th Minute		16th to 20th Minute		Total number
		Number	Percentage	Number	Percentage	Number	Percentage	Number	Percentage	
Normal	Approaches	19	41.8	16	35.2	6	13.2	5	11.0	46
	Contacts	9	29.9	9	29.9	8	26.6	4	13.3	30
	Crossings	76	28.2	63	23.4	60	22.3	70	26.0	269
Control	Approaches	57	45.2	27	21.4	25	19.9	17	13.5	126
	Contracts	25	47.1	11	20.8	11	20.8	6	11.3	53
	Crossings	79	27.7	61	21.4	73	25.6	72	25.3	285
Castrated at 5 mos.	Approaches	52	38.5	31	23.0	28	20.7	24	17.8	135
	Contracts	17	37.8	11	24.4	10	22.2	7	15.6	45
	Crossings	71	34.6	47	22.9	42	20.5	45	22.0	205
Castrated at 3 mos.	Approaches	44	40.0	26	23.6	22	20.0	18	16.4	110
	Contacts	26	48.1	17	31.5	4	7.4	7	13.0	54
	Crossings	48	39.3	39	32.0	22	18.0	13	10.7	122
Castrated at 3 mos., inj. placenta	Approaches	23	34.8	17	25.8	15	22.7	11	16.7	66
	Contacts	8	53.3	4	26.7	1	6.7	2	13.3	15
	Crossings	41	29.3	31	22.1	33	23.6	35	25.0	140
Castrated at 3 mos., inj. orchic 1	Approaches	26	55.3	8	17.0	7	14.9	6	12.8	47
	Contacts	16	76.2	3	14.3	0	0.0	2	9.5	21
	Crossings	33	45.8	17	23.6	12	16.7	10	13.9	72

TABLE 12 (*Continued*)

Group	Activity	0 to 5th Minute		6th to 10th Minute		11th to 15th Minute		16th to 20th Minute		Total number
		Number	Percentage	Number	Percentage	Number	Percentage	Number	Percentage	
Castrated at 3 mos. inj. orchic 2	Approaches	27	48.2	18	32.2	6	10.7	5	8.9	56
	Contacts	15	57.7	9	34.7	1	3.8	1	3.8	26
	Crossings	26	57.8	10	22.2	5	11.1	4	8.9	45
Castrated at 3 mos. inj. orchic 3	Approaches	14	32.6	12	27.9	10	23.2	7	16.3	43
	Contacts	8	53.3	4	26.7	3	20.0	0	0.0	15
	Crossings	31	46.3	23	34.3	7	10.4	6	9.0	67
Normal, inj. placenta 6 hours	Approaches	22	31.9	16	23.2	21	30.4	10	14.5	69
	Contacts	10	30.3	14	42.4	5	15.2	4	12.1	33
	Crossings	32	32.3	23	23.2	24	24.3	20	20.2	99
Normal, inj. placenta 2-3 days	Approaches	11	21.2	21	40.3	11	21.2	9	17.3	52
	Contacts	3	13.7	17	77.3	1	4.5	1	4.5	22
	Crossings	43	28.7	29	19.3	42	28.0	36	24.0	150
Senile, normal	Approaches	18	36.0	17	34.0	10	20.0	5	10.0	50
	Contacts	10	52.7	4	21.0	1	5.3	4	21.0	19
	Crossings	50	42.4	26	22.0	21	17.8	21	17.8	118
Senile, vasotomized	Approaches	12	28.6	11	26.2	7	16.6	12	28.6	42
	Contacts	14	66.7	3	14.3	0	0.0	4	19.0	21
	Crossings	35	41.2	25	29.4	15	17.6	10	11.8	85

TABLE 13

Reliability of the differences between the average number of crossings of the male groups

Groups	Difference	S. D. of the difference	Difference / S.D. of Diff.	Chances in 100 of a true diff. greater than 0
Normal—control	2.89	1.29	2.24	98.8
Control—castrated at 5 mos.	2.36	1.29	1.83	96.6
Control—castrated at 3 mos.	4.46	1.34	3.33	100.0
Castrated at 5 mos.—castrated at 3 mos.	2.10	1.32	1.59	94.4
Castrated at 3 mos.—castrated at 3 mos., inj. placenta	3.23	1.51	2.14	98.4
Control—castrated at 3 mos., inj. placenta	1.23	1.49	.83	79.7
Castrated at 3 mos.—castrated at 3 mos., inj. orchic 1	1.10	2.08	.53	70.2
Castrated at 3 mos.—castrated at 3 mos., inj. orchic 2	1.60	1.34	1.19	88.3
Castrated at 3 mos.—castrated at 3 mos., inj. orchic 3	0.60	1.62	.37	64.4
Normal—normal, inj. placenta 6 hrs.	3.55	1.68	2.11	98.3
Control—normal, inj. placenta 6 hrs.	0.66	1.70	.39	65.2
Normal—normal, inj. placenta 2-3 days	0.19	1.91	.10	53.9
Control—normal, inj. placenta 2-3 days	3.08	1.92	1.60	94.5
Normal—seniles normal	6.07	1.77	3.42	100.0
Seniles normal—senile vastomized	1.71	1.88	.91	83.9

several groups. Figure 4 shows graphically the effect of the various conditions imposed on the average scores. Figure 5 presents the percentages of Table 12 in diagrammatic form.

It appears from consideration of the average number of crossings that the *control operation* had a marked effect on the animals of this group; there is a difference of almost three points between the scores of the normal and control groups. This finding is especially surprising in view of the fact that the control operation performed on the females, which is apparently much more severe than is the corresponding male operation, caused only a very slight diminution in crossings. It is true, of course, that the scrotal wound is subjected to more friction and other irritations than is the dorsal wound in the female. Thus the males may have become more sensitive to traumatic stimuli of all kinds, and especially to those mediated through the scrotum; it seems quite possible that the animals felt the electric shock in the apparatus through this part of the body. Such an interpretation of the decreased number of crossings by the control males gains support from the fact that the approaches and contacts of this group are more numerous than are those of the normal animals; it may be supposed that the drive was equally great in both groups but that in virtue of the scrotal wound the physically constant shock had a more intense effect on the controls. Futhermore, it is seen from Figure 5 that the diminished number of crossings is not paralleled, as usual, by less persistence. In view of what has been said above (page 289), this fact mitigates the difference between the two groups.

There is, however, another possible explanation for the relatively low score of the controls. It was originally planned to have 20 animals each in the control group and in the group castrated one month before the test. The unexpected result in regard to the former group caused the writer to add 7 animals to the controls and 5 animals to the second group; these 12 animals were tested in the summer of 1928. The average of the first 7 animals, computed separately, is 12.0 (compared to an average of 10.56 for the entire 27 rats), and the average of the 5 animals added to the "castrated at 5 months" group is .6 higher than the average for the entire group of 25. This suggests that there was some factor operative in

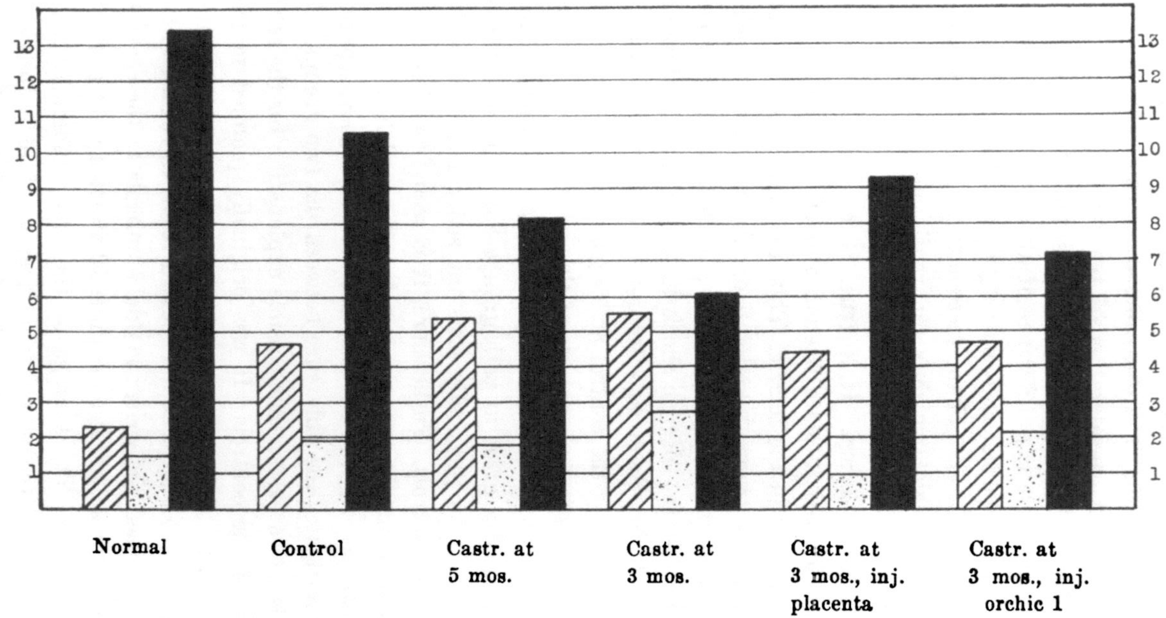

FIG. 4. Average number of approaches (hatched), contacts (stippled), and crossings (solid) for the male groups.

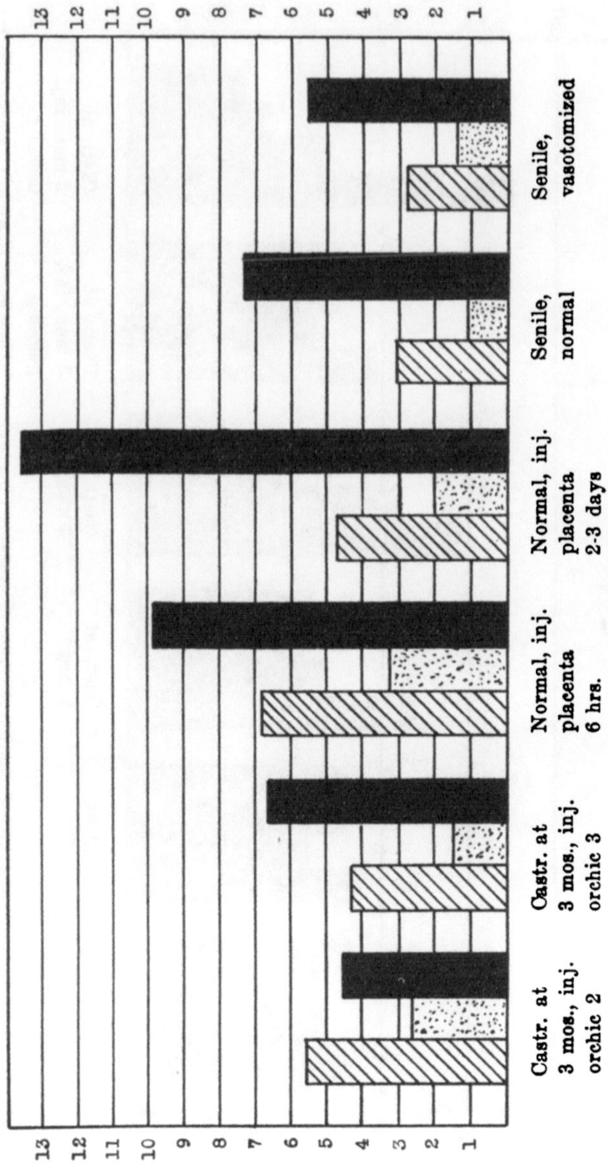

FIG. 4 (*Continued*)

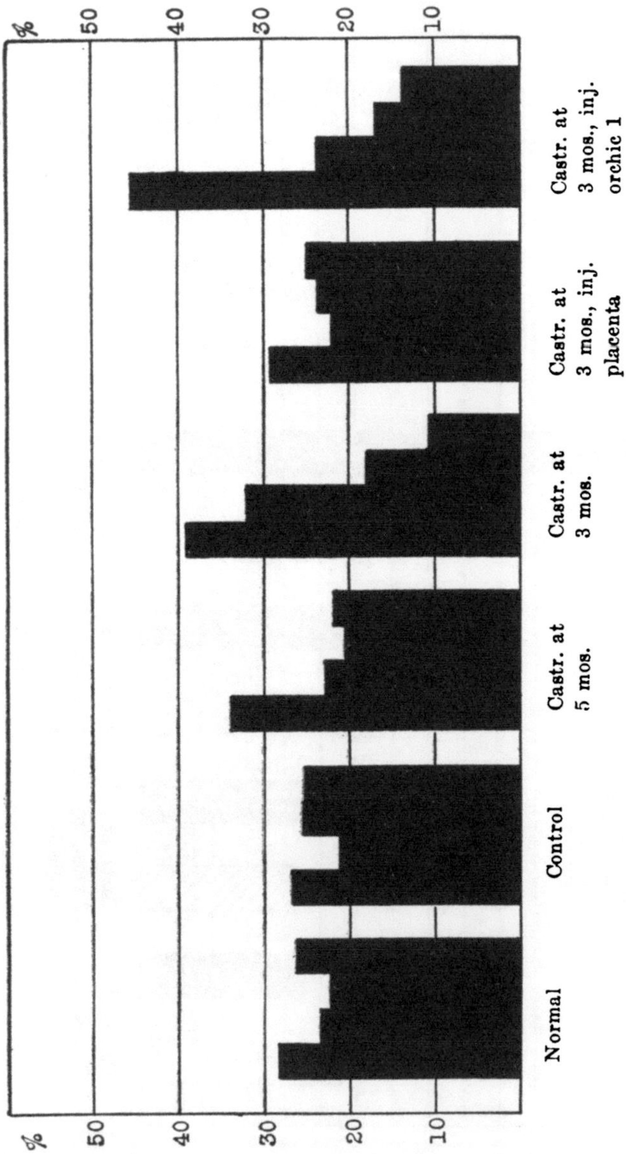

FIG. 5. Column diagram showing the percentage of crossings occurring during each 5-minute interval in the 20-minute test period for the male groups.

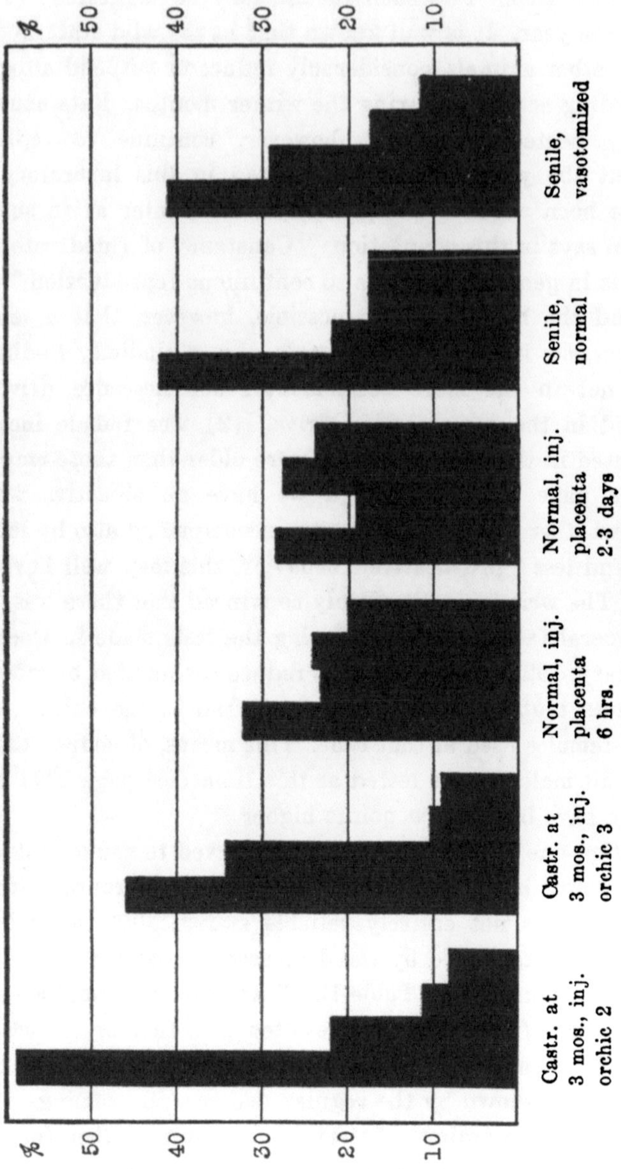

FIG. 5 (*Continued*)

the winter of 1927 which tended to reduce the crossings of animals tested at that time. Two such factors may be suggested: (1) the season of the year. It is well known that in the wild state, rodents, like many other animals, considerably reduce or suspend altogether their breeding activities during the winter months. Rats housed in an evenly heated laboratory, however, continue to reproduce throughout the year. It has been noted in this laboratory that litters are born about as frequently in the winter as in summer. Donaldson says in this connection: "Constancy of (moderate) temperature is in general favorable to continuous reproduction." It is not beyond the bounds of the possible, however, that a seasonal rhythm persists in laboratory animals, this periodicity finding expression not in the more fundamental detumescence drive but manifested in the contrectation drive. (2) The female incentive animals used in the winter of 1927 were older than those employed later (see page 279). Although we have no objective way of knowing whether this greater age was accompanied also by lowered activity and less "provocative" behavior, this may well have been the case. The writer is quite firmly convinced that there was, actually, a general factor operative during the tests made in December and January, 1927-1928, tending to reduce the number of crossings, and believes that this factor is to be found in the nature of the incentive females used at that time. This means, of course, that the scores of all male groups tested at this time (see page 279) would ordinarily have been a few points higher.

Castration one month before the test served to reduce the crossings 2.36 points below the average of the control group. Although this difference is not entirely reliable statistically, the indicated tendency is substantiated by the decreased persistence of the drive as shown in Figure 5 and Table 12. *Castration* two months earlier, i.e., *at the age of three months*, resulted in a further reduction of 2.1 points in the average number of crossings; the persistence of the drive is low, as shown by the regular decrease in crossings during each successive five-minute interval. It appears, therefore, that castration has a detrimental effect on the strength or intensity of the contrectation drive and that this effect is greater the longer the interval between the operation and the test. It should be noted, however, that extirpation of the gonads does not reduce the drive as quickly and completely in the male as it does in the female. Table 1 shows that copulations also decrease in number, but do not entirely cease, after castration. The latter finding is in accordance with the results reported by Steinach (119), Stone (134), and others.

Six possible explanations for the slow extinction of the male sex drive after gonadectomy may be suggested:

1) The stimulating hormone may persist in the blood stream for a long period of time, becoming gradually less and less efficacious. This hypothesis is made less probable by studies showing that the effects of injected testicular extract are very transitory.

2) Castration diminishes the stimulation of other glands, especially of the thyroid; the decrease in sex drive is only one manifestation of a generally lowered activity level following thyroid hypo-function. This hypothesis seems very improbable in view of the fact that other performances, also depending on the general activity of the organism, do not seem to suffer nearly as much as the sex drive in particular.

3) The testicular hormone originally stimulates sex behavior, but after the nervous system has mediated such activity for a long period under its influence, the nervous patterns involved continue to function for a time without the facilitation of the hormone. This is probably the most difficult of all the hypotheses to test and can be accepted only after the others have been proven inadequate.

4) The hormone stimulating the nerve centers controlling sex behavior issues from some gland other than the testis, but this "other gland" is in hormonal or other physiological dependence on the gonads. The prostate, composed of several parts and atrophying after castration, may be suggested as the source of the directly efficient hormone. Several lines of evidence, both for and against this view, have been briefly mentioned in Section I.

5) Tarchanoff (137) suggested that the clasping reflex in the male frog is stimulated (or facilitated) by the afferent impulses initiated in the distending seminal vesicles. Allport (5, p. 69) has applied this hypothesis to the human male and states that in man "the original stimulus for sex responses . . . is the gradual dis-

tension of the seminal vesicles, a condition requiring a fairly periodic discharge of their contents." This explanation is attractive in several respects, but is not consistent with the experimental findings of Steinach (120) who has shown that the clasping reflex is present in the frog before the vesicles are distended with fluid, and that extirpation of the gland does not result in disappearance of the reflex. The same author reports that removal of the seminal vesicles has no effect on sex behavior in rats.

6) The writer suggests the following generalized statement of the fundamental principle on which the hypothesis of Tarchanoff and of Allport is based.[6] Some organ, other than the gonads, but in structural and functional dependence on the testes, effects a tumescence or tension in itself or in another tissue or organ; this tension or tumescence initiates the afferent impulses which stimulate sex activity. This hypothesis shares, with the one outlined in (5) above, the advantage of supposing that the stimulation of the central nervous system is in the form of sensory nervous impulses. It is true that certain hormones as well as various foreign chemical agents (narcotics, stimulants, poisons) are known to exert a selective influence on different parts of the nervous system, but quite generally this selectivity follows the lines of more or less well-defined neuroanatomical divisions. While it is quite impossible to disprove the possibility of an hormonal action affecting only those parts of the nervous system which together control a given behavior pattern, it would appear equally or more probable that the nervous system is directly affected through afferent channels, mechanically stimulated. Any attempt at particularizing our hypothesis is clearly a speculative procedure, but since it may prove directive or provocative of further research, we may suggest the following: The penis is the organ whose tumescence or turgor initiates the afferent impulses in question. It will be recalled that reports of the "black eunuchs" of the Orient, who have been deprived of both testes and penis, indicate that the latter is of considerable influence on the sex drive. Since it is known that the penis atrophies after castration, two alternative explanations of the sex-stimulating mechanism present themselves: (1) The normally developed intromissive organ

[6] Compare, also, the similar viewpoint of Havelock Ellis (30).

responds to generally accelerated physiological vigor or to specific erotic exteroceptive stimuli by turgescing. After gonadectomy, this organ, having depended for its full development on the testicular hormone, becomes gradually less capable of tumescing and so of initiating the sensory impulses resulting in sex behavior. (2) Some other gland, such as the prostate, which maintains its normal structure and function in virtue of the gonadal influence stimulates the tumescence of the penis.[7]

The complications arising out of the fact that sex behavior apparently is often initiated by an external stimulus must be "explained," provisionally, by the laws of learning or by the assumption that this form of activity occurs normally in response to a proper combination of external and internal stimuli.

Placental extract injected two and three days before the test *into castrated males* caused an increase of 3.23 points in average number of crossings as compared to the score of castrated non-injected animals. As shown in Figure 5, the persistence of the drive, also, was greater in the injected group. Since it has been shown (144) that transplantation of ovaries into castrated male rats raises the spontaneous activity level of such animals considerably above the pre-grafting level, the present finding may be interpreted as the manifestation of augmented motility produced by the female hormone. This explanation is possible in spite of the fact that crossings are indicative of a definite drive rather than of non-specific random movement, for it is obvious that the measure of any specific drive includes always a certain value contributed by the general activity level of the test animal.

If, on the other hand, we wish to interpret the above result as indicative of a facilitating effect of the female hormone on the male sex drive, a synthesis of the results and theories of other investigators and the reconciliation of a number of apparently contradictory facts will have to be attempted. The resulting viewpoint is confessedly speculative but has the virtue of correlating a con-

[7] Professor R. S. Woodworth has suggested that if we consider the stimulation for penis turgescence as being primarily external (mediated through the exteroceptors), the function of the prostate—or its equivalent in this scheme—may be that of raising the irritability or lowering the threshold of the receptors in this organ.

siderable array of otherwise disconnected data. The points to be considered may be listed as follows:

(1) From what has already been said, it appears very probable that the testicular hormone does not directly influence the nervous system but exerts this effect through an interpolated mechanism. The arguments in favor of this viewpoint have already been discussed (page 313) and need not be repeated here. That the ovaries also control sex behavior indirectly seems probable when we consider the fact that the follicular hormone is present in the blood stream for some time preceding oestrous (77) and that injections of placental extract manifest their action several days later, rather than immediately.

2) The interpolated mechanism is undoubtedly under the control of the internal gonadal secretions. All other secondary sex characters depending on the sex glands for their existence, there is no reason for supposing that these particular mechanisms form an exception.

3) The soma is apparently equipotential as regards its secondary sex characteristics. The work of Steinach (123) and others in experimentally producing sex reversals demonstrates clearly that the gonads determine the distinguishing sex structures.

4) It follows, then, that the specific sex behavior expressed by an animal is under the direct control not of the gonads themselves but of the interpolated mechanism above mentioned. The sex glands determine sex behavior indirectly by dictating the nature of the development of the mediating mechanism.

5) There is no crucial evidence, so far as the writer is aware, against the assumption that the *activating* function of the gonads is non-specific as to sex. Halban's theory of the complete non-specificity of the gonads is no longer tenable; the present suggestion is merely that a given structure or mechanism — whose function is sex-specific — may be stimulated to its characteristic activity by either the ovarian or testicular hormone.

6) In the work on sex reversals and experimental hermaphroditism with alternations of male and female behavior hitherto reported, the effects were always produced by a transplanted gland or by a long-continued series of injections. Opportunity was

thereby given for the heterologous sex hormone to develop its sex-specific interpolated structure, through which the gonadal stimulation effected its corresponding behavior.

7) In the present work the mediating mechanism of the castrated males probably atrophied to some extent, but the corresponding structure of the opposite sex was not stimulated to development. When, three months after castration, the placental extract was injected, it exerted its stimulating, activating influence on the only mechanism present, that of the male.

To summarize: The gonads exert their influence on sex behavior through the mediation of an interpolated mechanism[8] which is sex-specific in its function but whose structure is in the first place determined by the specificity of the gonadal hormone; this mechanism may be stimulated to its characteristic function by either sex-gland.

Two groups of *normal* (uncastrated) *males* were given *injections of placental extract.* If we compare the results of these groups with normal animals, it appears that one injection six hours before the test diminished the drive, whereas two injections, given several days before the test, had no appreciable effect. The former result might be attributed either to an emotional disturbance resulting from the injection a few hours before the test or to a rapidly effective inhibitory influence of the female hormone on the male drive.

In view of what has been said (pages 278-279) of the incentive animals used when these two groups were tested, and especially if the evidence of a general factor operating to diminish crossings at this time is considered, it seems more fair to compare the results of the groups with those of the control animals. If this is done, it is seen that the injection six hours before the test caused a slight reduction in the intensity of the drive, whereas injections two and three days preceding the test produced an increase in the number

[8] Such a mechanism for the male has been already suggested (pages 313-14); testis hormone (prostrate function) — penis tumescence — afferent impulses stimulating sex behavior. In view of the more immediate dependence of female behavior on gonadal influence, we may very tentatively propose the following as the female mechanism: follicular hormone — tension in uterine or vaginal epithelium — sensory impulses initiating sex behavior.

of crossings. Thus the findings on the normal, injected males corroborate the view presented that either male or female hormone can activate the mechanism stimulating the sex behavior particular to that mechanism, and that the hormonal influence is indirect, requiring several days before it manifests itself.

The *orchic extracts* injected into *castrated males* apparently had no material effect on the drive, as shown either in crossings or in persistence. Each group tested with a given orchic extract was divided into two sub-groups, 5 animals receiving the entire quantity of extract on the day of the test, the other 5 receiving injections beginning 48 hours before the test. To see whether the temporal arrangement of the injections had any effect, averages and indices of variation and persistence were calculated separately for each sub-group. No significant differences were found in the case of the animals tested with orchic extracts "2" and "3." The five animals given orchic extract "1," beginning 48 hours before the test, averaged 9.8 crossings, the other sub-group, 4.6. The number of animals used is obviously too small to allow any definite conclusions to be drawn, but the results suggest, first, that type "1" of orchic extract alone has a facilitating effect on the male drive, and secondly, that this extract does not exert its effect immediately after injection. Thus our hypothesis of the indirect effect of the gonadal hormone receives additional evidence in its favor. It is curious to observe that the female hormone is found in the lipoid fraction of placental extract whereas the male hormone is apparently present in the non-lipoid, water soluable part of the testicular substance.

The *unoperated* group of *senile* males were a few days over 16 months old when tested. As seen from Table 1, the copulatory activity of the rats is still about as frequent and vigorous as in the younger animals.[9] The number of crossings in the Obstruction Apparatus, however, is considerably decreased when compared to normal rats of 185 days and is even slightly lower than the score for the younger animals castrated a month before the test. On the whole, these findings are quite in harmony with the observations

[9] In this connection it is necessary to bear in mind that selective factors, operative between the sixth and sixteenth month of life, may have eliminated (by death) more of the sexually weak than of the sexually strong animals.

made during the two-hour copulatory interval. On this occasion the old males paid slight attention to the female but, when they did occupy themselves with the latter, the activity usually terminated in copulation. The low number of approaches and contacts shown by this group is indicative of depressed activity and lack of motivation. The percentages of crossings occurring during each successive five-minute interval indicate lack of persistence of the drive. In male rats, therefore, the contrectation drive as measured by the Obstruction Method considerably decreases in intensity with increased age. Whether this diminution is relative as well as absolute, that is, whether the sex drive in relation to other drives is weaker in the senile animals than in males six months old, is not shown by our results. Richter (107) and others have found that the spontaneous activity of rats is at a much higher level when the animals are approximately 6 months old than it is 10 months later.

The *vasotomized seniles* scored 1.71 points less than the unoperated males of the same age. The difference is unreliable and is not supported (nor disproved) by the data on the persistence of the drive. It is clear, however, that the occlusion of the vas deferens did not serve to increase the contrectation drive in these animals; the slight decrease found may be attributable to the greater sensitization of the animals after the operation. Our results, therefore, do not agree with those of Steinach (127) and his followers. The negative findings obtained here have four possible explanations: (1) Vasotomy is not followed by interstitial cell hypertrophy. (2) The interstitial tissue does not produce the male hormone. (3) The testicular hormone follows the "all-or-none" principle, and the increase of its production above the necessary minimum is without effect. (4) It may be that the sex drive of our vasotomized animals increased shortly after the operation but that it had lapsed to a lower level of intensity by the time the test was made. Futakawa (36) found an accelerated desire for sexual intercourse in albino rats soon after vaso-ligation was performed, but this effect passed away very rapidly. Macht and Teagarden (85) report a transitory improvement in muscular coordination and in appearance following vasectomy.

IV. Conclusion

The sex drive of albino rats, and probably of other higher vertebrates as well, includes two fairly distinct behavior sequences or patterns, each of which has its own end or consummatory reaction. In the male rat there seems to be no correlation between the contrectation drive which leads to various forms of sex play and the detumescence drive which culminates in copulation. In the female the correlation between the two drives is high.

In the present study, quantitative indices of the influence of gonadal factors on the contrectation drive of white rats have been obtained by use of the Obstruction Method.

A. Summary of results

1) Female rats spayed when five months old manifest complete absence of a detumescence drive and only a very weak contrectation drive when tested one month after the operation.

2) Ovariectomized females show a slightly greater contrectation drive than do normal animals during dioestrum.

3) Two injections of placental extract (cow), prepared for the follicular hormone, given 72 and 48 hours, respectively, before the test, serve to restore the contrectation drive of ovariectomized females almost to normal.

4) Castration of male rats results in a slow and gradual diminution in the intensity of the contrectation drive. The score of a control group was 10.56; of a group castrated when five months old, 8.20; and of a group castrated when three months old, 6.10; all of these animals were tested at the age of six months.

5) Injections of placental extract apparently increase the contrectation drive — to the usual sex-object — of normal and castrated males. This result is obtained when two injections are made, 72 and 48 hours, respectively, before the test.

6) Injections of the lipoid fraction of testis-epididymis extract into castrated males are without effect. There is some indication of a facilitating influence on the sex drive of gonadectomized males by injections of the water-soluble residue of the testis-epididymis substance.

7) There is no significant difference between the detumscence drive of males 16 months old and those 6 months old. The contrectation drive, on the other hand, is much more intense in the younger animals.

8) Vasotomy performed on males 14 months old had no appreciable effect on the contrectation drive measured 60 days later.

9) Individual differences in sex behavior observed during the two-hour "satiation periods" are described (page 281).

B. Summary of discussion

1) The fact that spayed females show a very weak contrectation drive favors the view that the lack of sex activity in normal animals during dioestrum is caused by the absence of a stimulative factor rather than by the presence of an inhibitory ovarian hormone.

2) The following explanation is suggested for the finding that the contrectation drive is somewhat more intense in ovariectomized females than in normal animals during dioestrum: In normal animals the great stimulation, during oestrum, of the nervous mechanisms involved in sex activity is followed by a period of depressed functioning; in spayed animals the ovarian stimulation is absent, and sex activity remains on a level which is low, but not as low as that of normal animals in dioestrum. The fact that there is a sex drive at all after extirpation of the ovaries may be explained on the basis of nervous conditioning.

3) It is suggested that the observed delay of two or three days between the injection of placental hormone into spayed females and the manifestation of an effect on behavior indicates that the female hormone does not influence the nervous system directly but exerts its effect through a mediating mechanism.

4) The writer offers the following hypothesis for explaining the results reported in paragraphs 4 and 5 of the preceding section: The gonadal hormone does not directly stimulate nor facilitate sex behavior but controls the nature of the development of an interpolated mechanism or structure which is sex specific in function — that function being the stimulation of male *or* female sex behavior — but which may be activated by both ovarian and testicular hormones. Applied to the male, two alternative particularizations

of the hypothesis are tentatively suggested. (*a*) The penis, which has developed normally under the influence of the testicular hormone, becomes partially tumescent in response either to general physiological vigor or to specific external erotic stimuli. This tumescence initiates afferent nervous impulses which stimulate sex behavior. (*b*) The testicular hormone dictates the development of the prostate gland and penis and ordinarily stimulates the prostate to its characteristic activity. The latter causes a tumescence in the penis and thereby initiates the afferent impulses resulting in male sex behavior. After castration the prostate gradually atrophies but it may be reactivated by either male or female hormone.

Bibliography

(1) Albers-Schönberg, —. 1903. Ueber eine bisher unbekannte Wirkung der Röntgenstrahlen auf den Organismus. *Münch. med. Woch.*, 50, 1859.

(2) Allen, E., and E. Doisy. 1923. An ovarian hormone. *J. Amer. Med. Asso.*, 81, 819-21.

(3) Allen, E., E. Doisy, B. F. Francis, L. L. Robertson, C. E. Colgate, C. G. Johnston, W. B. Kountz, and H. V. Gibson. 1924. The hormone of the ovarian follicle; its localization and action in test animals, and additional points bearing upon the internal secretion of the ovary. *Amer. J. Anat.*, 34, 133-68.

(4) Allen, E., J. W. Whitsett, J. W. Hardy, and F. L. Kneibert. 1924. The follicular hormone of the hen ovary. *Proc. Soc. Exper. Biol. and Med.*, 21, 500-3.

(5) Allport, F. H. 1924. Social psychology. Boston: Houghton Mifflin, 453 pp.

(6) Aristotle. Historia animalium. (Trans. by D. W. Thompson.) 1909. The works of Aristotle, vol. 4. Oxford: Clarendon Press.

(7) Athanasiu, I., and A. Pézard. 1924. Influence de la castration sur l'energie nerveuse motrice. *Compt. rend. Acad. sci.*, 178, 874-76.

(8) Battelli, F., and J. Martin. 1922. La production du liquide des vesicules seminales en rapport avec la secretion interne des testicules. *Compt. rend. Soc. biol.*, 87, 429-31.

(9) Benoit, J. 1922. Sur les conditions physiologiques relatives à la parure nuptiale periodique chez les oiseaux. *Compt. rend. Acad. Sci.*, 174, 701-4.

(10) Berthold, A. A. 1849. Transplantation der Hoden. *Müller's Arch. Anat. Physiol.*, 42-46. (Cited by Vincent, S. 1924. An introduction to the study of secretion. London: Arnold.) 168 pp.

(11) Bertschi, H. 1920. Beiträge zur Physiologie der Drüsen. Untersuchungen über den respiratorischen Stoffwechsel kastrierter Kaninchen. *Biochem. Zentbl.*, 106, 37-55.

(12) Biedl, A. 1913. Innere Sekretion. Berlin: Urban & Schwarzenberg.

(13) Bouin, P., and P. Ancel. 1903. Recherches sur les cellules interstitielles du testicule des mammifères. *Arch. zool. exper.*, 1, 437-523.

(14) ———1906. Sur l'effet des injections d'extrait de glande interstielle sur la croissance. *Compt. rend. Acad. sci.*, 142, 298.

(15) ———1906. Action de l'extrait de glande interstitielle sur le développement du squellette et des organes génitaux. *Compt. rend. Acad. sci.*, 142, 232.

(16) Boukalik, W. F. and R. G. Hoskins. 1927. Further studies on testicular grafting. *Endocrin.*, 11, 335-37.

(17) Brambell, F. W. R., and A. S. Parkes. 1927. Changes in the ovary of the mouse following exposure to X-rays. III. Irradiation of non-parous adult. *Proc. Roy. Soc. London,* (*B*), 101, 316-28.

(18) Brandes, —. 1914. *Berlin Tagebl.*, June 7. (Quoted by Sharpey-Schafer, 114, p. 391.)

(19) Brown-Séquard, C. E. 1889. The effects produced on man by subcutaneous injections of a liquod obtained from the testicles of animals. *Lancet*, 2, 105-7.

(20) Busquet, H. 1910. La fonction sexuelle. Paris, 380 pp.

(21) ———1927. Determination ou retour des caractères de masculinite, chez les chapons et les viex coqs, par le serum de jeunes animaux males. *Compt. rend. Soc. biol.*, 97, 1463-1465.

(22) Ceni, C. 1922. Der Einfluss der Sehkraft auf die Funktion des Hodens und auf die äuszeren Geschlechtscharaktere. *Arch. f. Entwickmech.*, 51, 504-8.

(23) Champy, C. 1924. Les caractères sexuels considérés comme phénomènes de développement et dans leurs rapports avec l'hormone sexuelle. Paris, Doin, 376 pp.

(24) Courrier, R. 1921. Glande interstitielle du testicule et caractères sexuels secondaires chez les poissons. *Compt. rend. Soc. biol.*, 85, 42-43.

(25) ———1923. Sur le cycle de la glande interstitielle et l'evolution des caractères sexuels secondaires chez les mammifères à spermatogénèse periodique. *Compt. rend. Soc. biol.*, 89, 1131.

(26) ———1926. Sur l'action quantitative de l'hormone folliculaire. *Compt. rend. Acad. sci.*, 182, 1492-94.

(27) Doisy, E., E. Allen, J. O. Ralls, and C. S. Johnson. 1924. Preparation and properties of an ovarian hormone. *J. Biol. Chem.*, 59 43-44.

(28) DOISY, E., J. O. RALLS, AND C. N. JORDAN. 1926. Some chemical and physiological properties of the hormone of the liquor folliculi. Endocrin., 10, 273-85.

(29) DONALDSON, H. H., AND S. HATAI. 1911. Note on the influence of castration on the weight of the brain and spinal cord in albino rats and on the percentage of water in them. *J. Comp. Neur.*, 21, 155-60.

(30) ELLIS, H. 1905. Studies in the psychology of sex. Philadelphia; Davis. Vol. 3, 2d ed., 1913. Vol. 4.

(31) FALCONE, R. 1920. Sugli innesti della ghiandola interstiziale. *Rifor. med.*, 36, 1177-80.

(32) FELLMER, O. O. 1925. Ueber das Vorkommen des femininen Sexuallipoids in Vogeleiern und den Eirstöcken der Fische. *Klin. Woch.*, 4, 1651-52.

(33) FICHERA, G. 1905. Sur l'hypertrophie de la glande pituitaire consécutive a la castration. *Arch. ital. biol.*, 43, 405-26.

(34) FISHER, N. F. 1923. The influence of the gonad hormones on the seminal vesicles. *Amer. J. Physiol.*, 64, 244-51.

(35) FRANK, R. T., R. G. GUSTAVSON, J. HOLLOWAY, D. HYNDMAN, H. KRUEGER, AND J. WHITE. 1926. The occurrence and present chemical status of the female sex hormone. *Endocrin.*, 10, 260-72.

(36) FUTAKAWA, M. 1925. Influence of the ligation of the seminal ducts on the internal secretion of the testicles. *J. Okayama Med. Soc.*, 423.

(37) GANS, H. M. 1927. Effect of early castration on voluntary activity of male albino rats. *Endrocrin.*, 11, 141-4.

(38) ——— AND R. G. HOSKINS. 1926. Studies on vigor. VIII. The fatigability of castrated rats. *Endocrin.*, 10, 56-63.

(39) GERHARDT, U. 1924. Versuch einer vergleichenden Analyse des männlichen Geschlechtstriebes der Tiere. *Zsch. f. d. ges. Anat., Abt. III.*, 25, 661-95.

(40) GRIFFITHS, J. 1890. Observations on the function of the prostate gland in man and the lower animals. *J. Anat. Physiol.*, 24, 236-46.

(41) GUISY, B. 1912. Fernkomplikationen bei transvesikalen und perinealen Prostatektomien. Prä- und post-operative Geistes-störungen. *Zsch. Urol.*, 6, 124-30.

(42) GUTHRIE, C. C. 1908. Further results of transplantation of ovaries in chickens. *J. Exper. Zool.*, 5, 563-71.

(43) HALBAN, J. 1903. Die Entstehung der Geschlechtscharaktere. Eine Studie über den formativen Einfluss der Keimdrüse. *Arch. Gynäkol*, 70, 205-308.

(44) HALLION, L., L. MOREL, AND E. PAPIN. 1913. Action vaso-dilatrice pénienne de l'extrait prostatique. *Compt. rend. Soc. biol.*, 74.

(45) HAMMETT, F. S. 1920. Gynecomastia. *Endocrin*, 4, 205-20, 401-3.

(46) HARMS, W. 1910. Hoden- und Ovarialinjektionen bei Rana Fusca-kastraten. *Pflüg. Arch. f. d. ges. Physiol.*, 133, 27-44.

(47) ———1914. Experimentelle Untersuchungen über die innere Sekretion der Keimdrüsen. Jena: Fisher, 368 pp.

(48) ———1922. Keimdrüsen und Alterzustand. *Fortschr. Natur-wis. Forsch.*, 77, 189-298.

(49) ———1924. Morphologische und experimentelle Untersuchungen an alternden Hunden. *Zsch. f. Anat. u. Entwickmech.*, 71, 319-81.

(50) HARTMAN, C., C. DUPRE, AND E. ALLEN. 1926. The effect of follicular and placental hormones upon the mammary glands and genital tract of the opposum. *Endocrin.*, 10, 291-300.

(51) HATAI, S. 1913. The effect of castration, spaying or semi-spaying on the weight of the central nervous system and of the hypophysis of the albino rat; also the effect of semi-spaying on the remaining ovary. *J. Exper. Zool.*, 15, 296-314.

(52) ———1915. The growth of organs in the albino rat as affected by gonadectomy. *J. Exper. Zool.*, 18, 1-67.

(53) HOSKINS, R. G. 1925. Studies on vigor. II. The effect of castration on voluntary activity. *Amer. J. Physiol.*, 72, 324-30.

(54) ———1925. Studies on vigor. IV. The effect of testicle grafts on spontaneous activity. *Endocrin*, 9, 277-96.

(55) HUNT, H. L. 1922. Experiences in testicle transplantations. *Endocrin.*, 6, 652-54.

(56) ———1925. New theory of the function of the prostate deduced from gland transplantations in physicians. *Endocrin.*, 9, 479-89.

(57) HUNTER, J. 1928. (Quoted by C. R. Stockard in Chemistry in Medicine. New York: Chemical Foundation, 264 pp.)

(58) JÄGER, G.. 1880. Die Entdeckung der Seele. (Cited by Rieger, 108.)

(59) JENKINS, M. 1928. The effect of segregation on the sex behavior of the white rat as measured by the obstruction method. *Genet. Psychol. Monog.*, 3, 457-571.

(60) JENKINS, T. N., L. H. WARNER, AND C. J. WARDEN. 1926. Standard apparatus for the study of animal motivation. *J. Comp. Psychol.*, 6, 361-82.

(61) KINGSBURY, B. F. 1924. The endocrine organs: a point of view. *Endocrin.*, 8, 91-102.

(62) KNAUBER, E. 1896. Einige Versuche über Ovarientransplantation bei Kaninchen. *Zentbl. Gynäkol.*, 20, 524-28.

(63) KOCH, W. 1921. Ueber die Russisch-Romanische Kastratensekte der Skopzen. Jena: Fischer, 38 pp.

(64) KORENCHEVSKY, V. 1925. The sexual glands and metabolism. III. The influence of injections of testicular or ovarian emulsions

upon the nitrogen and gaseous metabolism of dogs and rabbits. *Brit. J. Exper. Path.*, 6, 158-71.

(65) Korenchevsky, V., and M. Carr. 1925. The sexual glands and metabolism. II. Influence of emulsions of testis and prostate upon the nitrogen metabolism of rabbits. *Brit. J. Exper. Path.*, 6, 74-83.

(66) Lee, M. O. 1927. Basal metabolism in the rat during the oestrous cycle. *Amer. J. Physiol.*, 81, 492-93.

(67) Lichtenstern, —. 1916. Untersuchungen über die Funktion der Prostata. *Zsch. Urol.*, 10, 1-24.

(68) Lillie, F. R. 1923. Supplementary note on twins in cattle. *Biol. Bull.*, 44, 47-78.

(69) ——— and K. F. Bascom. 1922. An early stage of the freemartin and the parallel history of the interstitial cells. *Science*, 55, 624-25.

(70) Lipschütz, A. 1917. On the internal secretion of the sexual glands. *J. Physiol.*, 51, 283-86.

(71) ———1918. Umwandlung der Clitoris in ein penisartiges Organ bei der experimentellen Maskulierung. *Arch. f. Entwickmech.*, 44, 196-206.

(72) ———1924. The internal secretions of the sex glands. Baltimore: Williams & Wilkins, 513 pp.

(73) ——— B. Ottow, and K. Wagner. 1922. Ueber Eunuchoidismus beim Kaninchen, bedingt durch Unterentwicklung des Hodens. *Arch. f. Entwickmech.*, 51, 66-78.

(74) Lipschütz, A., and E. Vinals. 1927. Fecondation par le cobaye experimentalement hermaphrodite. *Compt. rend. Soc. biol.*, 97, 1400-1.

(75) Livingston, A. E. 1916. The effect of castration on the weight of the pituitary body and other glands of internal secretion in the rabbit. *Amer. J. Physiol.*, 40, 153-85.

(76) Loeb, L. *Deutsch. med. Woch.* (Cited by Papanicolaou, 99)

(77) Loewe, S. 1925. Nachweis brunsterzeugender Stoffe im weiblichen Blute. *Klim. Woch.*, 4, 1407.

(78) Loewy, A. 1903. Neuere Untersuchungen zur Physiologie der Geschlechtsorgane. *Erg. d. Physiol.*, 2, 130-58.

(79) Loewy, A., and — Richter. 1902. Zur Frage nach dem Einfluss der Kastration auf den Stoffwechsel. *Zentbl. Physiol.*, 16, 449.

(80) Long, J. A., and H. M. Evans. 1922. The oestrous cycle in the rat and its associated phenomena. Berkeley: Univ. Calif. Press, 113 pp.

(81) Lydston, G. F. 1921. Two remarkable cases of testicle implantation. *N. Y. Med. J.*, 113, 232-33.

(82) McDougall, W. 1926. An introduction to social psychology. Boston: Luce, 513 pp.

(83) Macht, D. I. 1921. Endocrinological studies of the prostate. *Amer. J. Physiol.*, 55, 311-12.

(84) MACHT, D. I., AND J. L. ULRICH. 1922. Effects of prostatectomy on integration of muscular movements of the white rat. *Amer. J. Physiol.*, 59, 482.

(85) MACHT, D. I., AND E. J. TEAGARDEN. 1923. Rejuvenation experiments with vas ligation in rats. *J. Urol.*, 10, 407-13.

(86) MARSHALL, F. H. A. 1922. The physiology of reproduction. London: Longmans, Green, 770 pp.

(87) MEISENHEIMER, —. Experimentelle Studien zur Soma- und Geschlechtsdifferenzierung. Vol. 2. Jena: Fischer, 1912.

(88) MOLL, A. Untersuchungen über die Libido Sexualis. Berlin: Fischer, 1897. 310 pp.

(89) MOORE, C. R. 1919. On the physiological properties of the gonads as controllers of somatic and psychical characteristics. I. The rat. *J. Exper. Zool.*, 28, 137-60.

(90) ———1921. On the physiological properties of the gonads as controllers of somatic and psychical characteristics. IV. Gonad transplantations in the guinea pig. *J. Exper. Zool.*, 33, 365.

(91) ———1921. Sex gland transplantation and the modifying effect in rats and guinea pigs. *Anat. Rec.*, 20, 194.

(92) ———1924. The behavior of the testis in transplantation, experimental cryptorchidism, vasectomy, scrotal insulation and heat applications. *Endocrin.*, 8, 493-508.

(93) ———1926. On the physiological properties of the gonads as controllers of somatic and psychical characteristics. IX. Testis graft reactions in different environments. (Rat.) *Amer. J. Anat.*, 37, 351-416.

(94) MUHSAM, R. 1922. Endergebnisse der Hodenüberpflanzung. *Dtsch. med. Woch.*, 48, 1341-43.

(95) NAGEL, W. 1906. Physiologie der männlichen Geschlechtsorgane. Nagel's Handbuch der Physiologie des Menschen, Vol. XI. Braunschweig: Vieweg.

(96) OSLUND, R. M. 1926. Cryptorchid testes and testicular hormone production. *Amer. J. Physiol.*, 77, 76-82.

(97) ———1926. Ligation of vasa efferentia in rats. *Amer. J. Physiol.*, 77, 83-90.

(98) OTT, I., AND J. C. SCOTT. (Cited by Marshall, 86.)

(99) PAPANICOLAOU, G. N. 1926. A specific inhibitory hormone of corpus luteum. *J. Amer. Med. Asso.*, 86, 1422-24.

(100) PARKES, A. S. 1927. On the occurrence of the oestrous cycle after X-ray sterilization. II. Irradiation at or before birth. *Proc. Roy. Soc. London (B)*, 101, 71-95.

(101) PARKES, A. S., AND C. W. BELLERBY. 1927. Studies on the internal secretion of the ovary. IV. The significance of the occurrence of oestrin in the placenta. *J. Physiol.*, 62, 395-96.

(102) Pelikan, E. 1876. Gerichtlich-medizinische Untersuchungen über das Skopzenthum in Russland. (Russian text, xx, 1872.) German trans. by N. Iwanoff. Giessen.

(103) Pézard, A. 1911. Sur la détermination des caractères sexuels secondaires chez les Gallinacés. *Compt. rend. Acad. sci.*, 154, 1183.

(104) Reinke, F. 1896. Beiträge zur Histologie des Menschen. I. Ueber Krystalloidbildungen in den interstitiellen Zellen des menschlichen Hodens. *Arch. f. mikr. Anat.*, 47, 34-44.

(105) Retterer, E. 1927. Nouvelle observation d'un testicle de singe greffé à l'homme. *J. urol. méd. et chir.*, 24, 97-115.

(106) Retterer, E., and S. Woronoff. 1921. La glande génitale mâle et les glandes endocrines; études histophysiologique. Paris: Doin, 240 pp.

(107) Richter, C. P. 1922. A behavioristic study of the activity of the rat. *Comp. Psychol. Monographs*, 1, No. 2, 55 pp.

(108) Rieger, C. 1900. Die Castration in rechtlicher, socialer und vitaler Hinsicht. Jena: Fischer, 35 and 113 pp.

(109) Sand, K. 1919. Experiments on the internal secretion of the sexual glands, especially on experimental hermaphroditism. *J. Physiol.*, 53, 257-63.

(110) ———1921. Vasectomie pratiquée chez un chien dans un but de regeneration. *Compt. rend. Soc. biol.*, 85, 1201-5.

(111) ———1922. L'hermaphrodisme experimental. *J. physiol. path. gén.*, 20, 473-87.

(112) Schneidemühl, G. 1883. Vergleichende Untersuchungen über den feineren Bau der Cowperschen Drüse. *Dtsch. z. Tier-med.*, 6.

(113) Serralach, N., and N. Parés. 1907. Quelques données sur la physiologie de la prostate et du testicule. *Compt. rend. Soc. biol.*, 63, 790.

(114) Sharpey-Schafer, E. 1926. The endocrine organs. Part 2. London: Longmans, Green.

(115) Slonaker, J. R. 1912. The normal activity of the albino rat from birth to natural death, and its rate of growth and the duration of life. *J. Animal Behav.*, 2, 20-42.

(116) ———1927. The effect of the follicular hormone on old albino rats. *Amer. J. Physiol.*, 81, 325-35.

(117) Smith, P. E. 1926. Hastening the development of the female genital system by daily homoplastic pituitary transplants. *Proc. Soc. Exper. Biol. Med.*, 24, 131-32.

(118) Stanley, L. L. 1922. An analysis of one thousand testicular substance implantations. *Endocrin.*, 6, 787-94.

(119) Steinach, E. 1894. Untersuchungen zur vergleichenden Physiologie der männlichen Geschlechtsorgane insbesondere der accessorischen Geschlechtsdrüsen. *Pflüg. Arch. f. d. ges. Physiol.*, 56, 304-38.

(120) ———1910. Geschlechtsrieb und echte sekundäre Geschlechtsmerkmale als Folge der inneren sekretorischen Funktion der Keimdrüsen. *Zentbl. Physiol.*, 24, 540-66.

(121) ———1911 Umstimmung des Geschlechtscharakter bei Säugertieren durch Austausch der Pubertätsdrüsen. *Zentbl. Physiol.*, 25, 723-25.

(122) ———1912. Willkürliche Umwandlung von Säugetiermännchen in Tiere mit ausgeprägt weiblichen Geschlechtscharakteren der Keimdrüsen. *Pflüg. Arch. f. d. ges. Physiol.*, 144, 71-108.

(123) ———1913. Feminierung von Männchen und Maskulierung von Weibchen. *Zentbl. Physiol.*, 27, 717-23.

(124) ———1916. Experimentell erzeugte Zwitterbildung beim Säugetier. *Anz. Akad. Wiss., Wien*, No. 12.

(125) ———1916. Pubertätsdrüsen und Zwitterbildung. *Arch. f. Entwickmech.*, 42, 307-32.

(126) ———1920. Histologische Beschaffenheit der Keimdrüse bei homosexuellen Männern. *Arch. f. Entwickmech.*, 46, 29-37.

(127) ———1920. Verjüngung durch experimentelle Neubelebung der alternden Pubertätsdrüse. *Arch. f. Entwickmech*, 46, 557-618.

(128) Steinach, E., H. Heinlein, and B. P. Wiesner. 1925. Activation of the sexual cycle, development of sex characters, reactivating effect on the senile organism by ovarian and placenta extract. *Pflüg. Arch. f. d. ges. Physiol.*, 210, 598-611.

(129) Steinach, E., and H. Kun. 1928. Die entwicklungsmechanische Bedeutung der Hypophysis als Aktivator der Keimdrüseninkretion. *Med. Klin.*, 24, 524-29.

(130) Stockard, C. R., and G. N. Papanicolaou. 1917. The existence of a typical oestrous cycle in the guinea-pig, with a study of its histological and physiological change. *Amer. J. Anat.*, 22, 225-83.

(131) ———1919. The vaginal closure membrane, copulation, and the vaginal plug in the guinea-pig, with further consideration of the oestrous rhythm. *Biol. Bull.*, 37, 222-43.

(132) Stone, C. P. 1923. Experimental studies of two important factors underlying masculine sexual behavior: the nervous system and the internal secretion of the testis. *J. Exper. Psychol.*, 6, 85-106.

(133) ———1924. The awakening of the copulatory ability in the male albino rat. *Amer. J. Physiol.*, 68, 407-24.

(134) ———1927. The retention of copulatory ability in male rats following castration. *J. Comp. Psychol.*, 7, 369-87.

(135) Tandler, J., and S. Gross. Einfluss der Kastration auf den Organismus. II. Mitteilung; Die Skopzen. *Arch. f. Entwickmech.*, 30, 236-53.

(136) ———1913. Die biologischen Grundlagen der sekundären Geschlechtscharaktere. Berlin: Springer, 169 pp.

(137) Tarchanoff, J. R. 1887. Zur Physiologie des Geschlechtsapparatus des Frosches. *Pflüg. Arch. f. d. ges. Physiol.,* 40, 330-51.

(138) Thorek, M. 1924. Experimental investigations of the role of the Leydig, seminiferous and Sertoli cells and effects of testicular transplantation. *Endocrin.,* 8, 61-90.

(139) Tsubura, S. 1923. Beiträge zur Kenntnis der inneren Sekretion der Keimdrüsen. II. Keimdrüsen und respiratorischer Gaswechsel. *Biochem. Zentbl.,* 143, 291-322.

(140) Voronoff, S. 1923. Greffes testiculaires. Paris: Doin, 83 pp.

(141) ———1924. Quarante-trois greffes du singe à l'homme. Paris: Doin, 255 pp.

(142) Wallace, C. S. 1907. Prostatic enlargement. London: Frowde, 215 pp.

(143) Wang, G. H. 1923. The relation between "spontaneous" activity and oestrous cycle in the white rat. *Comp. Psychol. Monog.,* 2, No. 6, 27 pp.

(144) Wang, G. H., C. P. Richter, and A. F. Guttmacher. 1925. Activity studies on male castrated rats with ovarian transplants, and correlation of the activity with the histology of the grafts. *Amer. J. Physiol.,* 73, 581-99.

(145) Warden, C. J., and H. W. Nissen. 1928. An experimental analysis of the obstruction method of measuring animal drives. *J. Comp. Psychol.,* 8, 325-42.

(146) Warner, L. H. 1927. A study of sex behavior in the white rat by means of the obstruction method. *Comp. Psychol. Monographs,* 4, No. 22, 68 pp.

(147) Wheelon, H. 1914. Extirpation of testes and vaso-motor irritability. *Amer. J. Physiol.,* 35, 283-91.

(148) Wheelon, H., and J. L. Shipley. 1916. The effects of testicular transplants upon vasomotor irritability. *Amer. J. Physiol.,* 39, 394-400.

(149) Wilhelm, R. 1923. Beitrag zur histologisch-physiologischen Erforschung der sogenannten Erscheinungen der Verjüngung. *Rev. med. de Chile,* 51, Nos. 7 and 8. (Abstracted in *Ber. d. ges. Physiol.,* 1924, 23, 261

(150) Woodworth, R. S. 1921. Psychology. New York: Holt, 580 pp.

(151) Zondek, B., and S. Aschheim. 1926. Ovarialhormon, Wachstum der Genitalien, sexuelle Frühreife. *Klim. Woch.,* 5, 2199-202.

PART V

THE MATERNAL DRIVE

Part V

THE MATERNAL DRIVE

C. J. Warden

The limitations of the project allowed only a single study of the maternal drive. This is especially unforunate inasmuch as this drive offers so many interesting angles of approach. Among the many possible lines of investigation, a few of the more important are, (1) the effect of successive pregnancies on strength of drive, (2) the relative strength and persistance of the maternal drive from the birth of the young onward through the normal period of lactation and beyond, (3) the relation of this drive to age, (4) the analysis of the physiological factors underlying the maternal drive into glandular, neural, and other components, (5) the relation of suckling and other activities of the young to the development and maintenance of the normal drive, and (6) the effect on the drive of separating the young from the mother for various periods of time.

The study here reported deals with only certain of these lines of approach in a preliminary way. In order to carry out any general comparison of drives, it was necessary to test out a group of animals of standard age. In addition to this, a test of the incentive value of first litters at from 8 to 16 hours after birth was made. The effect of depriving the mother of the litter for some 3 or 4 hours before testing was also determined, as well as the variation in strength of drive with successive litters. The fact that the maternal drive proved to be so strong suggests the desirability of more systematic work than could be attempted at this time.

1. A STUDY OF MATERNAL BEHAVIOR IN THE WHITE RAT BY MEANS OF THE OBSTRUCTION METHOD [1]

H. W. NISSEN

I. INTRODUCTION

The maternal behavior of the white rat has been the subject of numerous observational studies, but of only a few quantitative investigations. The most commonly used criteria for the determination of the presence or absence of a maternal tendency or drive have been the following: (a) nursing activity; (b) general care of the young; (c) reactions of the mother animal to removal or presentation of the young. Such crude methods are obviously unsatisfactory and certainly cannot hope to give a reliable index of the effect of slight variations in conditions, such as size of litter, age of litter and of mother, and so forth. For the analysis of maternal behavior into its several component factors and for the determination of the relative importance of such conditions, more exact methods are required.

Szymanski (10) and Simmons (8) have compared the effectiveness of the litter as an incentive in learning situations with that of other incentives, such as various kinds of food, escape, return home, and sex object. Szymanski gave each of three female white rats of unknown age two trials per day in a maze, the goal box of which contained the litter in its living cage. Training was begun on the second day following parturition. Two of the animals did not learn the maze at all; they were at all times quite indifferent to their litters, the author reports. The third rat, which learned the problem quite rapidly, was much more concerned about her young and bestowed much care upon them. Szymanski continued giving this animal trials even after it had learned the maze and found that when the young had opened their eyes and moved about rather

[1] Reprinted from *The Journal of Genetic Psychology*, 1930, 37: 377-93.

independently (age about 20 days) the mother began to make more errors and require more time. He concludes that the maternal drive decreases in intensity as the litter grows older. It is interesting to note that the animal which succeeded in the maze had only six young, whereas those which failed had 7 and 12, respectively.

Simmons (8) used 12 female albino rats between 3 and 4 months old; presumably, therefore, they were all primiparous animals. Beginning from 12 to 18 hours after parturition massed trials in a simple maze were given until the animals had learned the problem or had conclusively demonstrated that they would not learn it. The litters were all born in the goal box of the apparatus so that, as in Szymanski's work, the incentive was really a double one: litter and return home. Six of the 12 animals did not learn the maze at all; the remaining members of the group made scores approximately as good as animals tested to a bread and milk incentive, although the variability of the former was higher.

The general disadvantages attendant on the learning method for measuring drive behavior have already been pointed out in a paper by Jenkins, Warner and Warden (2). The maternal tendency presents a special objection to this method for, unless an exceedingly simple learning situation be chosen, the time elapsing since parturition changes, and, as will be shown below, this factor seems to be of great importance in determining the intensity of the drive.

Wang (11) and Slonaker (9) have shown, according to Richter (7), that "pregnancy causes a 60 to 95 percent decrease in activity which lasts through the entire gestation and lactation period." The reference, here, is to "spontaneous" activity. The significance of this finding will be discussed in a later section.

Moss (4) tested five female rats to their "newly born litters" by noting whether they would cross an electrified grid in order to approach the young. One of the group crossed. Moss failed to control so many vital factors (2, 13), such as equalizing or compensating for individual variations in skin resistance, age of rats and period after parturition, that his data—aside from the essential differences between his "resistance" method and the Columbia Obstruction Method — are in no way comparable with those presented below.

The number of factors which materially influence the intensity of the maternal drive is probably very great; below are listed some of the more obvious ones:

(a) Age of test animal
(b) Length of interval between parturition and test period
(c) Serial order of litter used (primiparity vs. multiparity[2])
(d) Segregation of mother animal from litter
(e) Familiarity — or the opposite — of setting in which litter is found
(f) Time relations between test and nursing periods
(g) Size of litter (conditions *a* and *c* being constant)
(h) Weight of test animal; nutritional factors
(i) Tameness of animal and experimental conditions
(j) Paternity (?)
(k) Various combinations of preceding factors

In none of the studies reviewed above has the attempt been made to isolate any one of these factors, and in most cases many of them have been left uncontrolled. In the experiments described below the writer has endeavored to secure some preliminary data on the first five items of the list. The field is large and complex, and obviously the results yielded by the present work must be considered mainly suggestive and perhaps of chief value as directive for further, more systematic, experimentation.

II. Animals and Procedure

The data presented in this report are based on 34 tests conducted with 24 female albino rats of the same strain as the animals used in other studies of this project. It is unfortunate that so few animals could be brought into the study, but this circumstance was beyond the control of either the writer or of the Laboratory; we were obliged to make use of what time and materials were available. The rats were of the Wistar Institute "Experimental Colony" strain, although all of them had been born in this Laboratory. The diet used was McCollum's mixture, available in the cages at all times, supplemented by weekly rations of greens and occasional feedings of a meat and bone powder made into a pasty

[2] As here used this term refers not to number of young in the litter, but denotes that the female had had one or more litters previous to litter used in the test.

substance by the addition of water. Generally from three to four females were kept in a living cage together with an equal number of males. The pregnant animals were not removed from the regular living cages until from 1 to 7 days before parturition. At that time they were transferred to a specially constructed enclosure which we shall designate hereafter as the "maternity cage."[3] Three such cages were available, and each one housed, at a given time, only one animal — together with the litter which it delivered there. The maternity cages were of trapezoid shape, bases 4 and 7 inches respectively, altitude 9 inches, giving a floor area of 49½ square inches. They were 6 inches high. The floor was constructed of wood covered to a depth of one-half inch with wood shavings; the walls

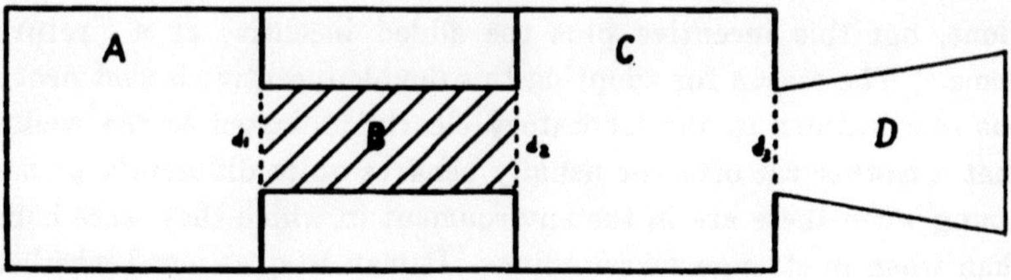

FIG. 1

were of ¼-inch wire mesh and the top was a piece of mathematical celluloid. The wire mesh wall arising from the 4-inch base was so arranged that it could be removed just before a test was to be made, and the cage could then be brought into the relation with the Obstruction Apparatus. (Fig. 1.) Before bringing the maternity cage into this position the mother rat was removed, being started, in the test, in compartment A. All of the details of procedure which were observed in other studies included in the project on animal drives of which this is one, were followed here. For a complete description of the apparatus and of the testing technique employed, the reader may be referred to earlier reports of the series, especially (1), (2) and (13). Briefly stated, the test animal was given 5 preliminary crossings from A to the incentive compartment D, four of these being without shock and the fifth with shock. That is, on the last of the preliminary crossings the grid, composing the floor of com-

[3] For one exception to this rule see Group C, Table 1.

partment B, was electrified. Immediately thereafter the test period proper, twenty minutes in duration, was begun. Scores were taken in terms of the number of times the test animal approached, touched (but retreated) and crossed the electrified grid during this period. Of these three types of scores, the latter is certainly the most significant (5).

It will be noted that our conditions permitted unrestricted mating and therefore provided as natural a situation as possible, in a laboratory setting, for the occurrence of pregnancies. No attempt was made to ascertain just when copulations occurred; daily macroscopic examination of the animals proved sufficiently trustworthy in determining an impending delivery.

The criticism may be raised that our incentive was not the litter alone, but this incentive plus the added incentive of a "return home." The reason for adopting this double incentive is that previous observations in the laboratory clearly indicated to the writer that a mother rat often or usually behaves quite differently to her young when these are in the environment in which they were born than when in strange surroundings. It may be questioned whether a litter *is* a litter, to the mother rat — or at least whether it has the same stimulative effect on the mother — outside of a certain setting. It will be recalled that Szymanski (10) and Simmons (8) also had their litter incentive in the enclosure in which delivery had taken place; in this connection Simmons says: "Observations of the maternal behavior of other rats had shown that when a litter is moved from its original place the mother often refuses to have anything more to do with it." In order to discover just what part the "return home" of the incentive situation played in influencing our results, a control group, consisting of animals which had delivered their young in the large living cages and which were tested to their litters in the maternity cages, was included.

Our animals were divided into five groups, the various conditions obtaining in each being summarized in Table 1. The letter "X" indicates variability in the particular condition so marked. Group A consisted of ten females approximately 185 days old (range 175 to 197 days) tested from 12 to 20 hours after parturition. All but one were multiparous, having had one or two litters before the one to which they were tested. None of the animals had had any previous contact with the Obstruction Apparatus (or with any other apparatus, for that matter). This group was formed in order to provide opportunity for comparing the intensity of the maternal drive with that of other drives, tested under strictly comparable conditions. One factor in our procedure, indeed, differed from that adopted in studying other forms of dynamic behavior: Our animals were not segregated (by sexes) 35 days before the test as had been done in previous studies. This variation, however, cannot be considered very consequential, since, in the first place, our animals were pregnant for about 22 days before the test and therefore had not copulated for that length of time and, secondly, because Jenkins (1) has shown the unimportance of a 35-day post-pubertal segregation period even when the sex drive itself is being measured.

TABLE 1

Summarizing the conditions under which each of the five groups of females were tested

GROUP	NUMBER OF ANIMALS IN GROUP	AGE IN DAYS	SERIAL NUMBER OF LITTER USED	TIME INTERVAL AFTER PARTURITION (HOURS)	SEGREGATED FROM LITTER	PREVIOUSLY TESTED	PLACE OF DELIVERY
A	10	185	X (1st-3rd)	15	No	No	Maternity cage
B	9	X (79-150)	1st	15	No	No	Maternity cage
C	5	X (about 185)	X (1st-3rd)	15	No	No	Large living cage
D	5	X (Avge. 195)	X (1st-4th)	168	No	Yes	Maternity cage
E	5	X (Avge. 188)	X (1st-3rd)	168	Yes	Yes	Maternity cage

The information supplied by Table 1 probably indicates sufficiently well the points of interest involved in each of the remaining four groups; further details will be supplied in the discussion of the section following.

III. Results and Discussion

The results for each of the five groups are presented in Tables 2, 3, 4 and 5. Each group is described (1) by a letter which corresponds to the one used in the summary of conditions found in Table 1, and (2) by three symbols which designate, in order, (a) age of test animals, (b) serial number of litter, i.e., whether the litter used in the test is the first, second, third or fourth which the animal delivered, (c) number of hours elapsing between parturition and test period. Again the letter "X" indicates that the corresponding factor varied among the individuals of the group. "Con" refers to the control condition of having the litters born in a cage other than the one used during the test. The letter "S" indicates that the mother rats were separated from their young for from 3 to 5 hours before the test period.

In Table 6 the averages of the several groups are compared and the reliabilities of the differences between these averages are shown. Objections may be raised against the first three comparisons made; it may well be argued that the intensity of one drive cannot be compared with that of another, even considering the scrupulous care which has been taken in the several studies of this series to keep all conditions rigidly constant. For how can we say that a bit of food is of the same relative incentive value in the hunger drive as is superficial contact with a sex object in the sex drive, or as a few seconds' contact with the litter is in the maternal drive? Indeed, this criticism cannot be completely refuted, but the following considerations should materially reduce its severity. Although it is true that the incentives used in various drives are "equated" only by common sense, it seems improbable that a more precise method for the purpose is, or will soon become, available. And probably no one will deny that such comparisons as we have made are of some value, at least from the "practical" viewpoint. Finally we may take refuge in the statement that the comparisons are valid "under the conditions of our experiment."

The average number of crossings of Group A, composed of animals about 185 days old, is higher than that of any group in any drive previously tested by the Columbia Obstruction Method. The difference is statistically reliable in only one case, however, because of the relatively small number of animals used and the variability in performance shown. Bimodality, such as we might expect in the light of the results of Simmons and Szymanski, is not indicated

TABLE 2

Distribution table covering approaches, contacts and crossings for each group

GROUP*	A 185-X-15			B X-1-15			C X-X-15, C			D X-X-168			E X-X-168, S		
Number of responses	Approaches	Contacts	Crossings	Approaches	Contacts	Crossings	Approaches	Contacts	Crossings	Approaches	Contacts	Crossings	Approaches	Contacts	Crossings
0	1	5		1	3			1	1		1			1	1
1	1	3			5		2	3		1	1	1	1	1	
2	3			3	1						1			2	1
3				1			1			1	1	1	1	1	
4	3							1					1		1
5		2		2			1		1	2			1		
6										1					
7			1	1			1				1				
8															
9				1									1		
10									1			1			1
11	2														
12															1
13						1									
14			2												
15			1												
16									1			1			
17															
18															
19															
20															
21															
22						2									
23			1												
24															
25			1												
26			1												
27						1			1						
28												1			
29															
30			1												
31															
32						3									
33			1												
34															
35						1									
36															
37			1												
38															
39															
40						1									

(Table 2) but the range in scores, from 7 to 37, is very large. It should be remembered that we are comparing here the maximum scores obtained in other drives with the single score — available for comparison — in the maternal drive. Were we to vary the several factors which influence the tendency, at the same time increasing

TABLE 3

Showing results for each of the five groups

ACTIVITY	APPROACHES						CONTACTS						CROSSINGS					
Group	Median	Range	Average	Average Deviation	Standard Deviation	Coefficient of Variation	Median	Range	Average	Average Deviation	Standard Deviation	Coefficient of Variation	Median	Range	Average	Average Deviation	Standard Deviation	Coefficient of Variation
A 125-X-15	3.0	0–11	4.10	2.76	3.67	89	0.5	0–5	1.30	1.48	1.90	146	24.0	7–37	22.4	7.92	9.14	41
B X-1-15	3.0	0–9	3.89	2.32	2.68	69	1.0	0–2	.78	.52	.63	81	32.0	13–40	28.33	6.52	7.73	27
C X-X-15, C	3.0	1–7	3.40	2.08	2.33	69	1.0	1–4	1.40	1.04	1.36	97	10.0	0–27	11.60	7.92	9.35	81
D X-X-168	5.0	1–6	4.0	1.60	1.79	45	2.0	0–7	2.60	1.92	2.42	93	10.0	1–28	11.60	8.32	9.77	84
E X-X-168, S	4.0	1–9	4.40	2.08	2.65	60	2.0	0–3	1.60	.88	1.02	64	4.0	0–12	5.60	4.32	4.63	83

TABLE 4

Showing temporal distribution of approaches, contacts and crossings during the twenty-minute test period in one-minute intervals

GROUP	ACTIVITY	1ST MINUTE	2ND MINUTE	3RD MINUTE	4TH MINUTE	5TH MINUTE	6TH MINUTE	7TH MINUTE	8TH MINUTE	9TH MINUTE	10TH MINUTE	11TH MINUTE	12TH MINUTE	13TH MINUTE	14TH MINUTE	15TH MINUTE	16TH MINUTE	17TH MINUTE	18TH MINUTE	19TH MINUTE	20TH MINUTE
A 185-X-15	Approaches	3	1	2	1	5	4	2	2	4	2	1	4		1	2	2	1	2		2
	Contacts		1				1	1	2	1			1	2	2					1	1
	Crossings	17	13	11	10	12	11	12	6	12	9	12	8	13	12	8	11	14	11	11	11
B X-1-15	Approaches	5	1	2	2	1	3	4	3	2	4	1	1	2		1	1	1			1
	Contacts		2	1	1					1						1	1				
	Crossings	12	15	11	13	13	13	12	12	9	13	12	14	12	15	12	11	11	17	13	15
C X-X-15, C	Approaches	2	1	2	2	2	2	1			1	2							2		
	Contacts		2			1		1	1				1				1				
	Crossings	5	3	1	4	2	4	2	1	4	2	2	2	5	2	3	2	2	5	3	4
D X-X-168	Approaches	2		2	1		1	1	2	1	2		2			1		1	3		1
	Contacts		2	2	1	1		3		1	1									2	
	Crossings	4	4	4	3	3	5	3	2	4	3	2	2	1	3	2	2	3	2	3	3
E X-X-168, S	Approaches	3	1	2	4	1				1			3		4			1	1	1	
	Contacts	1	1		1	1		1								1				1	1
	Crossings	4	3	1	1	2	1	2		1	1	2	1	2	1	2		1	2		1

TABLE 5

Showing temporal distribution of approaches, contacts and crossings during the twenty-minute test period in five-minute intervals

Group	Activity	0 to 5th minute		6th to 10th minute		11th to 15th minute		16th to 20th minute		Total number
		Number	Percent	Number	Percent	Number	Percent	Number	Percent	
A 185-X-15	Approaches	12	29	14	34	8	20	7	17	41
	Contacts	1	8	5	38	5	38	2	16	13
	Crossings	63	28	50	22	53	24	58	26	224
B X-1-15	Approaches	11	31	16	46	5	14	3	9	35
	Contacts	4	58	1	14	1	14	1	14	7
	Crossings	64	25	59	23	65	26	67	26	255
C X-X-15, C	Approaches	9	53	4	23	2	12	2	12	17
	Contacts	3	43	2	29	1	14	1	14	7
	Crossings	15	26	13	22	14	24	16	28	58
D X-X-168	Approaches	5	25	7	35	3	15	5	25	20
	Contacts	6	46	5	39	0	0	2	15	13
	Crossings	18	31	17	29	10	17	13	23	58
E X-X-168, S	Approaches	11	50	1	4	7	32	3	14	22
	Contacts	4	50	1	12	1	13	2	25	8
	Crossings	11	39	5	18	8	29	4	14	28

TABLE 6

Showing the reliability of the differences between the average number of crossings of the groups

GROUPS	DIFFERENCE BETWEEN THE AVERAGES	STANDARD DEVIATION OF THE DIFFERENCE	DIFFERENCE / S. D. OF DIFFERENCE	CHANCES IN 100 OF A TRUE DIFFERENCE GRATER THAN 0
A and Female sex drive at maximum*	8.26	3.09	2.67	99.6
A and Female hunger drive at maximum*	3.40	4.03	.84	80
A and Female thirst drive at maximum*	2.70	4.54	.59	72
A and *B*	5.93	3.87	1.53	93.7
A and *C*	10.80	5.07	2.13	98.3
A and *D*	10.80	5.23	2.07	98.1
A and *E*	16.80	3.55	4.73	100
D and *E*	6.00	4.65	1.29	90

* These three groups were tested by L. H. Warner (13, 14, 15). All three of the comparison groups were composed of female animals of the same age and strain as those of Group A.

the size of our groups, it is not at all improbable that the maternal drive, under the most favorable conditions, would prove to excel other drives in intensity or perseverance. That, however, is merely a possibility. What the data yielded by the experiment do indicate is that the maternal tendency is at least as strong, as measured by crossings in the Obstruction Apparatus, as are the hunger and thirst drives and rather definitely more so than the sex drive, when the latter are at their maximum.

It is interesting, in this connection, to consider the significance of the finding reported by Wang (11) and Slonaker(9), i.e., that "spontaneous" activity is diminished in amount during the time of lactation. It seems to the writer that the spontaneous activity method, insofar as it aims to give indices of the strength of specific drives is based on the assumption that any intra-organic stimulation tends to express itself in non-directed activity when the particular external stimulus situation which is related to it, is absent. To a limited extent this assumption seems quite justifiable, as is seen in the work of Richter, Slonaker, Hoskins and many others. The present study, however, considered in connection with the results of Wang and Slonaker, above mentioned, shows its limitations, for here we apparently have a specific drive correlated with a period of reduced spontaneous activity. It is true, of course, that both

Wang and Slonaker allowed the litters to remain with their mothers while the activity records of the latter during the lactation period were taken; under these conditions the external stimulus situation which "satisfies" the maternal drive was present, so that the intra-organic stimulation could find more or less complete release in directed behavior (nursing and general care of the young). If the litters had been removed soon after parturition the activity records of the mothers might have been quite different. One caution to be observed, then, in interpreting spontaneous activity scores is the control of the possibility of directed behavior. A second point to be considered is that the manifestation of one kind of intra-organic stimulation may be entirely obscured in spontaneous activity records by the influence of another physiological factor. Kinder (3), for instance, has shown that the tendency towards nest building (in the non-pregnant rat) is correlated with a period of greatly reduced spontaneous movement. The influence of the oestrous cycle, apparently, completely overshadows the effect — on spontaneous activity — of the nest-building tendency. The obstruction method obviously isolates to a far greater extent the operation of the several drives and can bring each to its maximum degree of measureable expression.

The animals of Group B were younger than those of A, ranging from 79 to 150 days at the time of the test. The average score of the former is 5.93 points higher, this difference being statistically unreliable. In variability Group A somewhat exceeds Group B. If the difference between the averages is considered significant — and after all it is rather large — the question may be raised as to what factor was responsible for the disparity, age or the matter of primiparity vs. multiparity. Obviously the data at hand do not make it possible to answer this question; either factor or a combination of the two may have determined the result. In order to solve this problem one could test another group of animals 185 days old reared, until about 160 days old, with vasotomized males. Checking through the individual records of all of the groups, no significant correlation was discovered, within a given group, between either age or number of litter and scores. Because of the small size of the groups such correlations, even if found, would not be very valuable.

The five animals comprising Group C delivered their young (first, second or third litter) in the large living cages and had their first contact with the maternity cages when they found their young in the latter during the test. In other respects (excepting a slightly wider age range) these rats were tested under conditions strictly comparable to those obtaining in Group A (Table 1). The results show rather definitely (Table 6) the influence of the variable factor. The fact that these animals were tested to a novel situation may have provided the incentive for an auxiliary drive: exploration. (For study of this drive see (6)). Thus the difference found between the averages of C and A — 10.8 points — might have been greater if this factor could have been properly controlled. Whether this finding means that the crossings of the other groups are to be attributed largely to the "return home" part of the double incentive, or that the mother rat does not "recognize" its litter in a novel setting, cannot be decided here. One who has observed the behavior of the A and B animals — how they dragged their young from one part of the cage to the other and how they struggled against removal from the maternity cage — and who has contrasted this with the "indifferent" behavior manifested by the C animals, certainly would be inclined to accept the second of these possibilities.[4] At first thought it might seem that a better control would have been given by a group of animals which delivered their young in the maternity cages and which were then tested to these cages minus the litters. The writer considered this possibility but rejected it because such a situation would doubtless have resulted in a large amount of "searching" behavior, thus giving deceivingly high scores.

The animals of Groups D and E were tested to their litters from 5 to 7 days after parturition; those of the latter group were separated from their young for about 4 hours immediately preceding the test period, whereas the D animals, as those of the other three groups, were tested within 5 minutes after separation from the

[4] On the record card of one of the animals of group *B* I find the following notation: "Lost considerable time trying to make mother leave young so as to replace her in entrance compartment. Score might have been higher if she had not been so determined to stick to litter."

litter. In age and number of litters delivered the animals of these two groups were closely similar to each other and to those of Group A (Table 1). The differences between the averages of these several groups are large (Table 6) but because of the great scatter in scores are not always reliable. If we take into consideration the facts (1) that the D and E animals had had previous contact with the apparatus (having been used in groups A or C), and (2) that Warden and Nissen (12) have shown that scores tend to increase materially as trials are repeated, even when the external and internal stimulating conditions are kept constant, it seems probable that with more carefully equalized conditions the differences between A and D or E scores would be even greater. The results indicate, at any rate, that the maternal tendency is more intense about 15 hours after parturition than it is after an interval of from 5 to 7 days, and that the drive diminishes in strength if the mother rat is separated from her young for several hours preceding the test. To a limited extent the former finding corroborates the conclusion drawn by Szymanski as noted above. It may be mentioned, incidentally, that within Group D a high correlation was found between the length of the interval: parturition-test (range from 5 to 7 days), and scores, the lower indices being associated with the longer intervals.

The fact that a segregation interval of several hours just before the test serves to decrease rather than to increase the number of crossings suggests that the internal stimulation for the maternal drive does not arise in afferent impulses initiated by the distending mammary glands. For it would seem reasonable to suppose that these glands had greater turgor after an enforced restraint from nursing (Group E) than if the young were allowed to suckle until just before the test (Group D). This is not a necessary interpretation, however, since it is possible that the interval of separation counteracted the glandular effect by weakening the recognition of the mother animal for her young.

IV. Summary

1. A study was made of the dynamic aspect of maternal behavior in twenty-four female albino rats by measuring the number of times

these animals crossed an electrified grid in order to approach their respective litters (Columbia Obstruction Method). Because of the relatively small number of animals used and the variability in performance found, the results have their chief value in suggesting systematic problems for future investigation.

2. The animals were divided into five groups corresponding to as many sets of conditions under which the tests were conducted. The results indicate that the intensity of the maternal drive

a) is slightly greater than that of the hunger and thirst drives at their maximum (80, 72);[5]
b) is greater than that of the sex drive at its maximum (99.6);[5]
c) decreases as the age of the animals increases, when litters are being dropped with normal frequency (94);[5]
d) decreases considerably as the age of the litter increases (98);[5]
e) decreases if the mother is separated from her litter for about four hours immediately preceding the test (90).[5]

Bibliography

(1) Jenkins, Marion. 1928. The effect of segregation on the sex behavior of the white rat as measured by the obstruction method. *Genetic Psychol. Monographs*, 3, 457-571.

(2) Jenkins, T. N., L. H. Warner, and C. J. Warden. 1926. Standard apparatus for the study of animal motivation. *J. Comp. Psychol.*, 6, 361-82.

(3) Kinder, Elaine Flitner. 1927. A study of the nest building activity of the albino rat. *J. Exp. Zool.*, 47, 117-61.

(4) Moss, F. A. 1924. Study of animal drives. *J. Exp. Psychol.*, 7, 165-85.

(5) Nissen, H. W. 1929. The effects of gonadectomy, vasotomy, and injections of placental and orchic extracts on the sex behavior of the white rat. *Genetic Psychol. Monographs.*

(6) Nissen, H. W. 1929. A study of exploratory behavior in the white rat by means of the obstruction method. *J. Genetic Psychol.*

(7) Richter, C. P. 1927. Animal behavior and internal drives. *Quart. Rev. Biol.*, 2, 307-43.

(8) Simmons, Rietta. 1924. The relative effectiveness of certain incentives in animal learning. *Comp. Psychol. Monog.*, 2, No. 7.

(9) Slonaker, J. R. 1925. The effect of copulation, pregnancy, pseudo-

[5] These figures represent the statistical reliability of the indications stated, in terms of the chances in 100 of a true difference (Table 6).

pregnancy, and lactation on the voluntary activity and food consumption of the albino rat. *Amer. J. Physiol.*, 71, 362-94.

(10) Szymanski, J. S. 1918. Versuche über die Wirkung der Faktoren, die als Antrieb zum Erlernen einer Handlung dienen können. *Pflüg. Archiv f. d. ges. Physiol.*, 171, 374-85.

(11) Wang, G. H. 1923. The relation between "spontaneous" activity and oestrous cycle in the white rat. *Comp. Psychol. Monographs*, 2, No. 6.

(12) Warden, C. J., and H. W. Nissen. 1928. An experimental analysis of the obstruction method of measuring animal drives. *J. Comp. Psychol.*, 8, 325-42.

(13) Warner, L. H. 1927. A study of sex behavior in the white rat by means of the obstruction method. *Comp. Psychol. Monographs*, 4, No. 22.

(14) Warner, L. H. 1928. A study of hunger behavior in the white rat by means of the obstruction method. *J. Comp. Psychol.*, 8, 273-99.

(15) Warner, L. H. 1928. A study of thirst behavior in the white rat by means of the obstruction method. *J. Genetic Psychol.*, 35, 178-92.

PART VI

THE EXPLORATORY DRIVE

Part VI

THE EXPLORATORY DRIVE

C. J. Warden

The question may arise as to whether or not the tendency to explore should be considered a specific drive, since the concept of drive as previously developed (Warden, The Columbia Obstruction Method) involves the notion of activity directed toward a specific incentive. Exploratory behavior is usually more or less random rather than so directed. In speaking of an exploratory drive, we must mean something more than a general tendency to be active, and the specific element may be introduced into the activity situation by providing an incentive which will induce exploratory behavior.

This condition seems to be fulfilled, at least in some degree, in the experimental conditions to be described in Part VI, 1. Instead of the small incentive compartment used in the measurement of the other drives, a large box designed especially to bring out exploratory behavior was provided. The results obtained under this arrangement have been directly compared with the various control tests of other drives in which compartment C was always empty. The difference in behavior under the two conditions would seem to be due to the lure of the exploratory box as thus utilized. All other drives were controlled in the same manner as in the previous work (see Warden: The Hunger Drive; The Thirst Drive; The Sex Drive; The Maternal Drive). The work could not be carried further than necessary in order to obtain an index of the exploratory drive in animals of the standard age selected for use in the project as a whole.

1. A STUDY OF EXPLORATORY BEHAVIOR IN THE WHITE RAT BY MEANS OF THE OBSTRUCTION METHOD [1]

H. W. Nissen

I. Introduction

The tremendous amount of activity expended by rats in "exploring" a novel situation or environment can hardly escape the notice of anyone working with these rodents in the laboratory. When one of these animals is placed into a new cage or when it is merely replaced into its old, unaltered cage after an absence of ten or fifteen minutes, it spends a considerable length of time in a hurried albeit rather thorough survey of the enclosure. Dashiell (1) has recorded the common observation that even a hungry animal will forego food until after it has completed its investigation of the renovated nest box. The writer (4), in the course of his study of the sex drive, found that a certain amount of exploratory activity—sometimes continuing for ten minutes or more — usually preceded any attempts at copulation on the part of sexually vigorous animals placed in an unfamiliar situation.

In spite of the frequency with which this characteristic form of behavior is encountered, few if any studies have been directed specifically towards its analysis or measurement. It is taken for granted and it is exploited; certainly the "curious behavior" of rats was a factor not of the least importance in determining the widespread adoption of these animals for laboratory use, especially in studies of learning.

Even the terminology to be employed in designating the behavior here under consideration presents difficulties, for it is not known exactly what intra-organic and environmental conditions are requisite for its manifestation. Apparently a novel situation — one which

[1] Reprinted from *The Journal of Genetic Psychology*, 1930, 37: 361-76.

does not frighten the animal — provokes exploration, the expression of other drives, as illustrated above, being temporarily deferred. Thus by a crude application of the method of choice we have some indication of the intensity of the exploratory tendency. Although such an uncontrolled observation cannot be taken too seriously, it suggests to the writer that exploration is more than a mere general activity drive which finds its outlet in the most common activities in the repertory of the animal, such as running, climbing, sniffing and moving the vibrissae. If exploration is in fact a dynamic form of behavior akin to the hunger, thirst, sex and maternal drives, it seems that when put to the test the animal would overcome a certain obstruction in order to explore . If, on the other hand, exploratory behavior is nothing more than the chance manifestation of a tendency towards the activity or exercise of the effector mechanisms, it does not seem reasonable to suppose that the animal would repeatedly overcome its negative reaction to such an obstacle or obstruction in order to reach an external situation which is especially favorable to exploration. In the latter case, indeed, the term "exploratory activity" would prove to be a misnomer, a result of misinterpretation.

It was in the light of these considerations that the present experiment was planned. The Columbia Obstruction Method was employed; a group of animals was given the opportunity to cross an electrified grid in order to reach a novel and "interesting" situation, and the scores of this group were compared with those of other animals, in a similar physiological condition, tested to a situation in which the novelty factor of the incentive was relatively insignificant. A further discussion of the theoretical aspects of the problem will be deferred until after the details of procedure and of results have been presented.

This study is one of a series of investigations of drive behavior in the white rat using the Columbia Obstruction Method. As far as possible all factors, with the exception of the variable being tested, have been kept constant throughout this larger project. The apparatus and testing technique employed have been fully described in earlier reports of the series, (2, 3, 5,) hence only a short account of the more important features will be given here, together with a description of certain modifications introduced because of the special nature of the present problem.

In the accompanying diagram (Fig. 1), the Obstruction Apparatus proper is indicated by the portions lettered A, B, and C. The test animal must pass from the entrance compartment A, through B, which contains the electric grid, to C, the incentive reaction compartment. Instead of the usual compartment D in which such incentives as food, water, etc., are placed, a new sort of compartment

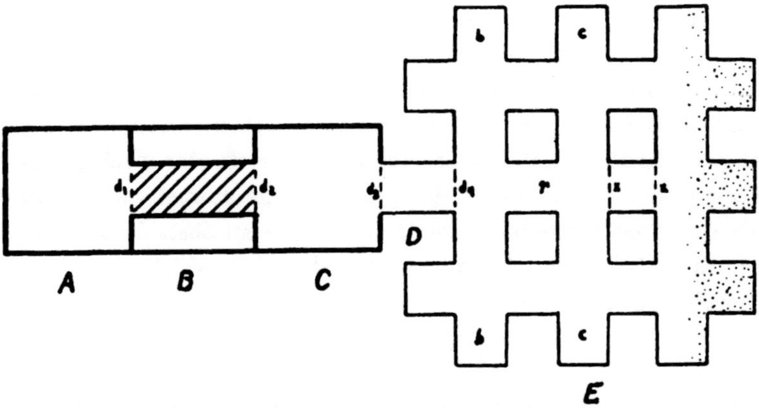

FIG. 1. A, B, C, comprise the Obstruction Apparatus proper. D, is the exploratory incentive compartment, which contained sawdust and objects as detailed in the text.

D, designed to stimulate exploratory behavior was used in this experiment. This compartment was constructed after a plan by Dashiell (1) of an apparatus by which to demonstrate spontaneous activity in the white rat in a simple manner. Several modifications of Dashiell's design were introduced. The walls (solid lines of inner portion of D) were made of galvanized sheet iron rising 5½ inches above the pine wood base. All pathways were 4 inches wide, the floor consisting of 32 squares each having an area of 16 square inches. In this way the total area of the apparatus to which the test animal had access after crossing the grid was approximately six times as great as it was in previous studies of the series. The entire top of the incentive compartment was covered by a large sheet of rigid wire mesh which was so hinged that it could be easily lifted up by the experimenter for removal of the test animal. Small pads of felt were fastened to the underside of the mesh cover so that its movement did not cause any appreciable amount of noise.

Preliminary observations with several rats confirmed the writer's guess that the many pathways and corners in compartment D would provoke a great amount of exploratory activity. Several features were added to further enhance the "interest value" of the situation: Wood shavings were piled up to the top of the walls in the back part of the incentive chamber (shown by stippling in Figure 1). In two of the square areas, blocks of wood (b) were placed, and a cork (c) in each of two others. There was a small rubber mat (r) in one alley-way; two walls (x) were formed of wire mesh. In the experience of the writer all of these items were of the nature to stimulate exploratory behavior in rats.

The automatic device (see Part I, 1) for the control of the door (d) leading to compartment D was used throughout the tests — not because it was necessary for our purposes but in order to keep whatever effect this mechanism may have the same as in the other drive studies in which this general method has been used.

II. Animals and Procedure

Twenty male albino rats about 185 days old (range: 179 to 191 days) were used in the experiment. All of them were born in this Laboratory, being the descendents (first and second generations) of animals procured from the Wistar Institute of Anatomy, Philadelphia. Males were used exclusively in order to avoid the complications introduced by the oestrous cycle of female animals, but also because it was planned to use all females available in the Laboratory at the time this study was made, for another investigation. The living conditions of our animals, described below, were planned with two requirements in view: (1) Strict comparability with the groups to be used as controls for our results; (2) Reduction to a minimum of the operation of other drives (hunger, thirst, sex, general activity) in the test situation. The success of our attempt to meet these demands can be judged from the following account and by reference to previous reports in which are described the conditions of testing the control groups (5, 6, 7).

The animals were not weaned (artificially, at least) until about 60 days old; from that time until about the 150th day of life they were kept with an equal number of females in cages containing 6 to 8 animals each. Beginning approximately 35 days before the test the males were separated from animals of the other sex; Warner (5) has shown that the sex drive of male albino rats is at a fairly low level of intensity after such a period of segregation.

The cages were supplied at all times with an abundance of the standard food, McCollum's mixture. Greens were given once a week and, until the animals were 135 days old, a meat and bone powder was fed about every ten days. Plenty of fresh water was always available in the cages. Since the animals were taken directly from their living cages to the testing apparatus, it would seem that the hunger and thirst drives did not enter as contributing factors into our results. Any social drive which may be operative in these animals in addition to the specific sex drive was probably kept at a low degree of intensity by the living conditions above described. The cages, each housing about 7 animals, were large enough (5900 cubic inches capacity) to insure ample opportunity for exercise so that any activity drive which may (and I believe does) exist in addition to the other, more specific tendencies, should have been pretty well "satisfied" at the time of the test.

Excepting the fact that care was taken not to clean the living cages or to otherwise disturb the animals for a period of two days immediately preceding the test, no special arrangements were taken to control the little known intra-organic conditions influencing the exploratory tendency. It is quite possible, of course, that the past history of the animal — the opportunities which it has had for exploration, the kind of environment in which it has lived, and so on — is an important influence in determining the intensity of this drive, but this is a matter for future work to decide. Regarding the present experiment we can say that the test animals had lived in a highly uniform environment giving little chance for unusual exploration and affording no definite rewards or incentives for such activity. It is probable, therefore, that our pre-testing conditions were generally unfavorable to the drive being measured.

All tests were conducted in the evening, usually between 9 and 11 o'clock and never before 8 o'clock. The same technique of handling the animals and the same precautions to avoid emotional disturbances were observed as in previous studies (2, 5) of this general project.

Preliminary crossings. In order (1) to adapt the test animal to the experimental situation, (2) to get the connection, other-side-of-the-grid-incentive, established, and (3) to exhaust as far as possible the influence of a conceivable auxiliary drive (exploration), the usual technique of the Columbia Obstruction Method calls for five preliminary crossings from the entrance to the incentive compartment; four of these crossings are allowed without any current passing through the grid, and on the fifth the shock is introduced after the animal is already in the obstruction chamber. After each of these preliminary crossings the animal gets a certain amount of stimulation from the incentive, food, water, sex-object or litter, as the case may be. The analogous procedure in the present study would have been, perhaps, to allow the animal to advance a certain distance — or for a certain length of time — into compartment D, at the end of each preliminary crossing, as well as after each crossing during the test proper. The following considerations, however, caused the writer to modify the preliminary procedure to the extent of allowing the animal to proceed only as far as d_8, (see Figure 1) after each of the first three crossings and to enter *E* only after each of the last two preliminary runs: (a) The novelty of the incentive obviously decreases with each opportunity to explore it. The procedure adopted preserved the stimulating effectiveness of compartment D inasmuch as when the test proper began the animal had been in it only twice. (b) In order legitimately to compare our data with those of other studies (thus giving the only possible basis for evaluating our results) it was necessary to equate approximately the length of the preliminary period. If door d_8 had been left open throughout this period — so it was found in preliminary work — the animals would have gone right through to D, stopping in C hardly at all. Then it would have been necessary to remove the animals rather quickly in order to prevent too much exploration of D (reason (a) above) and this, of course, would have meant that the preliminary period of our experimental animals would have been much shorter than that of their controls. (c) Preliminary work had demonstrated that if the animal were given no access to D at any time *before* the test period proper, it would often not even attempt to cross the grid *during* the test. Naturally the animal would need to have been previously stimulated in D if the latter was to function as an incentive and induce crossings in the test. (d) It had been found in studies of other drives that during the first few preliminary crossings the animal paid attention to the specific incentive only after it had thoroughly investigated compartment C. It seems, therefore, that our modified procedure gave sufficient opportunity for the formation of the association between the further side of the grid and the incentive, and at the same time equated fairly well the preliminary practice given to the experimental and control groups.

At the end of each of the first three preliminary crossings the test animal was given 3 minutes in C. This was more time than was given to most of the animals of the control groups but the difference, if effective at all, influenced the results in a direction unfavorable to the exploratory drive. After each of the last two preliminary runs the animal was allowed to explore D for ten seconds. During the test proper the animal was removed after 8 to 10 seconds in D; during this time it was able to traverse from 5 to 15 of the square areas (see Figure 1). If the rat made no attempts to explore it was removed anyway at the end of 30 seconds (compare (2) page 485). It may be said that on the average the amount of time allowed the test animal in the incentive compartments after each crossing, both in the preliminary period and during the test itself, was about the same in this study as in previous investigations of this series.

III. Results

The results for the group of twenty male albino rats tested are given in Tables 1, 2, 3 and 4. Of the three types of scores yielded by the method used, crossings probably have the greatest significance; they show how many times the animals actually crossed the electrified grid and reached the incentive. Contacts and approaches, while probably indicating a tendency to cross, may be, conceivably, only the expression of spontaneous, non-directed activity. The major portion of our discussion, therefore, will be based on scores in crossings.

In Table 5 are shown the differences — and the reliabilities of these differences — between the average score of the exploratory drive group and the average of each of the following: (A) A group of ten males tested to a food incentive immediately after being removed from a living cage well supplied with food. (B) A group of ten males tested to water immediately after being taken from a cage in which an ample supply of fresh water was available. (C) A group of twenty males tested to a small empty incentive compart-

TABLE 1

Distribution table covering approaches, contacts and crossings

NUMBER OF RESPONSES	APPROACHES	CONTACTS	CROSSINGS
0	3	6	
1		4	2
2	2	4	1
3	3	2	3
4	2	2	5
5	1	2	1
6	4		2
7			1
8			1
9	2		1
10	1		
11			1
12	1		1
13			
14	1		
15			
16			
17			
18			
19			
20			
21			
22			
23			1

TABLE 2

Showing measures of central tendency and of variability

APPROACHES						CONTACTS						CROSSINGS					
Median	Range	Average	Average Deviation	Standard Deviation	Coefficient of Variation	Median	Range	Average	Average Deviation	Standard Deviation	Coefficient of Variation	Median	Range	Average	Average Deviation	Standard Deviation	Coefficient of Variation
4.5	0–14	5.20	3.12	3.85	74	1.5	0–5	1.80	1.40	1.66	92	4.0	1–23	6.00	3.40	4.89	81

TABLE 3

Showing temporal distribution of approaches, contacts and crossings during the twenty-minute test period in one-minute intervals

ACTIVITY	1ST MINUTE	2ND MINUTE	3RD MINUTE	4TH MINUTE	5TH MINUTE	6TH MINUTE	7TH MINUTE	8TH MINUTE	9TH MINUTE	10TH MINUTE	11TH MINUTE	12TH MINUTE	13TH MINUTE	14TH MINUTE	15TH MINUTE	16TH MINUTE	17TH MINUTE	18TH MINUTE	19TH MINUTE	20TH MINUTE
Approaches	12	7	9	1	9	6	2	11	5	5	2	4	4	5	4	2	4	2	4	6
Contacts		3	4	4	1	2	2	6	1	1	1		1		2	4	1	1	2	
Crossings	12	14	7	7	8	6	8	7	5	3	6	4	3	6	4	3	3	4	4	6

TABLE 4

Showing temporal distribution of approaches, contacts and crossings during the twenty-minute test period in five-minute intervals

ACTIVITY	0 TO 5TH MINUTE		6TH TO 10TH MINUTE		11TH TO 15TH MINUTE		16TH TO 20TH MINUTE		TOTAL NUMPER
	Number	Per cent	Number	Per cent	Number	Per cent	Number	Per cent	
Approaches	38	36.5	29	27.9	19	18.3	18	17.3	104
Contacts	12	33.3	12	33.3	4	11.1	8	22.3	36
Crossings	48	40.0	29	24.1	23	19.2	20	16.7	120

TABLE 5

Showing the reliability of the differences between the average number of crossings of the exploratory group and of the three comparison groups

EXPLORATORY GROUP AND COMPARISON GROUP	DIFFERENCE BETWEEN THE AVERAGES *	STANDARD DEVIATION OF THE DIFFERENCE	DIFFERENCE / S. D. OF DIFFERENCE	CHANCES IN 100 OF A TRUE DIFFERENCE GREATER THAN 0
A — Ten males tested to food without previous starvation period (6)	3.30	1.52	2.16	98.46
B — Ten males tested to water without previous water-deprivation interval (7)	2.30	1.67	1.38	91.62
C — Twenty males tested to small empty incentive compartment during period of maximum sexual vigor (5)	3.05	1.16	2.63	99.57

* In all cases the score of the exploratory drive group is higher than those of the several comparison groups.

ment at a period of maximum sexual vigor (24 hours after a two-hour copulatory period). In all cases the members of the comparison groups were of the same age and sex as were the animals tested to the exploratory incentive; the physiological condition of

TABLE 6

Showing the relative degree of various kinds of stimulation operative in the three comparison groups of male rats, using the exploratory drive group as the standard

COMPARISON GROUPS	INTRA-ORGANIC STIMULATION				EXTERNAL STIMULATION			
	Sex	Hunger	Thirst	Exploratory	Sex	Food	Water	Exploratory
A — Ten males tested to food without previous starvation period (6)	=	=	=	=	=	+	=	—
B — Ten males tested to water without previous water-deprivation interval (7)	=	=	=	=	=	=	+	—
C — Twenty males tested to small empty incentive compartment during period of maxim sexual vigor (5)	+	=	=	=	=	=	=	—

Tested by Dr. L. H. Warner (5, 6, 7).

the animals of the first two comparison groups was as similar to that of our animals as possible. A summary of the internal and external stimulating conditions pertaining to the several groups is presented in Table 6; a plus, minus or equal sign indicates that the given condition is presumably present in greater, lesser or equal measure, respectively, as compared to the exploratory group.

It appears from Table 6 that in each of the three principal comparisons to be made we are dealing with more than one variable. The difference in intra-organic stimulation found in comparison group C, indeed, makes any very significant conclusions impossible here; we can only say that the external factor: exploratory incentive, results in a greater number of crossings than the internal condition of sexual excitation. The two variables in comparison groups A and B, on the other hand, are both external factors, and in each of these groups one variable may be disregarded; food and water may probably be considered as of zero stimulating value to animals which are neither hungry nor thirsty. Arguments denying this point will, of course, strengthen the conclusion drawn by the writer below. The only significant difference to be considered between the exploratory and the first two comparison groups, then, is the incentive which presumably is related to the exploratory tendency. This one important disparity in the experimental conditions resulted in a difference in scores in both cases favorable to the group tested with the exploratory incentive. In other words, our animals overcame their negative reaction to an electrified grid more often, within a stated period of time, if by crossing that grid they reached a situation affording opportunity for exploration than if the further side of the obstruction compartment was less conducive to exploratory activity. This suggests — although it does not prove, since complete statistical reliability is lacking—that exploration is a definite form of dynamic behavior similar to the hunger, thirst, sex and maternal drives.

It should be remembered that the physiological-neurological condition related to exploration, whatever it may be, was not controlled in such a way as to certainly obtain a maximum expression of the drive. Probably the scores of the exploratory group would have been higher had we restrained the animals in a small opaque enclosure, giving a minimum of extero-ceptive stimulation, for some time before the test. The writer had considered this procedure but abandoned it because of the necessarily concomitant variation in the possible activity tendency — the drive towards the exercise of the neuro-muscular mechanisms — whose importance in influencing behavior in the obstruction apparatus is not known. It may be argued that even under our actual experimental conditions the incentive used was more favorable not only to exploration but also to muscular activity (running, etc.). This is true to a certain extent, but consideration of the following facts makes it appear improbable that the activity drive entered as an important factor into our results: (1) Our animals had ample opportunity for muscular exercise in their cages until just beore the test (see page 6); (2) the entrance compartment of the apparatus, while not as large as the living cages, is extensive enough (10x10x10 inches) to permit free movement. We discuss below the possibility of considering exploration as an activity drive of a special kind.

In behavior tendencies like those of hunger and sex there are certain fairly definite physiological conditions related to each specific drive. Is there a comparable intra-organic mechanism in the exploratory drive — assuming that our results justify the use of this term? The answer to this question is so utterly remote that a few speculations may be permissible. Just as the muscles apparently require a certain optimum amount of exercise, may not the receptor organs and their more central connections also require stimulation? If this be considered a reasonable assumption we may go a step further and suggest two possible modes of origin of this tendency or trend toward stimulation: (a) A fundamental tendency inherent in all living tissue toward the expression of its characteristic activity; a basic drive toward functioning. (b) A derivative tendency; in the early life history of the animal (or of the species) specific intra-organic stimuli (hunger contractions, for instance) led to random movements and these, in turn, to various kinds of external stimulation and to "satisfaction of internal needs." With repetition of such sequences the novel situation — i.e., novelty — became conditioned to the primary drives and thus became of itself efficient in arousing a new type of drive behavior.

It has been shown (2, 5) that sex behavior is dependent upon two factors: environmental and intra-organic stimulation. The same is true of hunger drive (6) and thirst drive (7) behavior. To a limited extent the intra-organic factor may be measured in the absence of the related external stimuli, since an internal tension (or condition of excitation) has the tendency to express itself in spontaneous, non-directed activity when the proper incentive object is not present or available. Without both factors in the experimental situation, however, drive behavior, capable of measurement in terms of an obstruction overcome, is apparently impossible. We cannot, therefore, agree with Warner (5, page 4) when he says, speaking of the physiological condition of the animal and of the related external stimuli, that "of these two factors implied by the term, drive, the first is essential, the second non-essential." The writer holds that both are essential.[2] Nevertheless it seems to be true that in some cases the external factor is of greater relative importance, in other cases the internal factor. Warner (page 65) appears to mean this when he says, "The behavior data indicate that sex activity is initiated rather more by the external stimulus situation in the male than in the female, and rather more by internal stimulation in the female than in the male." In the exploratory drive we have, perhaps, the case in which the external stimulus situation is of the greatest relative importance, whereas the activity drive (tendency toward the exercise of the neuro-muscular mechanisms) possibly offers one of the best illustrations of the predominance of the intra-organic factor.

Summary

1. The tendency of twenty male albino rats approximately 185 days old to explore a novel situation was measured in terms of the number of times they crossed an electrified grid in order to reach an incentive especially designed to furnish favorable opportunity for exploration. The results of this group were compared with

[2] I do not mean, of course, that the incentive object must be immediately present to the animal. It is sufficient if the immediately present situation is capable of initiating fairly specific "preparatory reactions" which lead to the necessary environmental factor.

those of other animals tested under conditions varying only in respect to the exploratory "value" of the incentive employed.

2. The results indicate that use of an incentive offering opportunity for exploration increases the tendency of the animals used to cross the standard shock employed in the Columbia Obstruction Method. This indication is not entirely reliable statistically, variability in performance being very great.

3. In so far as the finding reported in the preceding paragraph is considered reliable, we are justified in speaking of an exploratory drive in rats of the age, sex and strain used.

Bibliography

(1) Dashiell, J. F. 1925. A quantitative demonstration of animal drive. *J. Comp. Psychol.*, 5, 205-8.

(2) Jenkins, Marion. 1928. The effects of segregation on the sex behavior of the white rat as measured by the obstruction method. *Genetic Psychol. Monographs*, 3, 457-571.

(3) Jenkins, T. N., L. H. Warner, and C. J. Warden. 1926. Standard apparatus for the study of animal motivation. *J. Comp. Psychol.*, 6, 361-82.

(4) Nissen, H. W. 1929. The effects of gonadectomy, vasotomy, and injections of placental and orchic extracts on the sex behavior of the white rat. *Genetic Psychol. Monographs.*

(5) Warner, L. H. 1927. A study of sex behavior in the white rat by means of the obstruction method. *Comp. Psychol. Monographs*, 4, No. 22.

(6) Warner, L. H. 1928. A study of hunger behavior in the white rat by means of the obstruction method. *J. Comp. Psychol.*, 8, 273-99.

(7) Warner, L. H. 1928. A study of thirst behavior in the white rat by means of the obstruction method. *J. Genetic Psychol.*, 35, 178-92.

PART VII

A COMPARISON OF NORMAL DRIVES

Part VII

A COMPARISON OF NORMAL DRIVES

C. J. Warden

The primary purpose of the project as a whole was to make possible a direct comparison of the normal drives in the white rat. By the term "normal" we mean merely that the conditions so designated were as natural as is possible in a laboratory environment except for the fact that the animal was deprived of the specific incentive under investigation. This general situation is normal as contrasted with many other conditions tested, such as those involving segregation, gonadectomy, vasotomy, injections, delayed incentive and the like. These normal conditions are of special interest in connection with our main purpose in determining the relative influence of the different drives under natural conditions. It is clear that the maximum scores obtaied under normal conditions may be made the basis of a ranking of the drives for this species, within the limits of the particular test method employed. The comparison of normal drives (Part VII, 1) is based upon data drawn from Part II, 1; Part III, 1; Part IV, 1; Part V, 1; Part VI, 1; and covers the hunger, thirst, and sex drives, both male and female, and the maternal and exploratory drives.

No attempt will be made to summarize and compare the results reported in the other studies; since these are concerned with an extension of the test to special conditions of a given drive and have no general comparative significance. Whatever comparisons are possible were indicated in each paper, and a summary of these would be little more than a repetition of the main conclusions in each case.

In bringing together the results of the five studies cited for purposes of comparison, only such scores as seem relevant to the discussion of the rank order of the various drives have been tabulated and used. The main emphasis throughout Part VII, 1, will be upon the maximum scores for the normal drives, since the ranking of the drives must rest upon these scores.

1. THE RELATIVE STRENGTH AND PERSISTENCE OF THE NORMAL DRIVES IN THE WHITE RAT

C. J. Warden

The central purpose of the project, as stated in the preceding section, was to determine the relative strength and persistence of the more important drives in the white rat when tested under the most natural conditions, e.g., with all conditions normal except for some definite and measurable deprivation of the incentive. We shall speak of the drives as tested under these conditions as "normal" drives. The value of the present study over previous studies of dynamic behavior lies not merely in the fact that a more direct and controlled method (The Columbia Obstruction Method) was used in the measurement of the several drives, but even more in the fact that the project was so standardized throughout that the scores of the various drives may be directly compared with one another. This standardization makes possible a fairly definite ranking of the drives and thus gives us a picture of the natural dynamic behavior patterns of the white rat, within the limitations of a single test method.

Before entering upon an analysis and comparison of the actual scores, it will be necessary to direct attention to the extent to which standardization was carried in method and procedure — since the validity of comparing the scores directly from drive to drive must depend, in the last analysis, upon the thoroughness of such standardization. This aspect of the problem will be covered in the first sub-section, and this will be followed by a second sub-section dealing with the comparison of the scores indicating the ranking of the different drives.

Standardization of Experimental Conditions

Our interest here is not to establish the validity of the Columbia Obstruction Method as an instrument for the measurement of animal drives; for such a treatment the reader may be referred to Part I of this volume. Our present concern is merely to show that the method was adapted to the measurement of the various drives under such uniform conditions that the scores from drive to drive may be directly compared. For the sake of clearness the conditions of uniformity adopted will be presented, in so far as possible, in tabular form. The conditions enumerated below were uniform for all the drives except for certain necessary variations as indicated.

A. Apparatus conditions (See Part I, Fig. 1, for general diagram of apparatus)

1. Compartment A (Entrance) empty except for test animal.

2. Compartment B (Obstruction) grid giving standard shock; alternating current of 60 cycles, with terminal pressure of 475 volts, external resistance of 10,100,000 ohms, and current of 0.047 milliamperes.

3. Compartment C (Incentive response-chamber) empty except for test animal.

4. Compartment D (Incentive container). Identical box used in tests on hunger, thirst, and sex; for special maternity cages, see Fig. 1, Part V, 1; for exploration box, see Fig. 1, Part VI, 1.

5. Doors separating the different compartments, including the automatic release door between compartments C and D, were operated in the same manner in testing all drives, regardless of whether the incentive required the use of a given door or not.

B. Animals employed

1. Wistar Institute experimental colony strain throughout.

2. Animals reared at Wistar Institute until approximately 150 days of age, or first and second generation of this strain reared in our own laboratory.

3. Animals weaned at approximately 30 days; males and females reared together until 150 days of age, at which time they were segregated as to sex for 35 days (range, 33-39 days) to eliminate pregnancies. Exception: groups used in study of the maternal drive were not segregated until near term, normally the fact of pregnancy being sufficient to prevent copulation during this pe-

riod; this exception was necessary in order to make the primiparous and multiparous groups strictly comparable.

4. Standard age for testing was 185 days (range, 175-196 days) or the age reached at the end of the 35-day segregation period. Exceptions: (a) Male sex drive, 28-day group (Part IV, 1) were of standard age but for the longer period of sex deprivation (28 days) involved; this group can be eliminated without effecting the comparisons to be made in this section. (b) First litter, maternal drive group in which age of testing was necessarily determined by the factor of primaparity (range, 79-150 days).

5. Animals shipped to us from the Wistar Institute at approximately 150 days were thus allowed about 5 weeks to become accustomed to living conditions in our laboratory before reaching the standard test age. See Part IV, 1, sub-section on Method and Procedure, for detailed statement of general living conditions in our laboratory — conditions which were strictly maintained throughout the testing of all drives.

C. Incentive conditions

1. The general plan was to deprive the animal of some one of the factors (food, water, sex, litter, etc.) of the standardized set of living conditions, keeping all other factors of the normal living conditions constant and uniform from drive to drive.

2. A summary of incentive conditions is presented in Table I of this study. The incentives used in the apparatus were samples of usual food (McCollum's diet), water supply (metal nipple bottle system), sex object, etc., so that the incentive response was always to the sort of stimulus to which the animals had been accustomed in cage life.

3. Isolation of a given drive for testing was accomplished, in so far as isolation could be secured, by keeping drives, other than the one being tested quiescent, or at a physiological minimum during the test period. A summary of the controls utilized are shown in the last five columns of Table I of this study. The conditions for keeping any one of the given drives at a minimum while testing some other drive was constant and uniform throughout. In the testing of any drive, four of these controls were operative, so that,

TABLE 1

Summary of experimental and test conditions for normal drives

DRIVE CONDITION TESTED	SIZE OF GROUP			INCENTIVE (Compartment D)	INCENTIVE RESPONSE	METHOD OF KEEPING OTHER DRIVES QUIESCENT				
	M.	F.	Comb.			Hunger	Thirst	Sex	Maternal	Exploratory
(Hunger)				Sample of regular diet (McCollum's mixture) in compart.	Nibble of powdered food; 30 sec. limit	Drive being tested	Regular water supply to hour of test	Males seg. 35 days; Females in dioestrum	Only non-pregnant females without litters	Specific incentive objects present; no place to explore
0 days	10	10	20							
2 days	10	10	20							
3 days	10	10	20							
4 days	10	10	20							
6 days	10	10	20							
8 days	10	10	20							
(Thirst)				Regular water bottle in compart.	Moistening tongue on damp nipple; 30 sec. limit	Regular diet, except greens omitted	Drive being tested	Same as above	Same as above	Same as above
0 days	10	10	20							
1 day	10	10	20							
2 days	10	10	20							
4 days	10	10	20							
6 days	10	10	20							
(Sex — male)				Female in cornified stage, except compart. empty for control group	Nipping, biting, nosing genitalia, mounting, etc.; 30 sec. limit	Regular diet to hour of test	Same as thirst above	Drive being tested	Same as above	Same as above
0 hours	20									
6 hours	20									
12 hours	20									
1 day	20									
1 day (control)	20									
4 days	20									
7 days	20									
28 days	20									
(Sex — female)				Normal male, except compart. empty for control group	Nipping, biting, nosing genitalia, exposing vulva; 30 sec. limit	Same as above	Same as above	Drive being tested	Same as above	Same as above
Early congestive		7								
Oestrum (cornified)		21								
Same (control)		22								
Late cornified		3								
Post-ovulative		6								
Dioestrum (4 stages)		32								
(Maternal)				Litter in maternity cage	Attending to litter; 30 sec. limit	Same as above	Same as above	Females with litter	Drive being tested	Same as above
First litter		9								
Multiparous (stand. age)		10								
(Exploratory)				Exploration box	Random activity; 30 sec. limit	Same as above	Same as above	Males seg. 35 days	Males only	Drive being tested
Males (stand. age)	20									

even if the controls should be considered inadequate, they were uniform for all normal drive conditions investigated. The adequacy of the various controls will be discussed at some length later.

D. Procedure

1. The experimenter was eliminated by making the tests in a dark room with observation made possible by means of a specially constructed illumination hood (see Part I, 1) which operated to place the experimenter outside the visual field of the test animal.

2. All tests were made between the hours of 9 P.M. and 4 A.M., or during the period when the white rat is normally most active; it was felt that maximum drive scores were most likely to be secured at this time when diurnal activity is also at a maximum. This time of day also favored better general laboratory conditions for testing.

3. The test period proper was standardized at 20 minutes immediately following the preliminary period to be next described. The advantages of a temporal over other forms of limiting the testing will be shown in the later discussion.

4. During the preliminary period, the incentive was indicated to the animal by allowing four crossings from compartment A into compartment C, with access to the incentive object in compartment D, with no shock present in compartment B. This tended to make the specific incentive dominant in the stimulation conditions afforded by the apparatus. The presence and nature of the shock was indicated to the animal by a fifth crossing, with the standard current turned on in compartment B. The testing proper followed immediately.

Exception: In testing the exploratory drive, the animal was not allowed to proceed into compartment D, except on the fourth and fifth crossings of the preliminary period, in order not to lower the value of the exploration box as an incentive and to allow compartment C to be thoroughly explored in the preliminary period and thus enhance the incentive value of the exploration box in the test proper. This change in procedure favored the exploratory drive somewhat, perhaps, but the scores for this drive were extremely low anyway, so that the ranking of drives is not disturbed thereby.

5. The general method of handling the animals was kept uniform; the mode of lifting the animal from incentive compartment and the rate of movement in transferring it back into the entrance compartment was developed into an automatic habit by the experimenter before being allowed to take data in connection with the project.

6. Scores were taken during the test period, covering the following types of response to the obstruction-incentive situation (see Fig. 3, Part IV, 1, for copy of score card):

(a) Approaches — Tip of nose across threshold of door between compartments A and B, but without touching the grid, facing incentive.

(b) Contacts—Touching grid with nose or stepping upon it; partial crossings involving shock.

(c) Crossings—Complete passage over grid into incentive compartment.

7. Only one sex was tested on a given day and the apparatus washed out thoroughly at the close of each test period.

It will be noted, in checking over the above summary of conditions, that strict uniformity was had from drive to drive, with the following exceptions: (1) The maternal groups were not segregated as to sex 35 days previous to testing. This variation in conditions may well be ignored in view of the fact that they were pregnant for a period of 22 days of the 35-day period, on the average, during which time copulation would naturally be avoided. Furthermore, as Jenkins (Part IV, 2) has shown, a segregation period of 35 days is of slight consequence in the case of the adult white rat. (2) One group of males (28-day sex deprived) were, on account of the long period of sex deprivation beyond the standard age at the time of testing. We have eliminated this group from the comparisons of this section, so that this variation in conditions can be entirely ignored. (3) The first litter, maternal group was younger than the standard age at time of testing, since age was necessarily determined by the factor of primiparity. In ranking the drives, this group may also be eliminated, and the standard age, maternal group score be taken as the maximum score for the maternal drive, under our standard conditions. (4) Passing the threshold into compartment D was allowed only on the fourth and fifth crossings, in testing the exploratory drive. As we have elsewhere stated, this variation in conditions favored the exploratory drive, and since the score for this drive is extremely low even with this advantage, this variation does not influence the rank order of the drives at all, and so may be disregarded.

Much more important than the above mentioned slight variations in procedure certain problems raised by the methods of control utilized to keep other drives at a minimum while a given drive was being tested, the comparability of the incentive-response (Table 1) from drive to drive, and the validity of a test period of set length. The questions raised here do not arise in so far as the method is utilized in the measurement of a single drive, but are of major consequence in any attempt to compare the scores and rank the different drives. These points will now be discussed in order.

It is manifestly impossible to isolate, in any absolute sense, a single drive in a complex organism like the white rat. Obviously, the different physiological systems upon which drives are based are more or less interdependent and operate simultaneously when the organism is maintained in a normal or natural state. The best that can be done is to keep drives, other than the one being tested in a minimum status of quiesence, and such a status must be empirically determined for each drive and with reference to the general living conditions which prevail during the investigation. Standardization of method covering the manner of securing the quiescence of a given drive is of as much importance as standardization of test procedure. Furthermore, in a project such as the present one in which the central aim is the ranking of drives, it is necessary to make use of the same method of allaying a given drive throughout the entire investigation.

Such standardization was accomplished in the present case in the following manner: (1) Except when the hunger drive was itself being tested, the animal was taken from a cage which had been continuously supplied with the usual diet of the laboratory. (2) Except when the thirst drive was itself being tested, the animal was taken from a cage which had been continuously supplied with water. (3) Except when the male sex drive was itself being tested, the male had been segregated from females for 35 days, and at approximately this time the male sex drive is at a minimum within the limits of our tests on male sex deprivation. (See score for 28-day sex deprived group, Part IV, 1.) (4) Except when the female sex drive was itself being tested, only females in dioestrum, at which stage the female sex drive is at a minimum, were used. This does not apply, obviously, to the maternal groups, but the same principle holds here, since females are sexually inactive while carrying the young. (5) Except when the exploratory drive was itself being tested, the animals were prevented from exploratory excitation by the small size of compartment D, by the preliminary crossings into compartments C and D, and by the dominance of appropriate incentive objects in compartment D, whenever they entered these compartments. The chief value of the Columbia Obstruction Method as a means of measuring definite incenitve-drive conditions lies in the directness and dominance of the incentive-object stimulation. (See Warden, The Columbia Obstruction Method, and Part I, 2.) Even if these controls were inadequate in maintaining at an absolute minimum all drives, other than the one being tested, whatever lack of control there may have been operated uniformly throughout and hence does not argue against the comparability of the scores from drive to drive.

Some question may arise as to the comparability of the incentive-response from drive to drive. The response of an organism to food, water, sex object, etc., naturally differs considerably not only qualitatively, but also as to the directness with which it is called out when brought in contact with the incentive object, as in compartments C and D. The hungry or thirsty animal reacts promptly enough to food or water, and likewise the maternal and exploratory response is usually prompt; at least this was true under our conditions in which the animals had gotten over the normal tendency to investigate a novel place during the five preliminary crossings into these compartments. But the completed sex act is not prompt in either male or female and we found it necessary to use some latitude in deciding when the animal had made an appropriate sex response so that it might be returned to the entrance compartment again. Such preliminary sex activities as biting the sex object or nosing the external genitalia rather than the act of copulation were the criteria adopted. In the case of the female, the familiar response of elevating the vulva under the nose of the male was considered sufficient. Mounting, or attempted mounting, in the male, if it occurred before the usual nosing or biting was also checked as a response to the sex object. The white rat exhibits no small degree of individual difference in the matter of preliminary sex activities, and it was necessary to use common sense in the matter of a criterion in the case of both male and female sex drive. It might be argued that, since the complete sex act was not permitted, the test conditions for the sex drive differed in an important manner from that for the other drives. But, as a matter of fact, deprivation during the test is necessary was the rule for all the drives; the incentive object was utilized as a stimulus to excite the drives rather than as a means of satiation. In the case of hunger, only the merest nibble of powdered food was permitted upon entering the incentive compartment; only the damp nipple of the water bottle could be licked in testing the thirst drive; only the merest attention to the young was allowed in the maternal group, and only the slightest running about was permitted in the exploration box in testing the exploratory drive. We cannot, of course, contend that the stimulation conditions from drive to drive were of the same degree, in so far as the incentive-response in compartments C and D are concerned, but we made use of the best criteria of response to the different types of incentive that we could devise after a large amount of preliminary observation of the animals in the apparatus. Further observation during the progress of the testing confirms us in our opinion that, when taken in connection with our method of scoring, to be next discussed, the criteria were sufficiently comparable as not to involve any question regarding the validity of ranking the different drives. This view of the matter is further strengthened by the fact that the ranking of the drives, in the last analysis, will be based upon the maximum scores for each drive; but, when a given drive is strongly aroused, the response to the incentive tends to be definite and prompt as might be expected.

The advantages of a test period of definite length has many obvious advantages, and among these, the matter of ease in scoring is of considerable importance. The amount of work done in a given time, or the number of repetitions accomplished within a definite practice period are well established methods of scoring human performance. The only questions that might arise regarding the validity of such a set test period as we used in this project (20 min.) would be: (1) Was the test period long enough to include a sufficient amount of behavior characteristic of a given drive to yield an adequate index of the strength of the drive; (2) since the incentive responses varied in promptness from drive to drive, was a test period of the same length equally fair to the various drives investigated?

The first question can be answered clearly in the positive. Preliminary work indicated that a test period of only 10 minutes by this method would give valid indices of drive conditions, the curves following the same general lines as those obtained by our standard 20-minute test period. We have checked through the results of several drives and have found that the ranking of drive conditions would not be disturbed at all by discarding the results obtained during the last half (10 min.) of the standard test period. In the case of each drive, approaches, contacts, and crossings were checked on the record card (see Fig. 3, Part IV, 1) for each minute of the 20-minute test period so that this matter is open to the most complete analysis. The test period was set at 20 minutes so as to insure the inclusion of any drive behavior of significance in the score.

The second point cannot be so clearly disposed of. For, as we have said, the incentive response varied not only as to general type from drive to drive, but also as to promptness, and this latter factor would be expected to influence the score since the test period was limited. In order to equalize the value of a crossing to the incentive from drive to drive, the animal was allowed to remain in the incentive compartment, after crossing over the grid, for not longer than 30 seconds, regardless of whether he had met the criterion of an appropriate incentive response for the given drive or not. That is to say, the time allowed for an incentive response was limited just as was the total period of the test. This was done with a view to standardizing the method so as to take account of differences in promptness in responding to the incentive from drive to drive. Again we must suggest that, in ranking the drives, we are directly comparing the maximum scores only, and that when strongly aroused the response to the incentive tended to be prompt and uniform in the case of each of the drives. As a matter of fact, the act of crossing and making an appropriate response to the incentive did not seem to vary any more from drive to drive than from one condition to another within the same drive. In any case, the incentive response could not last longer than 30 seconds, and in most cases actually occurred much sooner than that in each of the drives, and especially so when these were at a maximum.

From the above discussion it will appear that our apparatus and method was sufficiently well standardized to warrant the use of the scores in ranking the normal drives as proposed.

Analysis of Results

Our main interest, in the present section, will be in a comparison of the maximum scores of the different drives with a view to ranking them in a definite order on the basis of strength and persistence. Nevertheless, it seems best as a matter of convenience to the reader to bring together in tabular form the results of each normal drive condition investigated, and so to present the maximum scores in their appropriate setting.

The averages for each of the normal groups, covering approaches, contacts, and crossings, are given in Table 2, together with the usual statistical values for the several averages. These scores represent the full 20-minute test period. Separate indices have been entered for the two sexes in the case of hunger and thirst, in addition to the combined scores for these two drives.

In Table 3 will be found a set of values representing a grouping of the scores on the basis of successive five-minute parts of the 20-minute test period, and percentage scores, following the same grouping comprise Table 4. A finer temporal distribution of the scores (minute by minute) is available in the original reports, but no summary of this analysis has been included at this point, since even the five-minute grouping seems to be of little significance. The data were taken and reported in these temporal groupings in the hope that such temporal distribution would throw some light upon the persistence of the drive during the test period. It was thought possible that some useful measure of the *persistence* of a given

drive might be thus secured, corresponding to the *strength* score based upon the total score. However, it appears to be very doubtful whether the distinction between strength and persistence of drive is defensible — at least in so far as these may be indicated by total

TABLE 2

Showing results for the various drive conditions

DRIVE CONDITION TESTED	SIZE OF GROUP	APPROACHES			CONTACTS			CROSSINGS		
		Average	Standard Deviation	Coefficient of Variation	Average	Standard Deviation	Coefficient of Variation	Average	Standard Deviation	Coefficient of Variation
(Hunger—male)										
0 period	10	1.5	1.34	89	2.2	1.20	55	2.7	3.37	125
2 days	10	3.3	1.47	44	4.9	1.59	32	16.1	6.56	41
3 days	10	5.0	1.41	28	5.7	1.58	28	18.0	7.52	42
4 days	10	3.6	2.20	61	2.8	1.17	42	19.1	5.87	31
6 days	10	3.6	1.59	44	3.0	1.55	52	14.2	5.98	42
8 days	10	4.6	1.79	39	5.5	2.18	40	9.8	5.33	54
(Hunger—female)										
0 period	10	2.0	1.34	67	3.5	1.55	44	2.1	2.07	98
2 days	10	3.4	1.56	46	6.9	1.87	27	19.0	8.91	47
3 days	10	5.2	1.25	24	8.0	1.41	76	18.4	7.63	41
4 days	10	5.2	2.04	39	5.7	1.10	19	17.0	5.92	35
6 days	10	6.1	1.77	29	5.3	1.80	34	14.0	7.18	51
8 days	10	6.8	1.69	25	7.3	2.29	31	6.0	4.69	78
(Hunger—combined)										
0 period	20	1.75	1.36	79	2.85	1.42	50	2.40	2.78	116
2 days	20	3.35	2.81	84	5.90	3.92	66	17.60	7.78	44
3 days	20	5.10	2.46	48	6.85	2.93	43	18.20	7.58	42
4 days	20	4.40	2.37	54	4.75	1.91	40	18.10	5.98	33
6 days	20	4.85	3.58	74	4.15	3.61	87	14.10	6.61	47
8 days	20	5.70	3.77	66	6.40	5.71	89	7.90	5.07	64
(Thirst—male)										
0 period	10	3.0	1.6	53	1.7	2.1	123	3.7	4.0	108
1 day	10	7.3	3.3	45	5.6	5.4	96	21.1	11.6	54
2 days	10	5.0	3.7	74	4.5	4.3	95	16.7	12.3	74
4 days	10	3.0	1.8	60	1.8	2.0	111	12.7	9.0	71
6 days	10	0.8	0.7	87	0.9	1.2	133	7.5	5.5	73
(Thirst—female)										
0 period	10	3.1	2.9	94	0.8	0.9	112	4.6	3.8	83
1 day	10	7.3	4.7	64	4.7	3.9	83	19.7	11.1	56
2 days	10	4.0	3.5	87	3.8	3.4	89	15.3	11.7	76
4 days	10	3.2	3.3	103	1.1	1.4	127	14.5	8.3	57
6 days	10	1.7	1.9	111	1.8	1.9	105	6.9	6.9	100
(Thirst—combined)										
0 period	20	3.05	2.3	75	1.25	1.3	104	4.15	3.9	94
1 day	20	7.30	4.1	56	5.15	4.7	91	20.40	11.4	56
2 days	20	4.50	3.6	80	4.15	3.9	94	16.0	12.0	75
4 days	20	3.10	2.7	87	1.45	1.7	117	13.6	8.7	64
6 days	20	1.25	1.5	120	1.35	1.6	119	7.2	6.0	83

TABLE 2 (*Continued*)

(Sex—male)										
0 period	20	1.2	1.36	113	0.35	0.7	200	3.6	4.08	114
6 hours	20	4.15	3.26	79	4.50	2.77	62	8.1	5.20	64
12 hours	20	2.70	2.03	75	2.65	1.98	75	12.2	4.80	39
1 day	20	2.30	2.35	102	1.5	1.5	100	13.45	4.03	30
1 day (control)	20	1.20	1.03	86	1.05	1.18	112	2.95	1.80	61
4 days	20	1.80	1.90	106	1.85	1.56	84	12.60	6.24	50
7 days	20	2.25	1.50	67	3.10	2.50	81	12.25	7.07	58
28 days	20	2.50	2.90	116	2.00	1.97	99	10.55	7.10	66
(Sex—female)										
Early congestive	7	6.71	5.39	80	4.86	2.31	48	11.14	6.08	55
Oestrum (cornified)	21	5.43	3.63	67	3.14	2.73	87	14.14	5.14	36
Same (control)	22	3.23	2.70	84	2.27	1.91	84	5.05	2.55	51
Late cornified	3	2.33	1.39	60	3.66	1.39	38	8.66	5.55	64
Post-ovulative	6	1.17	1.04	89	4.00	2.92	73	4.66	5.28	113
Dioestrum (4 stages)	32	1.16	1.42	117	1.47	1.27	86	1.34	1.57	117
(Maternal)										
First litter	9	3.89	2.68	69	0.78	0.63	81	28.33	7.73	27
Multiparous (stand. age)	10	4.10	3.67	89	1.30	1.90	146	22.40	9.14	41
(Exploratory)										
Males (stand. age)	20	5.20	3.85	74	1.8	1.66	92	6.00	4.89	81

score and temporal distribution respectively. It is true, as an examination of Tables 3 and 4 will show, that when a drive is at its maximum as indicated by the maximum score in crossings, the score tends to be uniformly maintained throughout the test period; the one exception to be noted is the maximum score for the thirst drive. Scores representing other than maximum conditions of the drives usually show either an increase or a decrease through successive five-minute parts of the test period. But what this may mean in terms of the loose concept of *persistence of drive* is not at all clear, and we shall omit further discussion of these data from this section.

In comparing the various normal drives we shall limit ourselves to the scores for crossings, in which a complete response to the obstruction-incentive situation occurred. As a matter of fact, this plan was followed in the original reports on the several drives, since the other two types of response (approaches, contacts) appeared to be of little or no significance as between different drive conditions. From the first, approaches and contacts were checked in the hope that these partial crossings would reveal something of importance regarding strength of drive, and the practice was continued in the interests of uniformity in all later studies. However,

TABLE 3

Showing temporal distribution in five-minute intervals of approaches, contacts, and crossings during the twenty-minute test period

DRIVE CONDITION TESTED	APPROACHES				CONTACTS				CROSSINGS			
	Minutes of test period				Minutes of test period				Minutes of test period			
	0–5	6–10	11–15	16–20	0–5	6–10	11–15	16–20	0–5	6–10	11–15	16–20
(Hunger — combined)												
0 days	15	10	7	3	28	18	4	7	31	9	5	3
2 days	17	8	18	22	16	23	42	37	130	109	86	33
3 days	42	25	17	18	54	37	28	18	75	91	100	98
4 days	23	34	10	21	41	22	8	14	94	89	87	91
6 days	28	20	18	31	18	14	13	33	64	65	77	76
8 days	30	28	16	40	27	29	29	46	31	40	42	45
(Thirst — combined)												
0 days	20	18	13	10	8	6	4	7	42	21	7	13
1 day	74	31	22	19	38	24	20	10	153	106	84	65
2 days	24	31	16	20	28	22	14	19	153	81	60	26
4 days	18	17	20	7	14	3	6	6	115	70	48	39
6 days	8	8	4	5	10	6	4	7	39	31	42	32
(Sex — male)												
0 hours	4	3	12	5	2	2	2	1	10	19	18	25
6 hours	26	17	24	16	31	22	19	18	52	61	28	21
12 hours	6	22	15	7	13	13	8	19	74	76	44	50
1 day	19	16	6	5	9	9	8	4	76	63	60	70
1 day (control)	3	4	9	9	8	7	3	3	28	17	9	5
4 days	15	8	9	4	12	5	8	3	62	55	65	70
7 days	12	11	11	9	17	9	24	12	47	33	58	107
28 days	9	8	15	18	4	9	22	5	16	40	54	101
(Sex — female)												
Early congestive	11	17	13	6	6	3	8	17	11	18	21	28
Oestrum (cornified)	49	29	23	13	8	16	16	26	59	45	81	22
Same (control)	19	32	14	6	17	13	15	5	48	36	17	10
Late cornified	3	1	1	2	4	1	3	3	5	3	7	11
Post-ovulative	3	2	1	1	5	9	9	1	9	10	3	5
Dioestrum	12	17	5	3	15	18	9	5	18	13	9	3
(Maternal)												
First litter	11	16	5	3	4	1	1	1	64	59	65	67
Multiparous (Standard age)	12	14	8	7	1	5	5	2	63	50	53	58
(Exploratory)												
Males (Standard age)	38	29	19	18	12	12	4	8	48	29	23	20

we can find no indication of any significance in these two types of partial crossing to the incentive. From an inspection of the values for approaches and contacts in Table 2, it appears that they exhibit relatively slight variations from one condition to another within the same drive, and from one drive to another, sharply contrasting with markedly variable crossing scores. There seems to be no legitimate method by which the three types of reaction to the obstruction-incentive situation might be combined into a single index of strength of drive. A crossing always includes, of course, both an approach to, and a contact with the obstruction (grid). It is logical to suppose that any type of score for partial crossings would be reduced by actual crossings, as when the drive is strong. There is

TABLE 4

Showing temporal distribution in five-minute intervals of approaches, contacts, and crossings, (percentages) during the twenty-minute test period

DRIVE CONDITION TESTED	APPROACHES Minutes of test period				CONTACTS Minutes of test period				CROSSINGS Minutes of test period			
	0–5	6–10	11–15	16–20	0–5	6–10	11–15	16–20	0–5	6–10	11–15	16–20
(Hunger — combined)												
0 days	43	28	20	9	49	32	7	12	65	19	10	6
2 days	26	12	28	34	14	19	36	31	37	29	25	9
3 days	41	24	17	18	39	27	21	13	21	25	27	27
4 days	26	38	12	24	48	26	9	17	26	25	24	25
6 days	29	21	18	32	23	18	17	42	23	23	27	27
8 days	26	25	14	35	21	22	22	35	20	25	27	28
(Thirst — combined)												
0 days	33	30	21	16	32	24	16	28	51	25	8	16
1 day	51	21	15	13	41	26	22	11	37	26	21	16
2 days	26	34	18	22	34	26	17	23	48	25	19	8
4 days	29	28	32	11	48	10	21	21	42	26	18	14
6 days	32	32	16	20	37	22	15	26	28	22	28	22
(Sex — male)												
0 hours	17	12	50	21	29	29	28	14	14	26	25	35
6 hours	31	21	29	19	35	24	21	20	32	38	17	13
12 hours	12	44	30	14	24	25	15	36	30	31	18	21
1 day	41	35	13	11	30	30	27	13	28	24	22	26
1 day (control)	12	16	36	36	38	34	14	14	48	29	15	8
4 days	42	22	25	11	43	18	28	11	24	22	26	28
7 days	28	26	25	21	27	15	39	19	19	13	24	44
28 days	18	16	30	36	10	22	55	13	8	19	25	48
(Sex — female)												
Early congestive	23	36	28	13	18	9	23	50	14	23	27	36
Oestrum (cornified)	43	26	20	11	12	24	24	40	19	15	26	40
Same (control)	27	45	20	8	34	26	30	10	43	32	15	10
Late cornified	43	14	14	29	36	10	27	27	19	12	27	42
Post-ovulative	43	29	14	14	21	38	37	4	32	36	11	21
Dioestrum (4 stages)	32	46	14	8	32	38	19	11	42	30	21	7
(Maternal)												
First litter	29	34	20	17	8	38	38	16	28	22	24	26
Multiparous (Standard age)	31	46	14	9	58	14	14	14	25	23	26	26
(Exploratory)												
Males (Standard age)	37	28	18	17	34	33	11	22	40	24	19	17

some evidence for an inverse relationship between scores for partial crossings, particularly approaches, and actual crossings, although this relationship is not very consistent. This argues against any method of weighting the three types of incentive response and resorting to pooling in order to reduce all to a common index of strength of drive, aside from the element of arbitrariness that would enter into any such weighting of scores that are so disparate in significance. We did pool the averages for approaches, contacts, and crossings, without weighting, as a matter of information to ourselves, comparing these pooled scores with the scores for crossings, and while such a pooling is certainly not legitimate, we may as well report the result. There was no displacement of rank among the group averages of a given drive, except in the case of the male and female sex drive, where the grouping represents finer divisions in experimental analysis, and even here the displacements were based upon slight differences. This comparison serves at least to show the neutral character of partial crossing scores and their consequent insignificance in any index of drive. We feel justified in suggesting that, in later studies of drive in the white rat by means of the Columbia Obstruction Method, scoring be limited to crossings only, since it seems a mere waste of time to take account of partial crossings of any sort.

Since we are interested at this point primarily in the maximum scores as these may be useful in ranking the normal drives, further discussion of Tables 2, 3, and 4 will be unnecessary. In so far as the values there summarized relate to a particular drive, detailed analyses and comparisons will be found in the full reports under the appropriate sections. We may turn at once to an examination of the maximum scores covering the number of crossings to the incentive, considering the same a common index for the ranking of the various drives. It will be convenient to deal separately with the group of male and of female drives, as well as with the group of drives in which the sexes are combined.

The rank order of the several drives, according to these groupings will be found in Table 5, together with the maximum score in crossings upon which the ranking was based. Since, as we shall see, the sex differences for the hunger, thirst, and sex drives are too small to be of much significance, the ranking of the drives in the last grouping of the table (male and female combined) are of most interest to us.

Before dealing with the reliability of the maximum scores in comparisons from drive to drive, it may be well to discuss briefly the significance of the maximum scores for a given drive. In each case these indicate the number of crossings to the incentive when the drive was most strongly aroused. Except for the maternal and the exploratory drives, no less than five conditions, representing systematic changes in the variable, were tested in the case of each drive. The curves covering the successive changes in the variable (water, food, sex deprivation, etc.) comprise Figures 1, 2, and 3, of this study. The first value on each curve represents an approximate physiological zero for the given drive. Thus hunger and thirst begin with zero deprivation; male sex with a zero established by copulating with a female in heat until completely satiated (see Part IV, 1), while dioestrum constituted a zero point for the female sex drive. In thirst, hunger, and male sex, deprivation was carried far beyond the point of the maximum drive condition as the curves clearly indicate; each of the eight recognized stages of the oestrus cycle were tested in the female. We are able to say, then, that for each of these four drives, we secured the true maximum score under our conditions, within the limits of the unit of change adopted for the several variables. A finer unit of change might show the true maximum point to lie somewhere near rather than precisely at the points we indicate in each instance. For example, it is possible that a finer working of the field might result in showing that the maximum point for thirst is somewhat less or somewhat more than the one-day point where our maximum score was found; and so with the other drives. But, in any case, it must be admitted that our maximum points are approximately true, and are clearly established as true by the general trend of the scores downward in either direction from these maximum points, as may be observed in Figures 1, 2, and 3.

Only a single condition was tested in the case of the maternal drive, and the exploratory drive, under our normal standardization, and therefore, we cannot insist that the scores included for

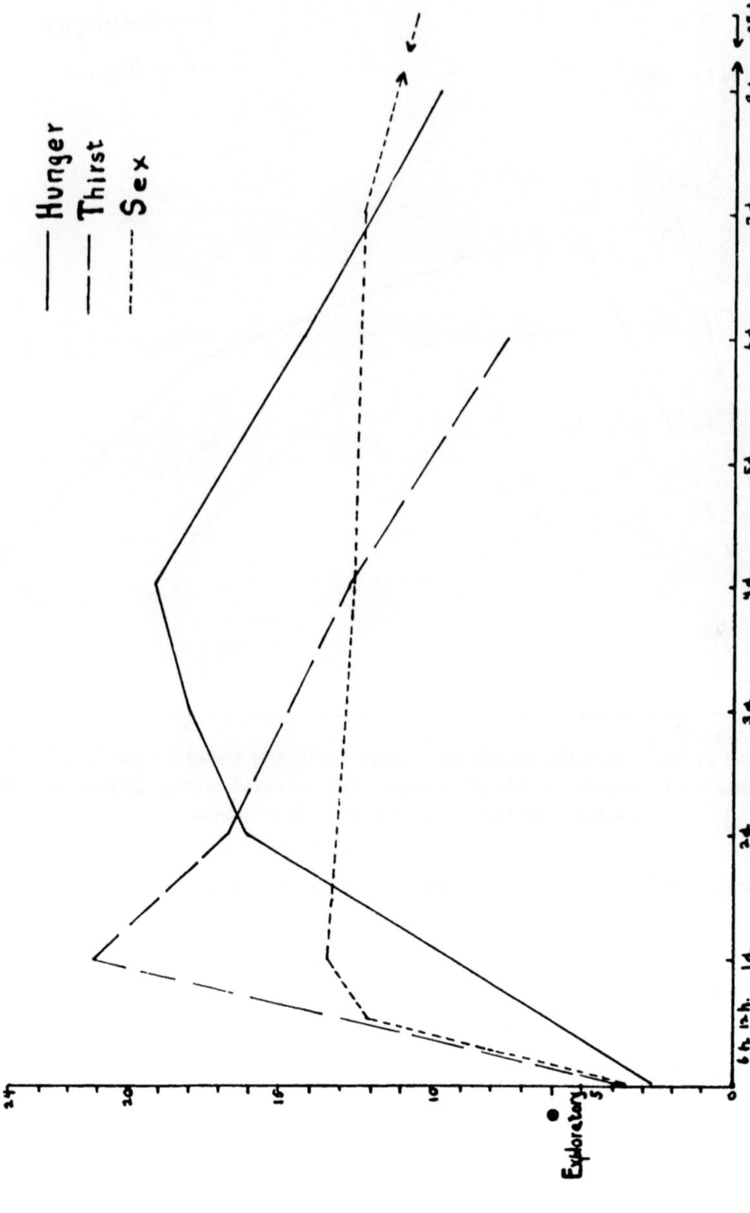

FIG. 1. The curves represent the different drive-conditions tested in the male. The several periods of deprivation are indicated on the abscissa and the average number of crossings on the ordinate.

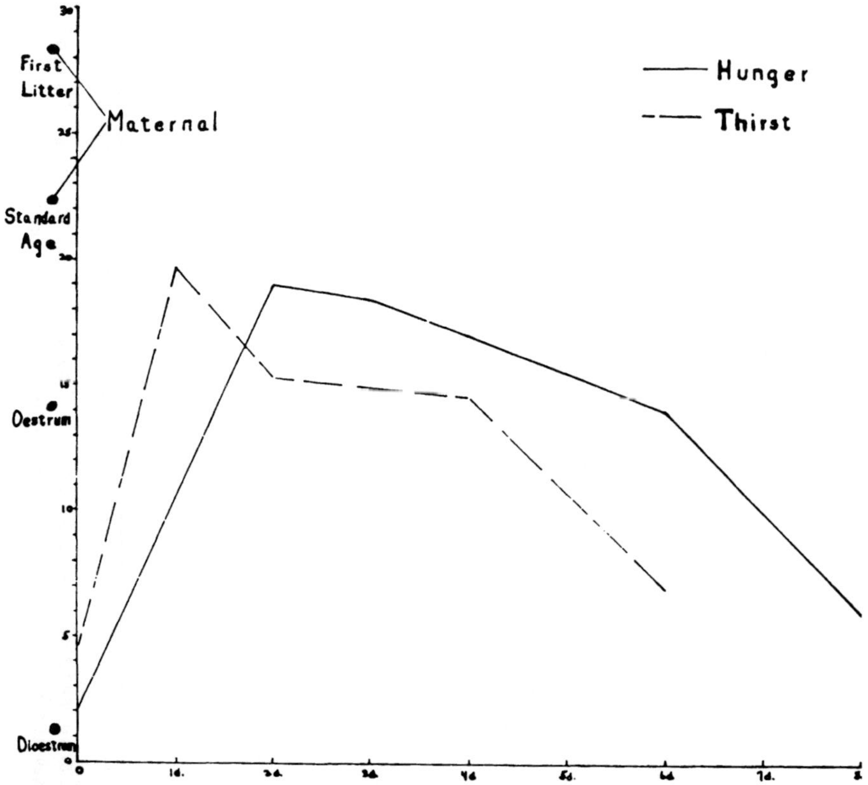

Fig. 2. The curves represent the different drive-conditions tested in the female. Th several periods of deprivation (thirst, hunger) are indicated on the abscissa and th average number of crossings on the ordinate.

these two drives are true maxima. So far as the maternal drive is concerned, it is quite doubtful whether the score in Table 5 is the highest possible maternal score, even under our standard conditions. The score for first litter is considerably higher (28.33) but this score could not be used here because the animals tested were naturally younger than our standard age. However, there is one important factor that might be varied in animals of standard age which might give a higher score than we secured, e.g., length of time between parturition and testing. Our standard age, maternal group were tested on the average 15 hours (range 12-20 hours) after casting the litter, and this may not have been the best time

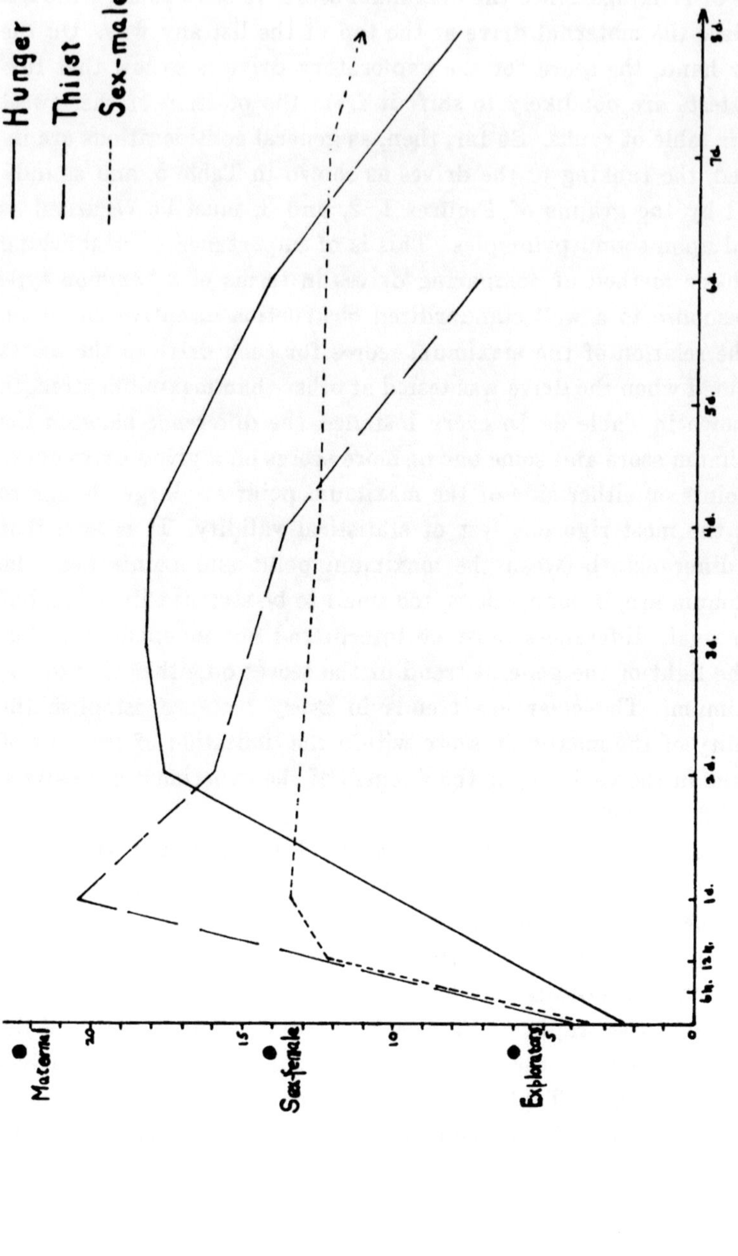

Fig. 3. The curves represent the different drive-conditions tested, with thirst and hunger scores for male and female combined. The several periods of deprivation are indicated on the abscissa and the average number of crossings on the ordinate.

to have tested them so far as strength of maternal drive is concerned. But, the admission of the possibility of an even higher score than we secured for the maternal drive does not disturb our table of rankings, since the maximum score we show is high enough to place the maternal drive at the top of the list any way. On the other hand, the score for the exploratory drive is so low that further tests are not likely to shift it from the position of the lowest in our table of ranks. So far, then, as general considerations are involved, the ranking of the drives as shown in Table 5, and as indicated by the graphs of Figures 1, 2, and 3, must be regarded as based upon sound principles. This is of importance as establishing the basic method of comparing drives in terms of a common type of response to a well standardized obstruction-incentive situation.

The relation of the maximum scores for each drive to the scores obtained when the drive was tested at other than maximum strength is shown in Table 6. In every instance, the difference between the maximum score and some one or more scores on a given drive curve at points on either side of the maximum point are large enough to pass the most rigorous test of statistical validity. It is true that the differences between the maximum point and points near the maximum are, in many cases, too small to be statistically valid, but such small differences must be interpreted not independently but in the light of the general trend of the scores on either side of the maximum. The clear cut trends in every instance establish the validity of the maximum score within the limitation of the unit of change in the variable, or the fineness of the experimental analysis, for a given drive.

Accepting the maximum scores as valid indices of the strength of the several drives, we must now inquire as to the significance of the differences between these scores from drive to drive. As previously mentioned, the ranking of the drives upon the basis of the maximum score is indicated in Table 5, and can also be seen by an inspection of the graphs of Figures 1, 2, and 3. The reliability of the differences between the maximum scores for the several male drives will be found in Table 7; similar values for the several female drives comprise Table 8, while Table 10 covers the same point with the drives grouped together without regard to sex. The relations of these reliabilities to rank order are summarized in Table 11.

Before discussing these data we shall examine Table 9 and dispose of the matter of sex differences in strength of drive. Separate scores for the sexes are available in the thirst, hunger, and sex drives only. The difference in both hunger and thirst favors the male, but is too small to be considered significant, while the sex drive is higher in the female by only a small margin of questionable validity. These findings refer only to the height of the curve for the drives and do not imply that the maximum point was identical

TABLE 5

Showing maximum scores (crossings) for the various drives, arranged in rank order

DRIVES TESTED	CONDITION OF MAXIMUM DRIVE	SIZE OF GROUP			AVERAGE	STANDARD DEVIATION	COEFFICIENT OF VARIATION
		M.	F.	Comb.			
(Male)							
Thirst	2nd day	10			21.10	11.60	54
Hunger	4th day	10			19.10	5.87	31
Sex	1st day	20			13.45	4.03	30
Exploratory	Only 1 tested	20			6.00	4.89	81
(Female)							
Maternal	Standard age		10		22.40	9.14	41
Thirst	1st day		10		19.70	11.10	56
Hunger	3rd day		10		19.00	8.91	47
Sex	Oestrum		21		14.14	5.14	36
(Combined)							
Maternal	Standard age		10		22.40	9.14	41
Thirst	1st day	10	10	20	20.40	11.40	56
Hunger	3rd day	10	10	20	18.20	7.58	42
Sex	1st day & Oestrum	20	21	41	13.80	4.64	34
Exploratory	Only 1 tested	20			6.00	4.89	81

for the two sexes. As will be seen by comparing the graphs of Figures 1 and 2, the peak of the thirst drive came at the same point (1 day) for both sexes, but this was not true in the case of the hunger drive. The peak for the females, in the latter drive, came after 2 days of starvation, while that for males came after 4 days of starvation. From the showing in Table 9, we may conclude that the hunger, thirst, and sex drives are about equally strong in the two sexes when at their maximum.

A detailed examination of Tables 7, 8, and 10 does not seem to be especially called for, since the main results in so far as the matter of the validity of the rank order of the various groupings of drives

TABLE 6

Showing the reliability of the difference between the maximum score (crossings) for each drive and all the other scores for the given drives

DRIVE CONDITIONS TESTED	S. D. OF DIFFERENCE	DIFFERENCE BETWEEN AVERAGES	DIFFERENCE / S. D. OF DIFFERENCE	CHANCES IN 100 OF A TRUE DIFFERENCE
(Hunger — male)				
4 days and 0 days	2.14	16.4	7.66	100
4 days and 2 days	2.78	3.0	1.08	86
4 days and 3 days	3.02	1.1	0.36	64
4 days and 6 days	2.65	4.9	1.85	97
4 days and 8 days	2.51	9.3	3.71	100
(Hunger — female)				
2 days and 0 days	2.89	16.9	5.84	100
2 days and 3 days	3.71	0.6	0.16	56
2 days and 4 days	3.38	2.0	0.59	72
2 days and 6 days	3,62	5.0	1.38	92
2 days and 8 days	3.18	13.0	4.08	100
(Hunger — combined)				
3 days and 0 days	1.81	15.8	8.73	100
3 days and 2 days	2.43	0.6	0.25	60
3 days and 4 days	2.16	0.1	0.05	52
3 days and 6 days	2.25	4.1	1.82	97
3 days and 8 days	2.04	10.3	5.05	100
(Thirst — male)				
1 day and 0 days	3.88	17.4	4.48	100
1 day and 2 days	5.35	4.4	0.82	80
1 day and 4 days	4.64	8.4	1.81	96
1 day and 6 days	4.06	13.6	3.35	100
(Thirst — female)				
1 day and 0 days	3.71	15.1	4.07	100
1 day and 2 days	5.10	4.4	0.86	81
1 day and 4 days	4.38	5.2	1.19	88
1 day and 6 days	4.13	12.8	3.10	100
Thirst — combined)				
1 day and 0 days	2.69	16.3	6.06	100
1 day and 2 days	3.70	4.4	1.19	88
1 day and 4 days	3.21	6.8	2.12	98
1 day and 6 days	2.88	13.2	4.58	100
(Sex — male)				
1 day and 0 hours	1.28	9.9	7.68	100
1 day and 6 hours	1.47	5.4	3.64	100
1 day and 12 hours	1.40	1.3	0.89	81
1 day and 4 days	1.65	0.9	0.51	69
1 day and 7 days	1.82	1.2	0.66	75
1 day and 28 days	1.83	2.9	1.60	95
1 day and control	0.99	10.5	10.61	100
(Sex — female)				
Cornified and control	1.25	9.1	7.28	100
Cornified and dioestrum	1.16	12.8	11.00	100
Control and dioestrum	0.61	3.7	6.08	100
(Maternal)				
First litter and standard age	3.87	5.9	1.53	94

have been summarized in Table 11. In column 1, of this table, the rank order for the several male drives tested, based upon the maximum scores is given; following this the rank order of the female drives, and finally the rank order for the drives as such without regard to sex. That is, the maximum scores for the two sexes covering the hunger, thirst, and sex drives are here combined; we have already found, in connection with the discussion of Table 9, that there is apparently no significant sex difference in maximum score,

TABLE 7

Showing the reliability of the difference between maximum scores (crossings) of the different drives in males

DRIVE CONDITIONS TESTED	S. D. OF DIFFERENCE	DIFFERENCE BETWEEN AVERAGES	DIFFERENCE / S. D. OF DIFFERENCE	CHANCES IN 100 OF A TRUE DIFFERENCE
Hunger and thirst	4.11	2.00	0.49	69
Hunger and sex	2.06	5.65+	2.74	99.7
Hunger and exploration	2.15	13.10+	6.08	100
Thirst and sex	3.78	7.65+	2.03	98
Thirst and exploration	3.83	15.10+	3.95	100
Sex and exploration	1.42	7.45+	5.26	100

Legend: The plus sign indicates that the first drive of the pair has the higher score.

TABLE 8

Showing the reliability of the difference between the maximum scores (crossings) of the different drives in females

DRIVE CONDITIONS TESTED	S. D. OF DIFFERENCE	DIFFERENCE BETWEEN AVERAGES	DIFFERENCE / S. D. OF DIFFERENCE	CHANCES IN 100 OF A TRUE DIFFERENCE
Hunger and thirst	4.50	0.70	0.16	56
Hunger and sex	3.03	4.86+	1.60	95
Hunger and maternal	4.04	3.40	0.84	80
Thirst and sex	3.69	5.56+	1.51	93
Thirst and maternal	4.55	2.70	0.59	72
Sex and maternal	3.10	8.26	2.66	99.6

Legend: The plus sign indicates that the score is higher than the female score.

TABLE 9

Showing the reliability of the difference between the maximum scores (crossings) of the male and female groups

DRIVE CONDITIONS TESTED	S. D. OF DIFFERENCE	DIFFERENCE BETWEEN AVERAGES	DIFFERENCE / S. D. OF DIFFERENCE	CHANCES IN 100 OF A TRUE DIFFERENCE
Hunger	3.37	0.10+	0.03	51
Thirst	5.08	1.40+	0.28	61
Sex	1.44	0.69	0.48	68

Legend: The plus sign indicates that the male score is higher than the female score.

so that such combination seems thoroughly justifiable. The reliability of the difference between a given drive score (maximum) and that for each drive of lower rank, taken singly, in terms of probability is shown for each drive in the next four columns of the table. The position of the sex drive, both male and female, as lower than either hunger or thirst, appears to be practically certain by this method of comparison. The exploratory drive is shown to be certainly lower than any of the other drives. The probability that thirst is higher than hunger is the least satisfactory, while the chances that maternal is higher than hunger and thirst, taken singly, are not any too good.

The problem of the validity of the rank order may be dealt with by the method of combining averages (see, Yule, pp. 115, and 142) rather than by the method of single comparisons. By this method the score of a given drive is compared, not with the score of the next lowest drive, but with the average score of all the drives of lower rank. This method is more pertinent to the present discussion because it deals simply with the factor of rank position within a group of values, such as we have here. The probability of the ordinal positions worked out by this method is much higher than that obtained by comparing the obtained differences singly, as will appear from an inspection of the last column of Table 11. The rank order as listed in this table amounts to practical certainty for the group of male drives, and for the group of drives in combination, while the chances that the listed rank order for the female drives is a true ranking is very high. The most general statement that can be made, then, is that for the white rat, tested under our conditions, and within the limitations of the scope of our investigation, the normal drives rank as follows in strength: Maternal, thirst, hunger, sex, exploratory.

The fact that the maternal drive ranks highest comes as something of a surprise, since several investigators using other methods have rated the maternal tendency rather low (see Part V, 1). In some cases, at least, the normal maternal drive has been more or less disturbed by the method employed. A common error in technique is to have the litter born in the living cage, and then transfer it for test to the apparatus used. We tested a group, following this procedure, and obtained a score only about half as high (11.6, see Table 3, Part V, 1) as in our regular group in which the small maternity cage in which the litter had been born was attached directly to the apparatus, as compartment D. It may be argued that, in such a case, we are testing the maternal drive plus the tendency to return to the home nest. But the truth seems to be that when the normal conditions of maternity are disturbed by transferring

TABLE 10

Showing the reliability of the difference between the maximum scores (crossings) of the different drives with sexes combined

DRIVE CONDITIONS TESTED	S. D. OF DIFFERENCE	DIFFERENCE BETWEEN AVERAGES	DIFFERENCE / S. D. OF DIFFERENCE	CHANCES IN 100 OF A TRUE DIFFERENCE
Hunger and thirst	3.06	2.20	0.72	76
Hunger and sex	1.84	4.40+	2.39	99
Hunger and maternal	3.35	4.20	1.25	89
Hunger and exploration	2.02	12.20+	6.03	100
Thirst and sex	2.65	6.60+	2.49	99.4
Thirst and maternal	3.85	2.00	0.52	70
Thirst and exploration	2.77	14.40+	5.19	100
Sex and maternal	2.98	8.60	2.89	99.8
Sex and exploration	1.31	7.80+	5.95	100
Maternal and exploration	3.09	16.40+	5.31	100

Legend: The plus sign indicates that the first drive of the pair has the higher score.

TABLE 11

Showing the reliability of the differences between maximum scores, as related to rank order. The difference between a given score and that for each drive of lower rank, taken singly, is treated in columns 2, 3, 4, and 5; in the last column the scores for all the drives of lower rank have been combined

DRIVE TESTED	RANK ORDER	CHANCES IN 100 OF A TRUE DIFFERENCE GREATER THAN 0				
		THIRST	HUNGER	SEX	EXPLORATORY	COMBINATION
(Male)						
Thirst	1.		69	98	100	99.4
Hunger	2.			99.7	100	100
Sex	3.				100	100
Exploratory	4.					
(Female)						
Maternal	1.	72	80	99.6		96
Thirst	2.		56	89		86
Hunger	3.			95		95
Sex	4.					
(Combined)						
Maternal	1.	70	89	99.8	100	99.6
Thirst	2.		76	99.4	100	99.7
Hunger	3.			99	100	100
Sex	4.				100	100
Exploratory	5.					

the litter to an unfamiliar apparatus, the tendency to explore the novel surroundings of the litter interferes with the maternal drive. Then, too, many animals ignore young that have been moved about — this is true of many species of birds such as the quail. Our purpose was to test the maternal drive, along with the other drives discussed in this section, under normal conditions, in order to secure maximum indices, and we feel that our ranking of the maternal drive is the only defensible one from this point of view. The fact that the score for first litter, maternal drive, was higher than the score for standard age, maternal is a further argument for the placing of the maternal drive at the top of the list. We have not used the score for first litter in the comparisons of this section relative to ranking the several drives, since the animals were younger than standard age.

The conclusions drawn above regarding the relative rank of the several normal drives in the white rat, must be, of course, qualified by the general and specific conditions representing the standardization of the project as a whole. In the strict sense, these conclusions apply only to the adult white rat, of the strain and age employed, and reared unsegregated as to sex until approximately the standard test age. The fact should be recognized that relative strength of drive is very likely a function of the age of the animal within wide limits, and there may be other factors also that enter here, such as strain, the precise incentive object used, and specific aspects of our method of testing and scoring. Our results, at any rate, justify the extension of the Columbia Obstruction Method, as a means of analyzing the dynamics of behavior, to other species, and further work in this direction has been planned by this laboratory for the immediate future. The importance of indices of motivation for typical animal forms, and secured by the same method of testing, can scarcely be overestimated for systematic comparative psychology.

APPENDICES

APPENDICES

INTRODUCTION

C. J. Warden

The following four papers have not been included in the main body of this report for the reason that the Columbia Obstruction Method as standardized for use in the general project was not employed. They are, however, studies of motivation in the white rat, and, except for the first study, were financed out of the regular project allotment.

As noted (see Warden, The Columbia Obstruction Method), the original obstruction apparatus, as devised by Jenkins and Warden was first used by Holden in 1924-25 in the study of the hunger drive, previous to the beginning of the project itself. Important changes in both apparatus and procedure were introduced later in further standardizing the method for use in the project. Nevertheless, this study of Holden was closely connected with the development of the project and belongs very properly in the present collection.

The paper by Warden and Hamilton (Appendix 2) and the later paper of Hamilton (Appendix 3) are concerned with the effect of delayed incentive in maze learning. The shorter paper may be considered as a preliminary attack upon this problem, which was more systematically and adequately dealt with in the longer report. The latter is of special interest inasmuch as it was done in connection with, and parallels the experiment on the effect of delayed incentive on the hunger drive as measured by the Columbia Obstruction Method (Part II, 2). This fact makes possible a comparison of these two essentially different methods, with respect to the influence of delay in the function tested.

The final paper is concerned with the relative value of reward and punishment as motivation situations in the Yerkes-Watson discrimination apparatus. This study was intended to be the beginning of a comprehensive comparison of these two opposite types of incentive, but this plan was later discontinued when it appeared desirable to concentrate the work along the main line of investigation.

1. A STUDY OF THE EFFECT OF STARVATION UPON BEHAVIOR BY MEANS OF THE OBSTRUCTION METHOD[1]

Frances Holden

I. Introduction

The present study is an attempt to measure the drive,[2] or motivation value, of starvation in the white rat in terms of objective behavior. For this purpose the Obstruction Method has been employed. This method involves the principle of placing an obstruction of some sort between the animal to be tested and an incentive stimulus, selected with reference to the drive under investigation. The animal is required to pass over the obstruction (in the present case an electric grill) in order to reach the incentive stimulus and thus satisfy the drive dominant at the moment. A quantitative determination of the strength of the drive can thus be secured in terms of the behavior of the animal toward the obstruction-incentive situation. By keeping the obstruction constant and varying systematically the physiological state of the organism from which the drive results (as would occur for example in the case of the hunger drive with a series of starvation periods of different length) it is possible to obtain indices of the motivation value of different organic states in relation to a specific stimulus. By keeping the obstruction constant it is also possilbe to obtain comparable data concerning the relative strength of different drives. A complete description of this method, including certain modifications intro-

[1] Reprinted with modifications from *Comparative Psychology Monographs*, 1926, 3: No. 17, 45 pp.

[2] The term hunger drive as used in this study, means nothing more than that the animal displayed a food searching response. The term carries no implications whatsoever concerning the presence or absence of so called "hunger contractions" of the stomach, or concerning the subjective state of the organism.

duced after my study was completed, has recently been published (6).

Very little work has been done up to the present in the field of animal motivation. The learning method has been employed for the most part. The recent work of Moss (7) in which he developed the Resistance Method bears the most direct relation to the present study. Both the Resistance Method of Moss, and the Columbia Obstruction Method, as devised by Jenkins and Warden for use in my work are similar in that both make use of the principle of placing an obstruction of some sort between the test animal and the incentive. The results of Moss are open to criticism not only because his work was not well controlled, but also because he used very small groups. He does not state the age of his animals, which is doubtless an important factor in the matter of strength of drive. Instead of using the method of equivalent groups, he tested the same animals over again, after increasing the length of the starvation period, so that his conclusions regarding the influence of this factor can hardly be considered valid. His investigation must be considered, however, very suggestive and valuable.

Moss also made use of the method of choice, in which food and a sex object were placed before the animal simultaneously, with free access to either allowed. This method has been further developed by Tsai (15) and used by him in comparing a 24-hour hunger period with the sex drive. The choice apparatus has been greatly improved by Jenkins of this laboratory, and the new design along with a criticism of the method has been recently published (6).

Moss further considered the relative effect of different drives on maze learning. This method had, however, been previously employed by Szymanski (13, 14) and Simmons (10). Szymanski used the maze both for the determination of the relative effects of different drives on learning and for the study of the effect of different hunger periods on learning. He found in the latter case that with a 30-minute, 4-hour, and 24-hour hunger period, the most rapid learning occurred with the 24-hour period, while the 30-minute period was least effective.

Simmons used both a simple and complex maze for a comparison of the effect of different food incentives and of different drives on learning. Since a constant hunger period was used throughout her study the results have little bearing on the present work.

Dodson (3) studied the effect of different periods of starvation on the development of a dark-light discrimination habit in the white rat. He used starvation periods of 24, 31, 41, and 48 hours and found that for rats 78 days old, the optimum starvation period for habit formation of this type was 41 hours; however for retention of the maze habit after 21 days, the 48-hour period appeared to be more effective.

Studies of the effect of different periods of starvation on general bodily activity as measured by an "activity cage" have been made by Richter and Nicholls. Richter (9) found that white rats lived 8 days when water but no food was provided, reaching a maximum degree of activity between 2 and 3 days. After three days the activity decreased gradually until death which occurred on the eighth day. When neither water nor food was provided the rats showed an immediate decrease in activity which was continuous until death which occurred on the fifth day. A similar study was made by Nicholls (8) upon the guinea pig. She found that the maximum degree of activity occurred on the second day of starvation and was followed by a marked decline on the third day.

Of the above investigators it appears that Szymanski, Dodson, Richter, Nicholls and Moss have each considered the effect of a series of starvation periods of different length upon behavior. Moss alone, however, used a method in any way comparable with the present one. The other investigators either measured general activity by means of some form of "activity cage" or investigated the relation of starvation to speed of habit formation. Essential differences in method make a comparison of these results impossible. From the small amount of work so far reported it is evident that the relative motivation value of different starvation periods has not been adequately determined.

The present investigation includes two main problems. The first is the determination of the effect of systematic variation of the period of starvation upon drive behavior when a constant amount of electrical stimulation is employed in the obstruction section. The second deals with the relative effect of three different degrees of electrical stimulation in the obstruction compartment upon the drive behavior resulting from a series of starvation periods.

II. Apparatus and Procedure

The apparatus used in this study consisted of a control box for testing the animal and a mechanism for the control of the current supplied to the grill. The general plan of this apparatus is shown

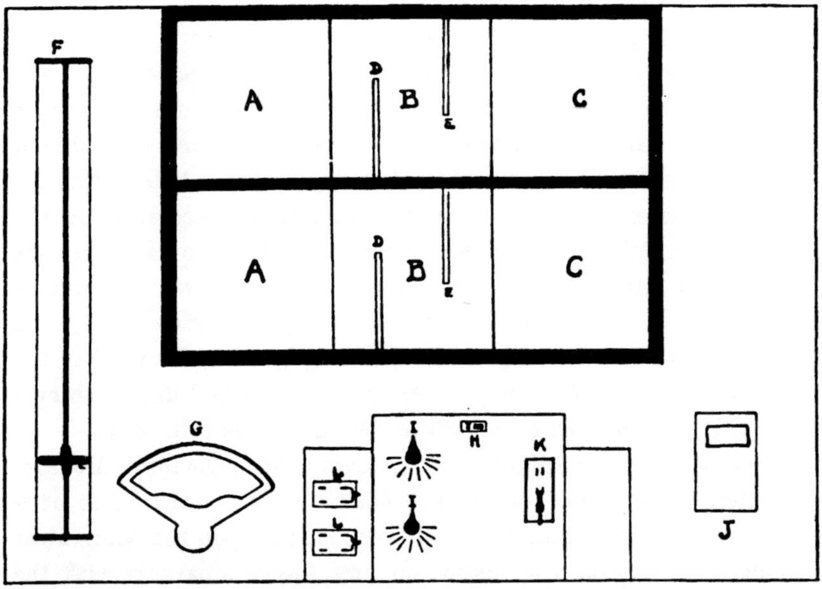

Fig. 1. Diagram of Control Box and Accessory Mechanisms (Wiring Omitted)

A, entrance compartment; *B*, obstruction compartment; *C*, incentive compartment; *D* and *E*, glass plates; *F*, potentiometer; *G*, voltmeter; *H*, switch controlling voltmeter; *I*, multipoint switches controlling each resistance unit; *J*, microammeter; *K*, intake switch; *L*, *L* switches throwing current into grills.

in figure 1. A second control box was also provided in order that two animals might be tested simultaneously. Each control box was divided into 3 sections, each of which was 10 inches wide, 10 inches long and 10 inches high. The entire box was painted a dull black. Sections *A* and *C* were made identical in every respect in order that any possible preference for either section might be obviated.

Section *A* was used throughout as the entrance compartment, section *C* as the incentive compartment. The floor of section *B* was made of a slab of bakelite 0.5 inches thick, wound with no. 18 copper wire at intervals of 0.25 inches in such a way that the wires from each terminal alternated, thus insuring a shock when any two adjacent wires were touched. The terminals of these wires were connected with the electric control apparatus. There were no divisions between the three sections (*A*, *B*, *C*) of the box but at *D* and *E* glass plates 6 inches wide and 12 inches high were placed as indicated in figure 1, in order to keep the animals from leaping across the grill. Section *A* was covered with a pane of glass to keep the animals from jumping out of the box.

In the box used by Moss the entry and incentive compartments were not similar. In the first place the entrance compartment was larger than the incentive compartment. In the second place the floor of the entrance compartment was covered with water, whereas the floor of the incentive compartment was dry. Since the degree of motivation resulting from the conditions of the box was not determined by a control group with which the food incentive was lacking, it cannot be ascertained in this case whether or not the food stimulus was the sole factor in causing the rats to cross the grill. Instead of a grill of the type described above, Moss used as a shocking device two brass plates both of which had to be touched simultaneously for the animals to experience a shock. A grill presenting a continuous stimulation surface, such as was used in the present study has certain obvious advantages over this simpler device.

The current supplied to the grill was controlled by means of a General Electric potential transformer and voltage changes were secured by means of the potentiometer, *F* (fig. 1). The electric pressure was measured at the beginning of each experiment by means of a high tension voltmeter, *G*, in order to obviate any deviations in the current supplied, adjustment being made by means of the potentiometer. The voltmeter when not in use was thrown out of the circuit by means of a snap switch, *H*. The current was regulated by means of variable resistance units each of which was controlled by means of a multipoint switch, *I*. Each resistance unit contained eight elements each having a resistance of approximately

1,250,000 ohms. The multipoint switches made possible a cumulative variation of the resistance in equal steps from 100,000 to 10,000,000 ohms. The current supplying the grills was measured by a microammeter, *J*. The method of wiring made possible, by means of the double throw knife switch, *K*, the measurement of the current in either grill (one for each control box) by one fixed measuring instrument. The current could be thrown into either or both of the grills by closing the switches, *L, L*. This apparatus makes possible an exact determination of the current supplied to the grills in terms of voltage and resistance and provides an adequate means of controlling the current and of varying it in systematic amounts. This control is a decided improvement over that of Moss, since the latter merely considers the current used in terms of voltage and disregards the resistance factor. Moreover, since the skin resistance of the animals undoubtedly varies greatly from rat to rat the amount of stimulation received must also vary, unless this factor is in some way eliminated. Moss attempted to overcome this difficulty by placing water in section *A* so that the feet of the animals were moistened on approaching the grill. This method is not entirely satisfactory however, as was indicated above. In the present study this factor was eliminated to as great an extent as possible by using a very large resistance within the electric circuit and thus rendering negligible the resistance of the rats. Since the resistance here used ranged from 3,850,000 to 10,100,000 ohms, it is believed that any variability in resistance in the rats is rendered practically negligible, thus insuring a constant degree of stimulation.

The following current values were used in this study:

	MILLIAMPERES	RESISTANCE (OHMS)	VOLTAGE
Shock 1	0.12	10,100,000	1,200
Shock 2	0.19	6,350,000	1,200
Shock 3	0.32	3,850,000	1,200

A total of 803 albino rats were used in this investigation.[3] The size of each of the 21 groups studied is indicated in table 1 which

[3] In view of the large number of rats used in this experiment an autopsy of the animals to determine the presence or absence of food in the stomach could not be made.

also shows the arrangement of starvation periods. With the 60- and 72-hour starvation periods many animals died due to the length of the starvation period, thus introducing a selective factor into these groups, which makes them not entirely comparable with the earlier groups. This factor is, however, not great except in the case of the 72-hour groups.

The rats used with shock 1 were highly uniform weighing between 80 and 100 grams and were approximately 8 weeks old. Those used with shock 3 were slightly more variable in size weighing between 80 and 110 grams but were of approximately the same age as those of shock 1. Those used with shock 2 were in general similar to those of shock 1 with the exception of half of the 12-, 24-, 48-, 60- and 72-hour groups which were similar to those of shock 3.

TABLE 1

Showing number of rats used in each starvation group

A indicates number of rats given preliminary training; B, number used in experiment proper; C, number that died during starvation period.

GROUP	SHOCK 1			SHOCK 2			SHOCK 3		
	A	B	C	A	B	C	A	B	C
Control	20	20	20	20	20	0	20	20	0
12-hour	37	37	0	40	40	0	40	40	0
24-hour	40	40	0	39	39	0	40	40	0
36-hour	40	40	0	40	40	0	40	40	0
48-hour	40	40	0	40	40	0	40	40	0
60-hour	41	35	6	40	40	0	53	40	13
72-hour	40	21	19	44	21	23	49	37	12

The rats were kept in the laboratory for one week before being given the preliminary training, in order that they might become adjusted to the laboratory conditions. During this period they were fed milk-soaked bread in their cages at noon each day.

The preliminary training consisted of placing the animals in groups of ten in section *A* and allowing them to run to the food in section *C* or explore freely within the box for 10 minutes. During this preliminary training there was of course no current in the grill. This procedure was repeated for three consecutive days, in order that the animals might become accustomed to the conditions of the control box.

In order to determine whether the food incentive was the sole factor causing the rats to cross the grill, a control group was used with each degree of shock. The preliminary training of these groups was the same as for the other groups except for the fact that section *C* contained no food while they were in the box. The control groups were fed immediately after being removed from the box.

In the experiment proper each rat was placed individually in section *A* with the current turned on in the grill and a dish of milk-soaked bread in section C. The behavior of each rat was recorded during the test period of 10 minutes. If any animal crossed the grill during the test period it was allowed a nibble of food and immediately returned to *A*. If a rat crossed the grill continuously (i.e., with only momentary pauses upon being returned to section *A*) a nibble was allowed after every third cross.

The procedure for the control group was the same as that for the starvation groups except for the fact that there was no food stimulus in section *C*. The experiment proper, occurred in the case of the control group, 12 hours after the last period of preliminary training. The control group is consequently equivalent to the 12-hour group in every condition but the absence of the food incentive.

The following types of behavior were recorded for each animal during the 10-minute test period: (*a*) crossings, (*b*) contacts, (*c*) jumps. The term crossing indicates that the animal actually crossed the grill and entered the incentive compartment. This measure appears to be the most significant of the three types, inasmuch as it represents the only reactions which resulted in satisfaction of the hunger drive.

By the term contact is meant that the animal merely touched the grill and withdrew upon being shocked, remaining in the entrance compartment. Partial crossings of the grill, followed by return to the entry compartment (instead of proceeding into the incentive compartment) were scored as contacts. Data concerning contacts are obviously important not only as indicating an attempt of the animal to react positively to the incentive stimulus but also as a measure of the conditioning process resulting in many cases in a negative reaction to the grill. Contacts are in reality frustrated crossings and thus are closely related to the latter type of behavior. It has, however, seemed advisable to consider the two kinds of activity separately.

The term jump refers to the behavior of the animal in the entrance compartment. A jump was checked only when the animal cleared the floor with all four feet. This type of activity seems to indicate an attempt to escape from the entrance compartment. This reaction was probably in most cases the result of the shock, since jumping did not occur until after the first experience with the grill.

The test period of every group in the experiment proper was so arranged as to fall between 3 and 6 p. m. The last of the three periods of preliminary training preceded the experiment proper by the number of hours of starvation requisite for each group. Consequently the preliminary training of the control, 12-, 36- and 60-hour groups was given as 4:30 a.m. while that of the 24-, 48- and 72-hour groups was given at 4:30 p.m.

III. Presentation of Results

The quantitative results of this study, covering both the effect of different periods of starvation and different degrees of shock on drive behavior are presented in tabular and graphic form as follows:

Tables 2, 3 and 4 are distribution tables for shocks 1, 2 and 3 respectively, covering in each case the three types of behavior — crossings, contacts, and jumps — for each starvation group.

Tables 5, 6 and 7 give the main results of shocks 1, 2 and 3 respectively, for each starvation period. Columns 2 to 5 of these tables give the measures of central tendency for crossings (median, average, S.D. and C.V.). Columns 6 to 10 give measures of the extreme tendencies for crossings. Column 6 indicates the range of crossings with each starvation group. Column 7 shows the per cent of rats that were immediately conditioned in each group (i.e., of those rats that did not cross the grill after their preliminary experience with it. The first experience with the grill may be either a crossing or a contact). This figure indicates the percentage of rats with which the resistance offered by the grill was at the

TABLE 2

Distribution table covering crossings, contacts and jumps under shock 1

DISTRI-BUTION	CROSSINGS							CONTACTS							JUMPS						
	C	12	24	36	48	60	72	C	12	24	36	48	60	72	C	12	24	36	48	60	72
0	6	8	4	1				2	0	6	7	7	2	1	15	8	7	8	20	9	8
1	6	5	1	1		3		8	7	15	6	8	9	7		8	4	6	6	4	6
2	6	3		2		5	1	7	7	7	3	9	6	6	2	6	7	3	3	2	1
3	1	2	2			4	1	2	3	2	6	4	10	6	1	3		4	1	4	1
4		1	3			3	2	1	5	3	6	3	2	1		2	2	4	2	1	1
5		3			2	1	1		2	4	7	4	3			5	6	1	3		1
6			1		1	2	4		3	1	1	3			1	3	3	3		4	
7					3	3			4		1	1	1				3	4	3	4	1
8				1	1	2			2		2		2				1	1	1	1	
9			1		2		1		1	1	1				1	1				2	
10		1				1				1							1				1
11				1		1			2			1					4			1	1
12	1			1					1									3			
13		1			2	2											1				
14																				1	
15						1	1											1			
16							1														
17					2													1			
18		1		1													1			2	
19																					
20		1	1	2		1															
21			1		1	1															
22			1		1														1		
23		1	1	1	2		1														
24				2	3		2														
25		1		1		1															
26																					
27			2				1														
28			3	1																	
29			1				1														
30					1		1														
31			1		1																
32				1												1					
33																					
34		1		1	1	1	1														
35					1																
36		1	1	2	1																
37			3				1														
38				1	1																
39					1																
40		1	4	1	2													1			
45		1	3	3	4	1															

TABLE 2 — *Continued*

DISTRIBUTION	CROSSINGS							CONTACTS							JUMPS						
	C	12	24	36	48	60	72	C	12	24	36	48	60	72	C	12	24	36	48	60	72
50			3	3	3	1															
55		2		2	1	1															
60		1	1	2																	
65				3	1		1														
70		1		1	1																
75		1	1	1																	
80				3	1																
85				1																	
90																					
95			1																		

The notation C refers to the control group, the numbers 12, 24, 36, 48, 60 and 72 to the series of starvation periods.

The interval of the distribution from 0 to 40 is 1, from 40 to the end of the scale, 5.

This table should read as follows: Control group: 6 rats crossed 0 times, 6 rats crossed 2 times, 1 rat crossed 3 times, etc.

outset stronger than the hunger drive. Column 8 gives the percentage of animals that crossed the grill only 5 times or less. The number 5 was chosen because the distributions of the groups indicated that the majority of rats with which the resistance soon proved stronger than the hunger drive, were conditioned by 5 crossings or less. Column 9 gives the precentage in each group that crossed 40 or more times. This number was arbitrarily chosen because few of the animals crossed more than 40 times within the test period. Column 10 shows the percentage of rats that failed to be conditioned against the grill (i.e., were still crossing the grill at the end of the 10-minute test period). This measure indicated the number of animals within each group with which the hunger drive was still stronger than the resistance offered by the grill after a 10-minute experience with it. Columns 11 to 14 give the measures of central tendency for contacts (median, average, S.D. and C.V.). Columns 15 to 18 give the same measures of central tendency for jumps.

Tables 8, 9 and 10 for shocks 1, 2 and 3 respectively, give the reliability of the difference between each two starvation groups with the same degree of shock. (Reliability is stated in terms of the ratio of the difference between two groups to the S.D. of that difference.)

TABLE 3

Distribution table covering crossings, contacts and jumps under shock 2

INTERVAL	CROSSINGS							CONTACTS							JUMPS						
	C	12	24	36	48	60	72	C	12	24	36	48	60	72	C	12	24	36	48	60	72
0		10	7	8	2	2	3	7	8	1	1	1	5	1	8	23	12	19	15	18	12
1	7	2	9	3	5	5	1	5	5	8	3	11	11	3	3	5	5	10	5	11	
2	8	7	6	6	1	8	2	5	8	12	12	5	7	4	3	3	5	4	6	1	1
3	2	4	3	8	1	6		2	5	9	12	7	7	4	1	4	1	3	2	4	2
4		2	1	3	5	3		1	7	3	4	6	6	1	2	1	3	1	2	1	1
5		1		2	1	3	3		2	1	6	2	2	1		2	4		1	1	1
6			1				1		4	1		4	1	3			4	1	2	2	1
7		1	1				1			2	1		1	3	1		3	1	1	1	1
8		2	1	1		1	1			1		1				1			1		
9					1							1			1				1		
10	1	1					1		1		1			1					1		1
11			1	1	1	1	1					1			1						
12						1										1					
13		2				2				1							1				
14		2																	2		1
15				1			2														
16							1													1	
17					2																
18		1																			
19					1												1				
20		1		1	1	1						1									
21				1	1	1	1														
22		1					1														
23																					
24																					
25						2															
26					1																
27			1		2		1														
28		1																1			
29	1				1																
30																					
31					1	1															
32																					
33			2		1																
34																					
35																					
36	1				1														1		
37																					
38																					
39				1	1																
40			1	2	3	2															

TABLE 3 — *Continued*

INTERVAL	CROSSINGS							CONTACTS							JUMPS						
	C	12	24	36	48	60	72	C	12	24	36	48	60	72	C	12	24	36	48	60	72
45		1	2		3	1															
50			1		2		1														
55				1	1																
60			1	1																	
65																					
70			1																		
75					1																
80																					
85																					
90																					
00																					
05		1																			

See legend accompanying table 2.

Table 11 indicates the relative effect of the 3 degrees of shock upon the various starvation groups.

Table 12 shows the reliability of the difference between any two shocks with a given starvation group.

Figures 2, 4 and 5 show the cumulative percentile distributions for crossings for shocks 1, 2 and 3 respectively. These figures make possible a point to point comparison of any decile in the distribution of one group with any decile of any other group with which the same degree of shock was used.

Figure 3 shows graphically the effect of the different starvation periods upon the central and extreme tendencies of each group with shock 1.

Figure 6 shows graphically the median number of crossings made by each starvation group with each degree of shock.

The results of this study will be considered in two sections. The first section is concerned with the effect of the series of different starvation periods upon behavior in the obstruction-incentive situation. The second section will consider the effect of different degrees of shock upon behavior when the preceding starvation period was the same for each degree of shock.

The effect of different periods of starvation

This section is concerned with the effect of a series of different

TABLE 4

Distribution table covering crossings, contacts and jumps under shock 3

INTERVAL	CROSSINGS							CONTACTS							JUMPS						
	C	12	24	36	48	60	72	C	12	24	36	48	60	72	C	12	24	36	48	60	72
0	2	3	4	2	6	4	2	5	6	6	1	6	4	3	7	9	14	11	12	9	13
1	7	4	4	7	7	4	3	7	10	9	4	11	6	7	2	7	8	4	6	6	5
2	4	12	3	4	6	9	3	4	15	6	10	5	8	4	2	5	4	2	4	5	3
3	1	5	4	4		1	6	2	3	6	10	7	10	8		3	4	5	1	1	3
4	1	5	2	2	4			2	1	4	4	6	2	8	1	2	1	2	4	4	3
5	1	1	1	2	4	1	1		3	8	5	2	4	2		5		3	2	1	
6	1		1	4		3	1		1	1	2	2	3	3	1		2	4	2	4	2
7		1	1	1	1	1	3		1			1	1	1	2	1	2	2	2	2	1
8	1		2		1	1					2		1			1	3	2		2	
9											1				2			2	2		2
10						1									1	2		1	1		1
11		1				2	2							1		1	2				
12					1	2					1					2		2	1	4	
13			1				2									1					
14		1	1	1			2									1			1		
15				1	1		1								1				1	1	3
16				1		1															
17		1	1			2															
18	1																				
19		1				1	1														
20			1	1	1		1														1
21																					
22					1	2															
23			1			1							1								
24			1																	1	
25			1																		
26		1				1															
27		1																			
28			1	1																	
29					1		1														
30					1																
31		1			1																
32				1			3								1				1		
33				1																	
34				1																	
35				1																	
36			1																		
37		1			1	2	1														
38																					
39				1			1														
40		1	1	1																	

TABLE 4 — *Continued*

INTERVAL	CROSSINGS							CONTACTS							JUMPS						
	C	12	24	36	48	60	72	C	12	24	36	48	60	72	C	12	24	36	48	60	72
45			2	1	1		2														
50	1		3	1																	
55					1																
60																					
65						1															
70					1																
75			2	1			1														
80																					
85																					
90			1																		
95																					

See legend accompanying table 2.

starvation periods upon the behavior of the white rat in the obstruction-incentive situation. The effect of the different starvation periods will be considered when the obstruction involves (*a*) a low degree of electrical stimulation — shock 1; (*b*) a medium degree — shock 2; and (*c*) a high degree — shock 3.

The effects of the different periods of starvation will be discussed in respect to five main types of data. First, in relation to the number of crossings made by each group, as indicated by the median,[4] or by the percentage of animals within each group that crossed the grill 40 or more times; or more fully by a comparison of the cumulative percentile graphs showing the distribution of each group in regard to the number of crossings. Second, by the per cent of animals within each starvation group that were immediately conditioned, i.e., that failed to cross the grill after the first experience with the shock. This figure probably indicates the number of animals in which the hunger drive was at the outset weaker than the resistance offered by the grill. Third, by the percentage of rats that failed to be conditioned within each group, i.e., were still crossing the grill at the end of the ten-minute test period. This figure probably indicates the number of animals in which the hunger drive was at the end of the test period still stronger than the

[4] The median rather than the average will be referred to throughout this discussion. Owing to the fact that the extreme deviations always occurred at the same end of the distribution scale, the average does not in this case give a reliable measure of central tendency.

TABLE 5

Results of different starvation groups under shock 1

STARVATION PERIOD	CROSSINGS									CONTACTS				JUMPS			
	Median	Average	Standard deviation	Coefficient of variability	Range	Per cent immediately conditioned	Per cent crossing 5 times or less	Per cent crossing 40 times or more	Per cent not conditioned	Median	Average	Standard deviation	Coefficient of variability	Median	Average	Standard deviation	Coefficient of variability
Control group..	1.6	1.7	2.5	153	0–12	60	95	0	0	2.0	1.6	0.97	61	0.66	1.1	2.34	213
12 hours........	4.5	17.2	23.3	135	0–77	33	55	19	11	4.6	4.5	3.10	69	2.4	3.3	5.28	160
24 hours........	28.6	29.1	22.3	77	0–98	13	25	33	18	1.9	2.3	2.32	101	5.0	4.7	4.26	91
36 hours........	37.0	41.6	25.4	61	0–89	5	10	50	50	3.7	3.2	2.42	76	3.8	5.2	7.01	136
48 hours........	27.0	30.7	19.6	64	5–80	0	5	33	30	2.6	2.7	2.37	89	1.0	2.2	3.97	179
60 hours........	6.7	11.8	14.7	124	1–58	9	46	9	23	3.1	2.8	1.99	71	3.6	4.7	4.89	104
*	4.6	10.1	14.2	140	0–58	22	54	7	19								
72 hours........	16.5	18.0	15.8	88	2–68	0	24	5	33	2.4	2.0	1.0	51	1.4	2.3	3.24	141
*	2.0	9.5	14.6	154	0–68	48	60	25	17								

A crossing indicates that the animal passed over the grill into the incentive compartment. A contact involves merely touching the grill and refusing to cross, or if partially across, turning back into the entrance compartment. Jumps occurred in the entrance compartment and involved clearing the floor with all four feet.

* Results in this row include the dead.

TABLE 6

Results of different starvation groups under shock 2

STARVATION PERIOD	CROSSINGS									CONTACTS				JUMPS			
	Median	Average	Standard deviation	Coefficient of variability	Range	Per cent immediately conditioned	Per cent crossing 5 times or less	Per cent crossing 40 times or more	Per cent not conditioned	Median	Average	Standard deviation	Coefficient of variability	Median	Average	Standard deviation	Coefficient of variability
Control group..	2.4	5.3	9.4	177	1–36	35	85	0	0	1.6	1.3	1.12	90	1.7	2.4	3.13	133
12 hours.......	3.3	9.3	18.4	198	0–108	30	65	5	5	2.9	2.4	2.17	90	0.87	1.4	2.47	174
24 hours.......	2.6	12.4	19.9	160	0–74	40	67	15	5	2.9	3.0	2.41	80	2.5	3.3	3.89	119
36 hours.......	3.5	9.4	15.6	166	0–60	28	75	10	8	3.3	3.2	1.77	56	1.1	1.8	4.49	249
48 hours.......	20.0	22.5	20.0	89	0–77	18	38	25	38	3.4	3.7	3.54	96	2.0	3.7	6.37	175
60 hours.......	4.1	9.4	12.2	131	0–46	18	68	2.5	0	2.6	2.4	1.60	66	1.2	1.8	2.96	169
72 hours.......	6.0	10.9	11.6	111	0–50	19	43	4.8	10	3.6	3.7	2.57	70	0.89	2.5	3.81	155
*	0.0	2.5	9.7	389	0–50	61	73	2.3	5								

See legend accompanying table 5.

TABLE 7

Results of different starvation groups under shock 3

STARVATION PERIOD	CROSSINGS									CONTACTS				JUMPS			
	Median	Average	Standard deviation	Coefficient of variability	Range	Per cent immediately conditioned	Per cent crossing 5 times or less	Per cent crossing 40 times or more	Per cent not conditioned	Median	Average	Standard deviation	Coefficient of variability	Median	Average	Standard deviation	Coefficient of variability
Control group..	2.3	5.6	11.8	210	0–54	45	80	5	5	1.7	1.4	1.14	81	2.5	5.2	7.49	144
12 hours........	3.2	6.8	10.8	159	0–44	18	75	2.5	0	2.3	2.0	1.80	89	2.8	4.2	5.08	122
24 hours........	8.0	20.0	24.3	122	0–93	20	25	23	0	2.8	2.5	1.82	72	1.8	2.6	3.22	124
36 hours........	5.5	14.1	18.2	129	0–79	23	53	10	0	3.5	3.6	2.40	67	3 6	3.8	3.51	92
48 hours........	4.3	11.0	16.6	151	0–70	33	68	7.5	0	2.6	2.4	1.84	77	2.5	4.3	6.01	141
60 hours........	6.3	10.6	13.5	127	0–69	20	48	2.5	0	3.2	3.4	3.69	109	3.0	4.5	5.13	114
*	2.6	8.0	12.6	158	0–69	40	60	2	0								
72 hours........	7.5	14.8	16.3	110	0–70	14	41	8.1	0	3.6	3.2	2.18	69	2.2	3.9	5.17	133
*	3.7	11.2	15.5	138	0–70	35	55	6.1	0								

See legend accompanying table 5.

TABLE 8

Showing the reliability of the differences between the various starvation periods (average crossings) under shock 1

STARVATION GROUP	12		24		36		48		60		72	
	SD_D	$\frac{D}{SD_D}$	SD_D	$\frac{D}{SD_D}$	SD_D	$\frac{D}{SD_D}$	SD_D	$\frac{D}{SD_D}$	SD_D	$\frac{D}{SD_D}$	SD_D	$\frac{D}{SD_D}$
Control	3.87	4.03	3.57	7.70	4.06	9.85	3.15	9.23	2.55	3.33	2.38	3.31
12			5.21	2.27	5.55	4.40	4.93	2.73	4.57	1.55	4.48	1.72
24					5.34	2.34	4.70	0.09	4.31	4.40	4.22	4.64
36							5.07	2.15	4.59	6.49	4.63	6.93
48									3.81	5 41	3.87	5.48
60											3.20	0.19

TABLE 9

Showing the reliability of the differences between the various starvation periods (average crossings) under shock 2

STARVATION GROUP	12		24		36		48		60		72	
	SD_D	$\frac{D}{SD_D}$	SD_D	$\frac{D}{SD_D}$	SD_D	$\frac{D}{SD_D}$	SD_D	$\frac{D}{SD_D}$	SD_D	$\frac{D}{SD_D}$	SD_D	$\frac{D}{SD_D}$
Control	3.59	0.90	3.82	1.83	3.23	1.27	3.80	4.52	2.85	1.43	2.56	1.09
12			4.31	0 72	3.80	0.03	4.30	3.07	3.49	0.03	3.26	2.08
24					4.02	0.75	4.49	2.24	3.72	0.81	3.51	2.82
36							4.00	3.28	3.12	0.00	2.85	2.42
48									3.70	3.54	3.48	5.74
60											2.42	2.85

TABLE 10

Showing the reliability of the differences between the various starvation periods (average crossings) under shock 3

STARVATION GROUP	12		24		36		48		60		72	
	SD_D	$\frac{D}{SD_D}$	SD_D	$\frac{D}{SD_D}$	SD_D	$\frac{D}{SD_D}$	SD_D	$\frac{D}{SD_D}$	SD_D	$\frac{D}{SD_D}$	SD_D	$\frac{D}{SD_D}$
Control	3.14	0.38	4.66	3.09	3.90	2.17	3.72	1.45	3.16	0.76	2.81	1.63
12			4.20	3.14	3.35	2.18	3.13	1.34	2.43	0.49	2.80	1.57
24					4.80	1.22	4.65	1.93	4.21	2.85	4.42	1.98
36							3.89	0.80	3.36	1.81	3.63	0.80
48									3.14	0.96	3.43	0.06
60											2.81	1.13

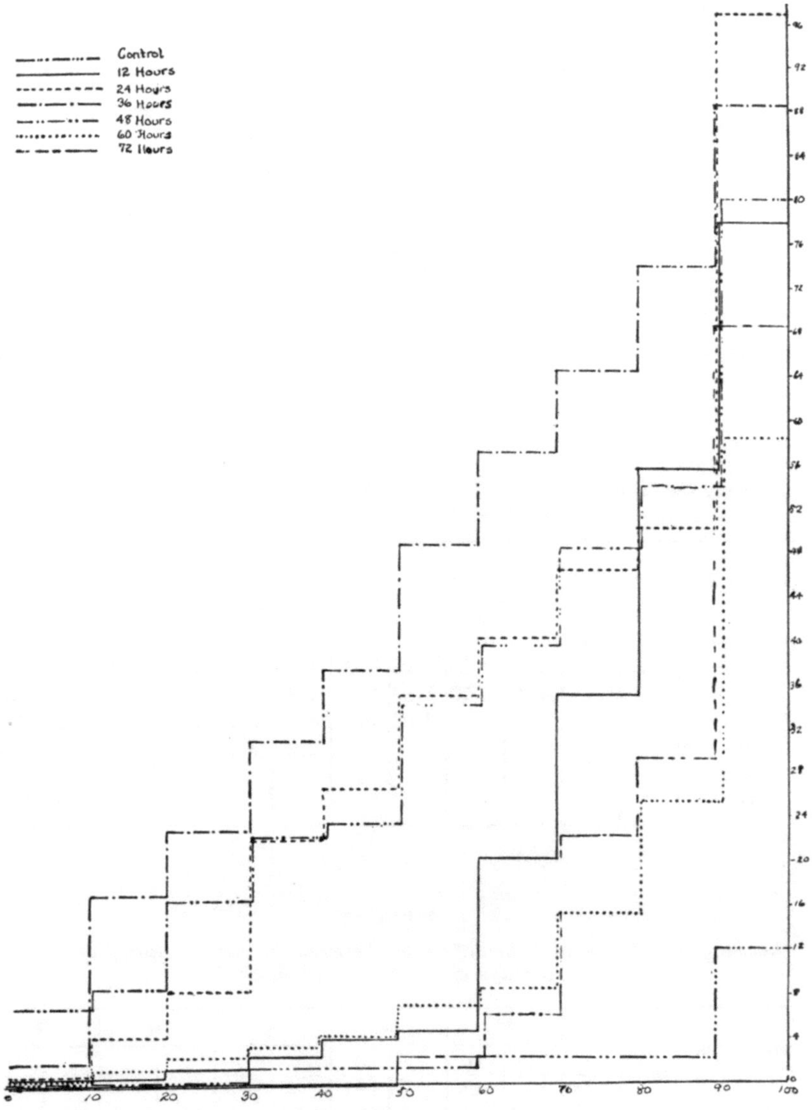

FIG. 2. PERCENTILE DISTRIBUTION OF EACH STARVATION GROUP COVERING CROSSINGS WITH SHOCK 1

Ordinates indicate the number of crossings, abscissas the deciles of each distribution.

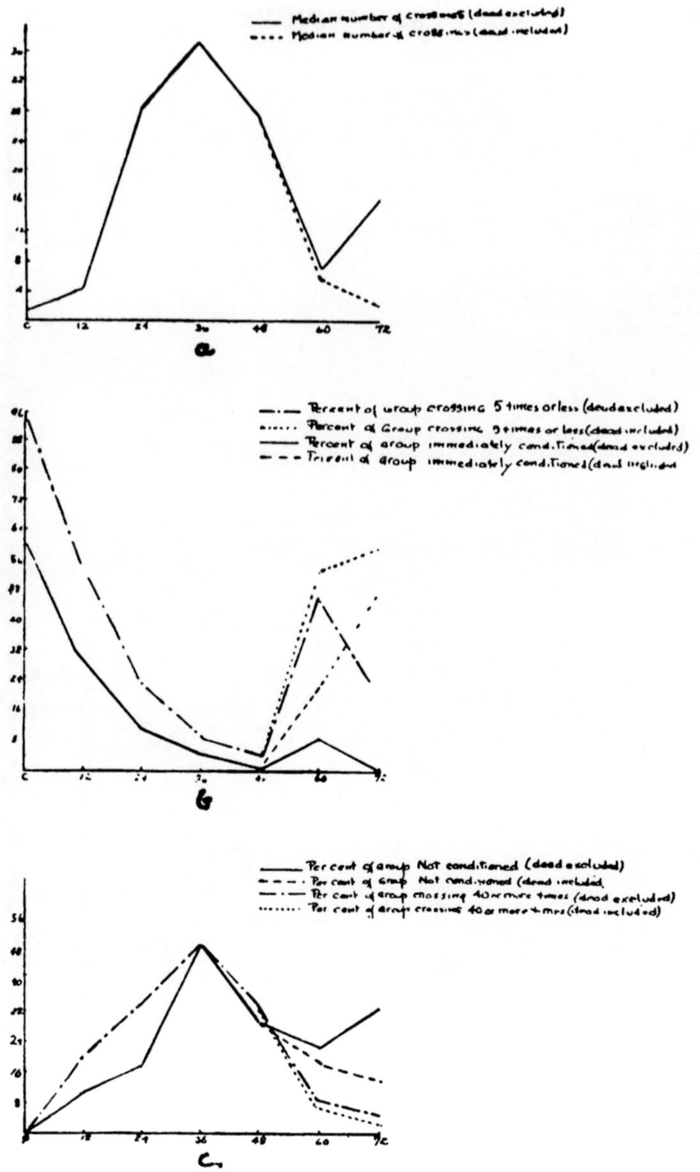

FIG. 3

a. Showing median number of crossings made by each starvation group. Ordinates indicate number of crossings, abscissas periods of starvation.

b. Showing percentage of animals immediately conditioned and percentage crossing 5 times or less in each starvation group. Ordinates indicate percentages, abscissas periods of starvation.

c. Showing percentage of animals crossing 40 times or more and percentage that failed to be conditioned in each starvation group. Ordinates indicate percentages, abscissas periods of starvation.

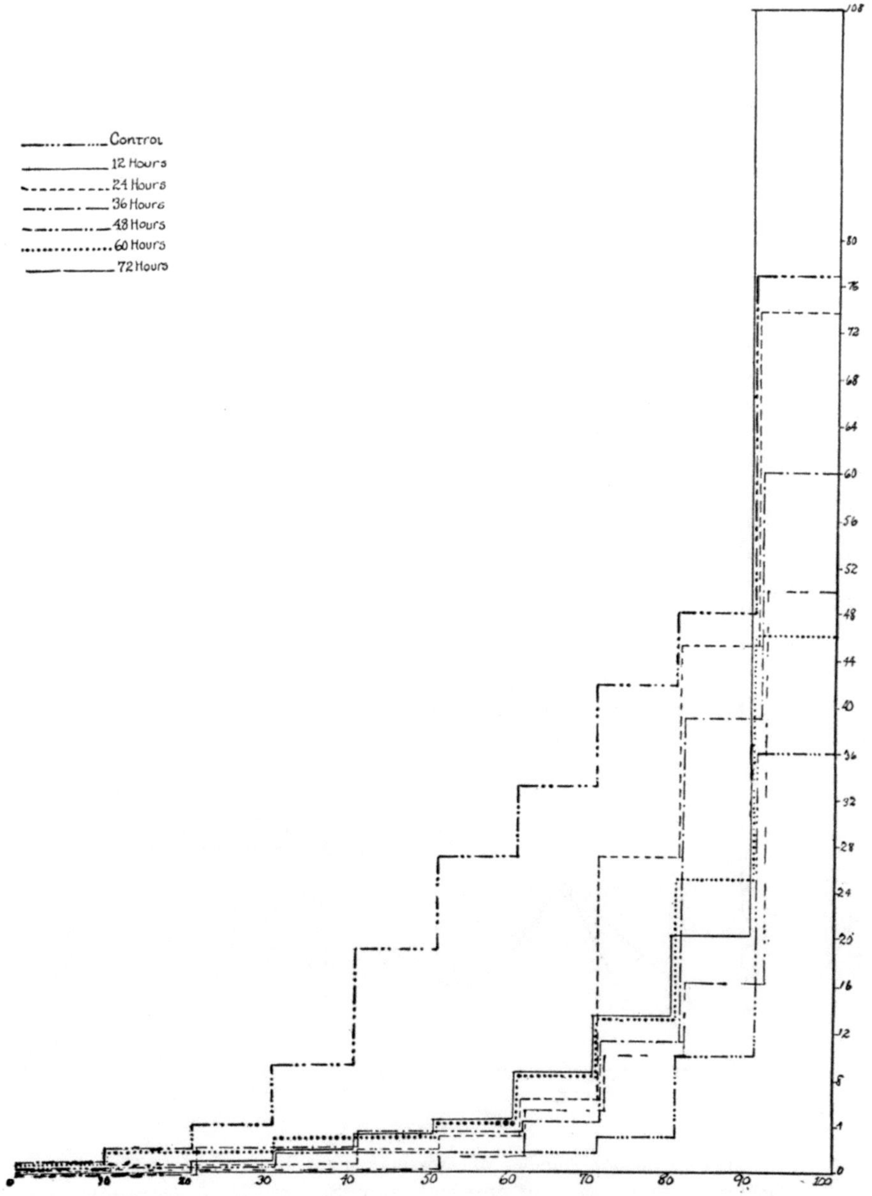

Fig. 4. Percentile Distribution of Each Starvation Group Covering Crossings with Shock 2

Ordinates indicate the number of crossings, abscissas the deciles of each distribution.

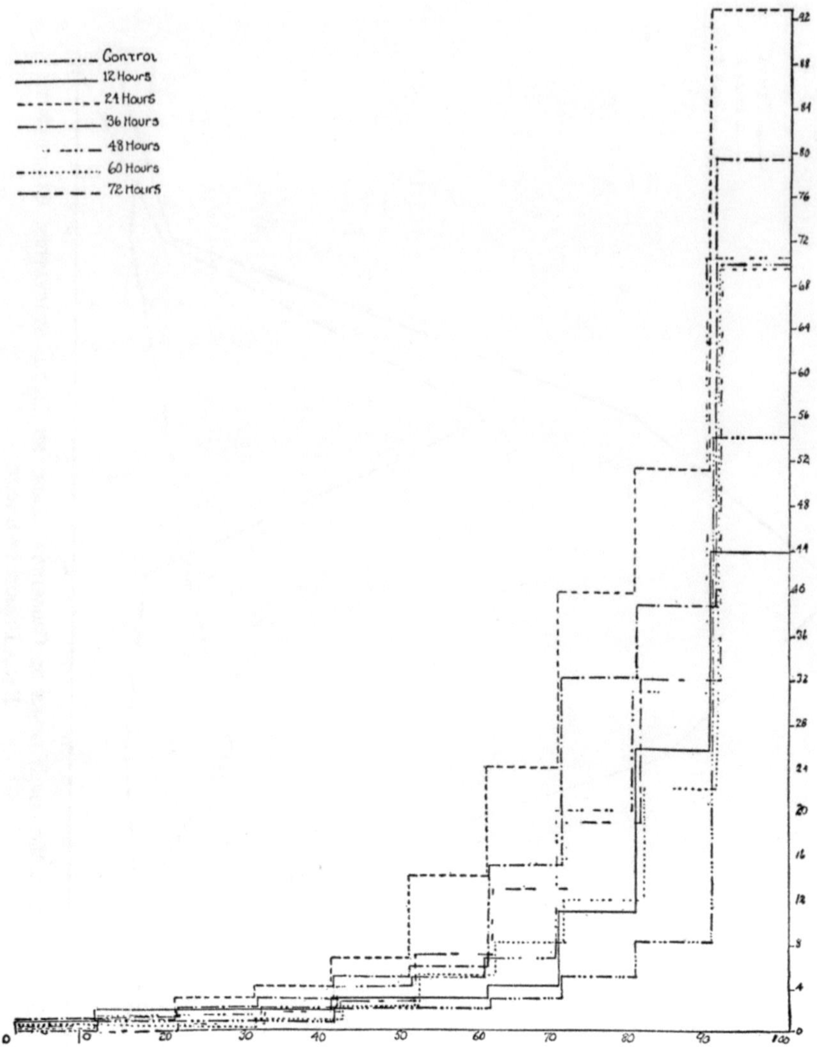

Fig. 5. Percentile Distribution of Each Starvation Group with Shock 3 Covering Crossings

Ordinates indicate the number of crossings, abscissas the deciles of each distribution.

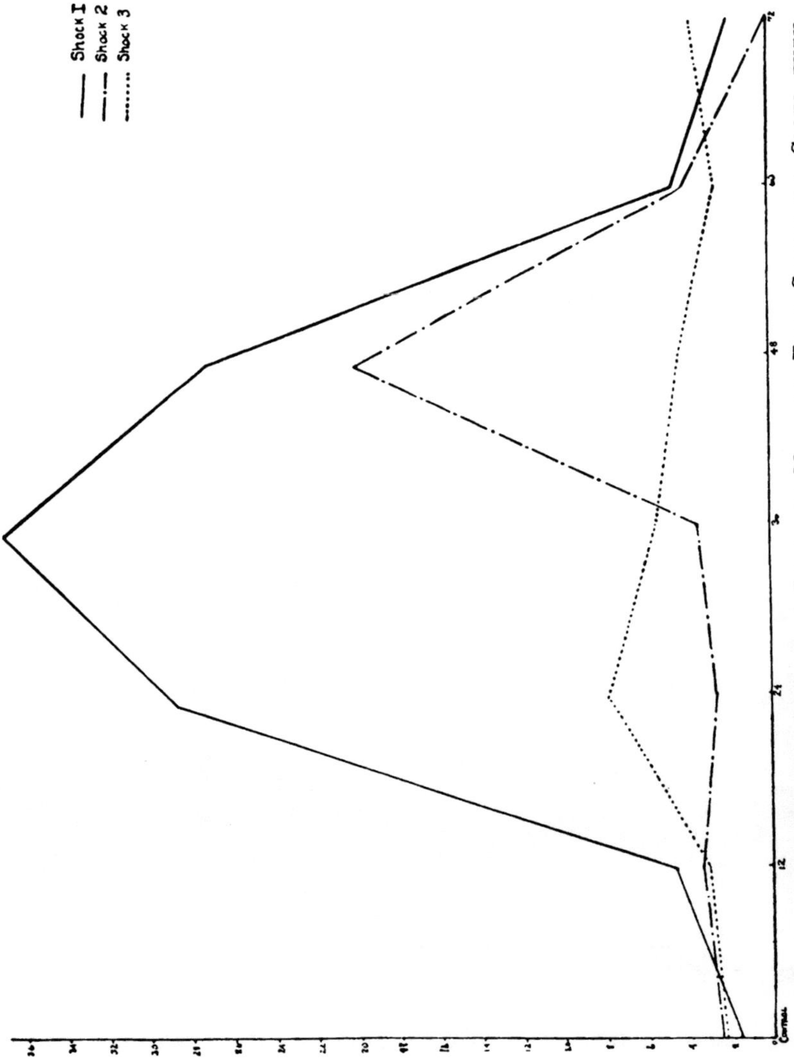

FIG. 6. SHOWING THE MEDIAN NUMBER OF CROSSINGS MADE BY EACH STARVATION GROUP WITH EACH DEGREE OF SHOCK

resistance offered by the grill. Fourth by the median number of contacts and fifth by the median number of jumps made by each starvation group.

Shock 1. The results of the control group considered by any of these measures indicate that the tendency to cross the grill with shock 1, in the absence of the food incentive is negligible, the median number of crossings in this case being 1.6, while only 1 rat crossed the grill more than 3 times (table 5, row 1). In general the tendency to touch the grill and withdraw is greater than that of crossing it.

The number of crossings for the 12-, 24- and 36-hour groups increases with the length of the starvation period. When this period is increased beyond 36 hours there is a consistent decrease in the number of crossings. This holds true of the 72-hour group only if the dead are included in the results as failing to cross the grill.[5] When the dead are excluded, however, the number of crossings made by the 72-hour group is greater than that for the 60-hour group. These results are clearly indicated by the medians of each group (table 5, column 1) and are shown graphically in figure 3, a. The results of the 24-hour group are in this respect approximately the same as those of the 48-hour group, while those of the 12-hour group are similar to those of the 60-hour group. The relative effect of the different starvation periods as shown by the percentage of rats crossing 40 or more times (table 5, column 9) is similar to that shown by the median number of crossings with the exception that the number of animals crossing very frequently is greater in the 12 than in the 60-hour group (see Fig. 3, c).

The variability of the different starvation groups (crossings — table 5, column 5) is in general inversely related to the strength of

[5] The results of the groups in which rats died of starvation are stated in Tables 5, 6 and 7 both excluding and including the dead. Neither of these measures are entirely satisfactory for a direct comparison with the results of the other groups, but it is probable that the results in which the dead are included give a better measure of the tendency of the group than those which exclude them. This is due to the fact that the dead in all probability correspond to the least active rats in the other groups. Consequently any results excluding them probably indicate only the tendencies of the upper end of the distribution scale.

the drive. In the control group in which the hunger drive appears to be a negligible factor, the variability is extremely large (C.V. 153). With the series of starvation groups the variability is in every case smaller than with the control group. In general the variability tends to diminish as the starvation period is increased to 36 hours and to increase with the longer periods.

The distribution covering the crossings for each of the starvation groups is shown in Figure 2 by means of cumulative percentile graphs. It is apparent from this figure that the differences between the starvation groups indicated above by the median and per cent crossing 40 or more times holds true between the corresponding deciles of each group. The differences between the control group and any of the starvation groups are immediately apparent. The fourth decile of the 12-hour, the second decile of the 24-hour, and the first decile of the 36-hour groups show as frequent crossings as the ninth decile of the control group. The 12- and 60-hour groups are simliar throughout the first 6 deciles of their distributions, but from the sixth through the tenth deciles the 12-hour group shows more frequent crossings than the 60-hour group. The distributions of the 24- and 48-hour groups are similar from the fourth through the ninth deciles, but from the first through the third decile the 48-hour group crossed considerably more frequently than the 24-hour group. In the tenth decile the 24 crossed more frequently than the 48-hour group. From the second through the ninth decile the 36-hour group crossed more frequently than any of the other groups. These results again indicate the similarity of the 12- and 60-hour groups but further emphasize the fact that the more active animals of the 12-hour group cross more frequently than those of the 60-hour group. It is also shown that the least active animals of the 48-hour group cross considerably more frequently than those of the 24-hour group although the results of these two groups are otherwise very similar.

The same results considered in relation to the percentage of rats that failed to cross the grill after their first experience with the shock, i.e., were immediately conditioned (table 5, column 7) indicate that increases in the period of starvation from 12 to 48 hours result in corresponding decreases in this type of behavior. After a starvation period of 48 hours all the rats crossed the grill at least 5 times. The results of further increases in the period of starvation, in this respect, cannot be determined because of the large proportion of animals that died in the 60- and 72-hour groups. Within these two groups all the living animals crossed at least once while in the 72-hour group no rat crossed less than 2 times. These results may indicate that the hunger drive of rats starved from 48 hours to the point of physical exhaustion is after the first experience with the shock still stronger than the resistance offered by the grill. The percentages of rats crossing the grill 5 times or less (table 5, column 8) give results similar to the percentages immediately conditioned for starvation periods of 12 to 48 hours (see Fig. 3, b). With starvation periods of 60 and 72 hours, however, this measure seems to indicate that in many cases the hunger drive tends to become weaker than the resistance offered by the grill after 5 experiences with the shock. This appears to be more often the case with the 60-hour than with the 24-, 36-, or 48-hour groups.

The percentage of rats that failed to be conditioned indicates that the hunger drive is strongest with the largest proportion of any group at 36 hours, since with this period of starvation 50 per cent of the group failed to be conditioned after 10 minutes experience with the shock, i.e., were still crossing the grill at the end of the test period. From this fact it might be concluded that the motivation resulting from 36 hours starvation is equivalent for this group to the resistance offered by 10 minutes experience with the grill under the conditions of this experiment. A further increase in the period of starvation to 48 hours reduced the number of animals that failed to be conditioned to 30 per cent. Further increases in the starvation period, however, do not appreciably alter this percentage. The number of rats failing to be conditioned in any group starved longer than 36 hours, is in every case, greater than that of any group starved less than 36 hours (see Fig. 3, c).

From these results it appears that, measured by the number of crossings, the hunger drive tends to increase in strength as the period of starvation is increased from 12 to 36 hours and to decrease as it is further increased from 36 to 72 hours. It is probable, however, that any determination of these results in terms of the number of times the grill was crossed measures not only the hunger drive but also the factor of general activity since physical exhaustion necessarily results from long starvation periods. As has been stated above the tendency to cross the grill very frequently is considerably greater with the 12- than with the 60-hour group, whereas the tendency to persist in attempts to reach the food stimulus, as shown by the per cent of each group that failed to be conditioned and the per cent immediately conditioned, is considerably greater with the 60- than with the 12-hour group. As indicated by the number of times the grill was crossed the hunger drive appears to diminish from a maximum at 36 hours to the point of physical exhaustion which occurred with the majority of these rats after 72 hours' starvation. However, a rat that was too weak to cross the grill still made attempts to do so, thus indicating that the hunger drive had not been entirely overcome, even though extreme physical exhaustion had set in. Consequently the number of crossings within the test period does not measure the hunger drive alone but also includes the factor of general bodily activity. Richter (9) found that the general activity of the white rat increases with the period of starvation, reaching a maximum at between 2 and 3 days, and then continuously decreases to the point of physical exhaustion at 8 days. Richter's results would appear similar to those obtained in the present study when measured in terms of the number of times the grill was crossed by each group. It is probable therefore that with the longer hunger periods (48 to 72 hours) this measure indicates the physical capacity of the rats to attain the incentive stimulus rather than the actual motization resulting from this stimulus. The latter factor is probably better indicated by the percentages of animals immediately conditioned and failing to be conditioned, since these measures are based upon persistence in reaching the food stimulus even when the physical capacity to do so is relatively slight.

Considering these results together it appears that the hunger drive reaches a maximum strength at 36 hours. After 48 hours this drive seems to be slightly diminished, but further increases in the length of the starvation period to 60 and 72 hours do not seem to result in any further weakening of the hunger drive, although the capacity to reach the stimulus is greatly reduced owing to the onset of physical exhaustion.

The reliability of the differences in crossings between any two groups in the shock 1 series will be found in table 8. It is apparent that the differences between any starvation group and the control group are completely reliable. The differences between the 12- and 36-hour groups and between the 36- and 60- or 72-hour groups are also extremely reliable. The differences between the 12- and 24-, or between the 24- and 36-hour groups are less reliable as are also the differences between the 36- and 48-, or that between the 48- and 60-hour groups.

The tendency to touch the grill without crossing as indicated by the median number of contacts made by each group (table 5, column 11) does not appear to be related directly to the length of the starvation period. With the 12- and 60-hour groups, in which the contacts tended to deter the rats from further crossings, the tendency to touch the grill is great. It was also large with the 36-hour group, but in this case the contacts generally immediately preceded the crossings and were not as effective in conditioning as in the other groups. The tendency to jump was in general greater with starvation periods of 36 hours or less than with the longer periods.

Shock 2. The results of the shock 2 series are difficult to interpret because of the lack of consistent tendencies. These results seem to indicate that the shock used in this series was so great as to arouse drive factors other than those of hunger. Some animals appeared to be reacting against the electric grill in section *B* rather than to the food stimulus in section *C*. After rushing across the grill into section *C* they did not eat but sat in a corner with sides palpitating as if in fear, or attempted to jump from the box. Apparently they did not discriminate between sections *A* and *B* for upon being replaced in *A* after crossing they would immediately dash across the grill.

The type of reaction thus described was evident in the control group, as shown by the fact that 15 per cent of the animals crossed between 10 and 36 times, in spite of the fact that section *C* contained no food. The median number of crossings (2.4), however, indicates that the majority of these animals were easily conditioned.

The differences between the control group and any of the starvation groups in this series, with the exception of the 48-hour group, are too small to be reliable. The median number of crossings of the control, 12-, 24-, 36-, and 60-hour groups all fall between 2.4 and 4.1. The median number of crossings is 6.0 for the 72-hour group excluding the dead and 0.0 when they are included. The percentage of animals crossing the grill 40 or more times is not large with shock 2; the largest number (25 per cent) occurred in the 48-hour group. The tendency to cross the grill continuously is slightly greater with the 24- and 36-hour than with the 60- and 72-hour groups. The greatest differentiation between the groups is shown by the percentage of animals that were immediately conditioned (table 6, column 7). The results of the control, 12-, and 24-hour groups are quite similar in this respect ranging from 30 to 40 per cent. There is a decrease, however, to 28 per cent with the 36-hour group and to 18 per cent with both the 48- and 60-hour groups and to 19 per cent with the 72-hour group. The number of animals that failed to be conditioned (table 6, column 10) never rises above 15 per cent in any but the 48-hour group in which 38 per cent were still crossing the grill at the end of the test period.

The variability (crossings) of the 12-, 24-, and 36-hour groups (table 6, column 5) is very great ranging from 160 to 198. With the 48-hour group it falls to 89 and with the 60-hour group to 131.

In general it may be said that the individual differences between the animals of each group are so great as to obliterate any group differences that might result from the various starvation periods. It was observed that each group included both animals that regularly ate after crossing and animals that failed to do so. In the latter case the escape reaction was evident. Upon being returned to section *A* these rats either withdrew as far as possible from the grill making no further attempt to cross, or immediately recrossed. In the latter case the crossings occurred in rapid succession and then ceased altogether. In the former case there were frequent pauses between crossings and the animals gave no indications of fear.

The cumulative percentile graph (Fig. 4) for shock 2 indicates clearly the difference between the 48-hour and the other groups of this series. The tendency to cross the grill in the 48-hour group is evident in the fourth decile and very marked from the fifth through the tenth decile. The majority of the animals in this group ate after crossing the grill and showed little evidence of fear. They seemed to be reacting directly to the food stimulus rather than attempting to escape. Pauses within section *A* and contacts followed by withdrawal without crossing were frequent in this group. This type of reaction, shown by the majority of the 48-hour group, occurred in only a few cases in the other groups of this series. The tendency to cross the grill is very slight with any other group. In fact from the first through the sixth decile the number of crossings made by any starvation group other than the 48-hour differs only slightly from that of the control group.

In the ninth and tenth deciles the tendency toward extreme reactions is considerable and greater in every case with the starvation groups than with the control group. This fact, however, does not warrant the conclusion that these extreme reactions were due to the hunger drive alone, since in many cases the animals failed to eat after each crossing, although given ample opportunity to do so.

It appears from these results that the 48-hour group is the only one of the shock 2 series in which the strength of the hunger drive is at all comparable with the resistance offered by the grill. Throughout the other groups this shock appears to disturb seriously the hunger drive of the majority of the animals. This disturbance seems to be greater with groups that have been starved less than 36 hours than with those starved more than 36 hours, inasmuch as the number of animals immediately conditiond is considerably less with the longer starvation periods. This fact considered in conjunction with the results of the 48-hour group above discussed suggests that the degree of shock sufficient to overcome the hunger drive (as indicated by a failure to react to the food stimulus when in the presence of it) is not identical for all periods of starvation but is dependent upon the length of the starvation period preceding the shock.

The tendency to touch the grill without crossing in the shock 2 series appears to bear some relation to the length of the starvation period although the differences between the groups are not large. The number of contacts made by each starvation group is in every case twice as great as the number made by the control group. The median number of contacts (table 3, column 11) increases from 2.9 with the 12-hour group to 3.4 with the 48-hour group and falls to 2.6 with the 60-hour group. Jumping behavior seems to bear no consistent relation to length of starvation period under the conditions of shock 2. The number of jumps (table 3, column 15) is greatest with the 24-hour group and slightly smaller with 48-hour group, but the other groups do not differ greatly from the control group in this respect.

Shock 3. The tendency to react against the grill rather than to the food stimulant is noticeable to a much greater extent with shock 3 than with shock 2. The results of the control group (crossings (of this series are extremely variable (C.V. 210), and seem to indicate that a strong degree of shock results with a few animals in a decided tendency to cross the grill. The only rat that failed to be conditioned by shock 3 occurred in the control group, in the absence of any food stimulus. This rat showed evidence of fear and seemed to be fleeing from section *A*.

The tendency to cross the grill with shock 3 is relatively slight except in a few extreme cases. Every rat in each starvation group was conditioned against the grill by shock 3 within the test period (table 7, column 10). The greatest tendency to cross the grill frequently, as shown by the percentage of animals crossing 40 or more times (table 7, column 9) occurred in the 12-and 24-hour groups. With the other groups this tendency does not differ noticeably from that of the control group. The median number of crossings (table 7, column 2) increases from 2.3 with the control group to 8.0 with the 24-hour group and then falls to 4.3 as the period of starvation is increased to 48 hours. The effect of a 60- and 72-hour starvation period upon the shock 3 groups cannot be determined owing to the fact that so many animals in these groups died. If the dead are excluded from the results these groups appear to cross the grill more frequently than the 48-hour group, whereas if they are included the medians are smaller than that of this group. The percentage of rats immediately conditioned and the percentage crossing the grill 5 times or less (table 7, columns 7 and 8) show the same trend as the median number of crossings.

The percentile graph for shock 3 (fig. 5) indicates that the difference between the various starvation groups from the first through the fifth decile is negligible. Moreover there is no apparent difference between the starvation and control groups within this range. From the sixth through the tenth decile, however, the 24-hour group crosses more frequently than any of the other groups, while from the seventh through the tenth decile the 36-hour rats cross more frequently than those of any group with the exception of the 24-hour. Since the differences between the control and starvation groups (excepting the 24-hour group) are very small and unreliable, and since animals usually failed to eat after crossing, shock 3 cannot be said to offer a measure of the hunger drive alone. That some factor other than the incentive stimulus operates with shock 3 is shown by the fact that animals crossed frequently in the absence of food and after refusing to eat. This additional factor appears to be the tendency to escape, which is stimulated by the shock itself and results from an association of the grill with section *A*. The fact that the number of crossings increases with an increase in the starvation period from 12 to 24 hours and decreases with further increases from 24 to 48 hours indicates that there is some relation between the tendency to cross the grill and the length of the starvation period. This relation is probably due to differences in the organic state of the animals resulting from the different starvation periods. Richter (supra) found that general activity bore a direct relation to differences in organic state resulting from different starvation periods. Consequently it is probable that shock 3 indicates more clearly the factor of general activity than the motivation resulting directly from the incentive stimulus.

The median number of contacts in this series increases from 2.3 with the 12-hour group to 3.5 with the 36-hour group and then decreases to 2.6 with the 48-hour group. There appears to be an increase in the number of contacts with the 60- and 72-hour groups but owing to the number of deaths in these groups, the results are not comparable with the others of this series. The jumping activity (table 7, column 15) shows no relation to the length of the starva-

TABLE 11

Showing the effect of different degrees of shock upon the same starvation groups

	CROSSINGS							CONTACTS (MEDIAN)	JUMPS (MEDIAN)
	Median	Coefficient of variability	Range	Per cent immediately conditioned	Per cent crossing 5 times or less	Per cent crossing 40 times or more	Per cent not conditioned		
Control group									
Shock 1......	1.6	153	0–12	60	95	0	0	2.0	0.66
Shock 2......	2.4	177	1–36	35	85	0	0	1.6	1.7
Shock 3......	2.3	210	0–54	45	80	5	5	1.7	2.5
12-hour group									
Shock 1......	4.5	135	0–77	33	55	19	11	4.6	2.4
Shock 2......	3.3	198	0–108	30	65	5	5	2.9	0.87
Shock 3......	3.2	159	0–44	18	75	2.5	0	2.3	2.8
24-hour group									
Shock 1......	28.6	77	0–98	13	25	33	18	1.9	5.0
Shock 2......	2.6	160	0–74	40	67	15	5	2.9	2.5
Shock 3......	8.0	122	0–93	20	25	23	0	2.8	1.8
36-hour group									
Shock 1......	37.0	61	0–89	5	10	50	50	3.7	3.8
Shock 2......	3.5	166	0–60	28	75	10	8	3.3	1.1
Shock 3......	5.5	129	0–79	23	53	10	0	3.5	3.6
48-hour group									
Shock 1......	27.0	64	5–80	0	5	33	30	2.6	1,0
Shock 2......	20.0	89	0–77	18	38	25	38	3.4	2.0
Shock 3......	4.3	151	0–70	33	68	7.5	0	2.6	2.5
60-hour group									
Shock 1......	6.7	120	1–58	9	46	9	23	3.1	3.6
	4.6	140	0–58	22	54	7	19.5	2.6	1.8
Shock 2......	4.1	131	0–46	18	68	2.5	0	2.6	1.2
Shock 3......	6.3	127	0–69	20	48	2.5	0	3.2	3.0
*	2.6	158	0–69	40	60	2	0	2.4	1.9
72-hour group									
Shock 1......	16.5	88	2–68	0	24	5	33	2.4	1.4
*	2.0	154	0–68	48	60	2.5	17.5	0.0	0.0
Shock 2......	6.0	111	0–50	19	43	4.8	10	3.6	0.89
*	0.0	389	0–50	61	73	2.3	5	0.0	0.0
Shock 3......	7.5	110	0–70	14	41	8.1	0	3.6	2.2
*	3.7	138	0–70	35	55	6.1	0	2.6	0.0

* Results in this row include the dead.

tion periods. It is rather large in every instance but that of the 24-hour group.

The effect of different degrees of electrical stimulation

This section in concerned with the effect of different degrees of electrical stimulation upon drive behavior resulting from a given starvation period. The relative effect of the three degrees of shock upon the control group and upon each of the six starvation periods will be considered, as measured by crossings, contacts, and jumps.

Control group. In the absence of a food stimulus in section *C*, the central tendencies of the groups indicated by the medians (table 11, column 1) are not materially affected by increasing the shock. The extreme tendencies of the groups are however, considerably affected. The percentage of animals immediately conditioned by shock 1 is greater than that by either shocks 2 or 3. Frequent crossings of the grill increase directly with an increase in shock as shown by the fact that only one rat crossed the grill more than 3 times, and that only 12 times, with shock 1, whereas 3 rats crossed between 10 and 36 times with shock 2, and 3 between 8 and 54 times with shock 3. Moreover the only animal that failed to be conditioned in any control group occurred with shock 3. As indicated by the coefficient of variability an increase in shock directly increases the variability of the groups (table 11, column 2).

The number of contacts made by the control groups was greater with shock 1 than with either shocks 2 or 3 (table 11, column 8). This appeared to be due to the fact that the rats moved much more rapidly and less cautiously with shocks 2 and 3 and consequently upon touching the grill did not stop but proceeded across the grill, whereas with shock 1 the animals approached the grill rather slowly and generally withdrew after experiencing the shock. The number of jumps made by the control groups appears to be directly increased by an increase in shock, the median number of jumps being 0.66 with shock 1, 1.7 with shock 2, and 2.5 with shock 3.

In brief, in the absence of the food stimulus an increase in shock while affecting the central tendencies (crossings) of the groups only slightly produces a marked effect upon the extreme tendencies of the groups. The lowest degree of shock appears to be most effective in immediately deterring the animals from crossing the grill, whereas a heavy shock may result in repeated crossings. Inasmuch as the degree of shock constitutes the only variable condition in the control groups, it must be assumed that the increased tendency to cross the grill is due to the shock itself. Observations of the behavior of the animals indicated that an increase in the degree of shock used resulted in an increase in the reactions indicating fear. Such reactions were not evident with shock 1 but occurred frequently with shock 2. In such cases the rats after crossing, ran wildly about section *C* or tried to jump from the box. Upon being replaced in section *A* they would dash immediately across the grill or bury their heads in a corner of the box, in which position they would remain, breathing very rapidly until finally removed. With shock 3 the number of animals showing these reactions was greater and the conditions described more exaggerated. Jumping also occurred more frequently with an increase in shock. From these facts it appears that in the absence of the incentive stimulus shocks 2 and 3 were both strong enough to frighten the animals. The increase in

TABLE 12

Showing the reliability of the difference between the various shocks (average crossings) for a given starvation group

STARVATION GROUP	SHOCKS 1 AND 2		SHOCKS 2 AND 3		SHOCKS 1 AND 3	
	SD_D	$\frac{D}{SD_D}$	SD_D	$\frac{D}{SD_D}$	SD_D	$\frac{D}{SD_D}$
Control	2.17	1.70	3.37	0.09	2.70	1.48
12	4.81	1.64	3.37	0.74	4.19	2.48
24	4.75	3.51	4.99	1.69	5.21	1.75
36	4.70	6.85	3.78	1.24	4.94	5.56
48	4.43	1.85	4.11	2.80	4.06	4.85
60	3.14	0.02	2.59	0.54	3.03	0.69
72	2.73	2.56	2.65	3.28	3.20	0.53

the number of crossings made by some rats with an increase in shock seemed to be due to the fact that fear aroused random movements resulting in a failure to discriminate between sections *A* and *B* so that *A* was reacted to in the same way as *B*.

Twelve-hour group. With the introduction of the incentive stimulus in section *C* the relative effect of the three degrees of shock upon the behavior of the rats is changed. The differences between the central tendencies of the groups are, as with the control group, small and unreliable (table 12). An increase in shock in this case, however, appears to decrease the number of times the rats will cross the grill, the medians being 4.5 with shock 1, 3.3 with shock 2, and 3.2 with shock 3.

The results of the three degrees of shock in regard to immediate conditioning indicate a somewhat different effect. Thirty-three per cent were immediately conditioned by shock 1, 30 per cent by shock 2, and 18 per cent by shock 3. Animals reacting to shock 3 are thus more likely to cross the grill at least twice than when the lighter shocks were used. On the other hand shock 3 is the most effective of the three shocks in finally conditioning the animals since no rats failed to be conditioned in this case, whereas 2.5 per cent remained unconditioned with shock 2 and 11 per cent with shock 1. Moreover only 25 per cent of the group crossed more than 5 times with shock 3, while 35 per cent did so with shock 2 and 45 per cent with shock 1. The number of rats crossing 40 or more times was also directly decreased by an increase in shock.

The number of contacts in the 12-hour group decreased with an increase in shock (table 11, column 8). The tendency to jump does not vary consistently with the degree of shock as will be seen from table 11.

While the control and 12-hour groups had been without food exactly the same length of time, the results are quite opposed in respect to the relative effect of the three degrees of shock. It will be recalled that in the control group the animals reacting to shock 1 showed the least tendency to cross the grill, while in the 12-hour group they show the greatest tendency to do so. Moreover whereas the greatest tendency to cross in the control group occurred with shock 3, in the 12-hour group this shock is most effective in deterring the rats from crossing.

Twenty-four-hour group. The difference between the results (crossings) with shocks 1 and 2 after 24-hour period of starvation is large and reliable (table 12). The number of crossings made with shock 1 is considerable as shown by a median of 28.6 and the fact that 75 per cent of the group crossed more than 5 times. The tendency to cross with shock 2 is, on the other hand, very slight as shown by a median of 2.6 and the fact that only 23 per cent of this group crossed more than 5 times. Only 13 per cent of the animals used with shock 1 were conditioned immediately, while 40 per cent of the group with shock 2 were immediately conditioned.

The number of crossings made with shock 3 is in this case greater than with shock 2, but considerably less than with shock 1, the median number of crosses being 8.0. The tendency to cross 40 or more times with shock 3 was not quite as marked as with shock 1 as shown by the fact that 23 per cent of the group with shock 3 crossed 40 or more times whereas 33 per cent did so with shock 1. The number of rats immediately conditioned by shock 3 is only half as great as that by shock 2.

The results of the 24-hour group are least variable with shock 1 and most variable with shock 2 (table 11, column 2).

The number of contacts made by the 24-hour groups was smallest with shock 1. The number of jumps in this case decreased directly with an increase in shock.

The behavior of the rats in response to the food after crossing to section *C* varied considerably with the different shocks. With shock 1, the rats ran immedately to the food and started eating with all possible speed. With shock 2 some rats reacted in like manner showing practically no evidence of fear, while others were obviously frightened and paid no attention to the food, but tried to jump out of section *C*. With shock 3 hardly any of the animals noticed the food. They either cowered in a corner of section *C* or attempted to jump from the box. The speed in crossing the grill when replaced in section *A* also differed with the different shocks. With shock 1 hesitation for several seconds at the edge of the grill followed by a crossing, frequently occurred, whereas with shock 3, if they crossed at all, the tendency was to dash immediately across the grill after being placed in *A*. Very little hesitation at the edge of the grill occurred with shock 2 although the temporal interval between crossings varied considerably from rat to rat.

Thirty-six-hour group. The relative effect of the three degrees of shock upon crossings with the 36-hour group is in general similar to that noted with the 24-hour group. The largest number of crossings was made with shock 1 (median 37.0) and the smallest number with shock 2 (median 3.5). The difference between the results with shocks 1 and 2 is in this case extremely reliable (table 12). The number of crossings made with shock 3 (median 5.5) is slightly greater than that with shock 2.

The number of animals that failed to be conditioned is largest with shock 1 (50 per cent), while none failed to be conditioned by shock 3. The number of rats immediately conditioned was again smallest with shock 1 (5 per cent) and greatest with shock 2 (28 per cent).

The number of contacts made with shock 1 was rather large (median 3.7). In this case, however, a contact was generally a preliminary reaction immediately preceding a crossing. With shocks 2 and 3 the contacts did not bear as close a relation to the crossings but were only slightly less frequent than with shock 1. Jumping activity was considerable with shocks 1 and 3 but was relatively slight with shock 2.

Forty-eight-hour group. With the 48-hour group the tendency to cross the grill is directly related to the degree of shock. The group with shock 1 crossed most frequently (median 27.0) and that with shock 3 least frequently (median 4.3). The number of crossings with shock 2 was in this case only slightly less (median 20.0) than with shock 1. With shock 1, however, only 5 per cent crossed 5 times or less whereas 38 per cent did so with shock 2 and 68 per cent with shock 3. The number of rats failing to be conditioned was slightly greater with shock 2 than with shock 1, but the number of crossings 40 or more times was slightly less than with shock 1.

The number of contacts was approximately the same with shocks 1 and 3 (medians 2.6) but slightly greater with shock 2 (median 3.4). The jumping activity increased directly with an increase in shock.

The variability (crossings) of the 48-hour group is least with shock 1 and greatest with shock 3 (table 11).

Sixty-hour group. Due to the fact that some of the 60-hour rats, with shocks 1 and 3, died, the results for the three degrees of shock are not entirely comparable. The general effect of an increased shock appears, however, to be very slight with the 60-hour group. Considering these results in terms of the medians the differences between the groups are negligible and have no reliability. The variability of the three groups is approximately the same (table 11, column 2).

If the results are considered in respect to the percentage of animals that were immediately conditioned, there appears to be a slightly greater tendency for all the rats to cross with shock 1 than with shocks 2 or 3. More rats crossed 5 times or less with shock 2 than with shocks 1 and 3. The chief difference between the 3 degrees of shock occurs at the upper end of the scale as 9 per cent of the group with shock 1 crossed 40 or more times, whereas only 2.5 per cent did so with shocks 2 and 3. 23 per cent of the group with shock 1 failed to be conditioned while none failed to be with shocks 2 and 3.

The number of contacts is relatively large with shocks 1 and 3 (medians 3.1 and 3.2, respectively) but is somewhat less with shock 2 (median 2.6). The tendency to jump is also considerably greater with shocks 1 and 3, (medians 3.6 and 3.0) than with shock (median 1.2).

It appears from these results that an increase in shock with rats that have been starved 60 hours was only effective in respect to the animals that crossed the grill most frequently, since an increase in shock diminished the number of extreme cases but did not noticeably affect the central tendency. In general, the activity shown by touching the grill and jumping was greater with shocks 1 and 3 than with shock 2.

Seventy-two-hour group. With these groups the factor of selection is again operative so that the results are not entirely comparable. In general the tendency to cross the grill is greatest with shock 1 since no animals were immediately conditioned in this case whereas 19 per cent were by shock 2 and 14 per cent by shock 3. The percentage of rats crossing 5 times or less is also smaller with shock 1 than with shocks 2 or 3. The number crossing 40 or more times is approximately the same with the three degrees of shock, but the number failing to be conditioned decreased directly with an increase of shock. As shown by the median the number of crossings was greatest with shock 1 and considerably less with shocks 2 and 3 (table 11).

In general the number of contacts is greater with shocks 2 and 3 than with shock 1. The number of jumps in this case seems to bear no relation to the degree of shock.

Summary. The results of the three degrees of shock upon the same starvation periods show that, in the absence of the incentive stimulus, the tendency to cross the grill is relatively slight with any degree of shock. In this case, however, the general tendency of an increased shock, as has been stated in the discussion of the control group, is to increase the tendency to cross the grill. This fact indicates that a strong shock in itself produces a certain degree of motivation.

These results show that when the food stimulus is introduced into the otherwise identical situation the relative effect of the three degrees of shock was dependent upon the length of the starvation period that had preceded the shock. The relative effect of the three degrees of shock on each starvation group can be clearly observed in figure 6. With either a short (12 hours) or long (60 and 72 hours) starvation period the tendency to cross the grill is not markedly different with the three degrees of shock, although there is a slight tendency to cross more frequently with shock 1 than with shocks 2 or 3, except in the case of the 72-hour groups where the results of the three degrees of shock are not comparable owing to the deaths occurring in these groups. With a median period of starvation (24 or 36 hours), however, the number of crossings made with shock 1 is very great, but is relatively slight with shocks 2 and 3. With these shocks the tendency to cross the grill in response to the food stimulus does not appear to differ radically from that observable when the food stimulus is absent. This fact indicates that shocks 2 and 3 do not give adequate measures of the relative effect of different periods of starvation. Shock 2 appears to divide the animals into two groups, those motivated largely by the food stimulus and those motivated chiefly by fear. With shock 3 the fear motivation appeared to become dominant in practically all cases. With shock 1, however, all the animals seemed to be motivated by the food stimulus. Consequently the data of shock 1 alone, can be considered as affordig measures of the hunger drive.

IV. Summary and Conclusions

1. Of the three degrees of electrical stimulation used in compartment *B* as an obstruction against which to measure the hunger drive, only the lowest proved to be suitable for use with this method. Both of the higher shocks were severe enough to disturb the normal food response. Shock 2, in general, had the effect of separating into two classes the animals of each starvation group: in one case the animals appeared to be motivated mainly by hunger and in the other mainly by fear. Shock 3 was so severe that it seemed to introduce an additional source of motivation which disturbed and obscured the normal hunger drive.

2. The lowest degree of shock furnished a degree of stimulation adequate for the measurement of the hunger drive by the Obstruction Method, in that it was high enough to bring out differentiation in drive behavior resulting from different periods of starvation, and low enough not to arouse other conflicting tendencies.

Inasmuch as shock 1 alone proved suitable with this method for the measurement of the hunger drive the following statements are based solely on data obtained with this shock.

3. As measured by the number of crossings within the test period, the hunger drive appears to increase from 12 to 36 hours and to decrease with further increases in the length of the starvation period.

4. The hunger drive as indicated by the percentage of animals immediately conditioned and by that failing to be conditioned, apparently reaches its maximum at 36 to 48 hours and remains fairly constant from this point to 72 hours.

5. As measured by the number of crossings the results of the 24- and 48-hour drives appear to agree, while the 12- and 60-hour drives have highly similar values. As indicated by the per cent that failed to be conditioned, however, the hunger drive, is stronger with a 48- and 60-hour starvation period than with 12- and 24-hour periods. The difference between the results of these two measures is probably in large part due to the factor of physical exhaustion occurring with the longer starvation periods.

6. The number of contacts and jumps do not appear to bear any consistent relation to the length of the starvation period.

Bibliography

(1) Carlson, A. J. 1919. Control of Hunger in Health and Disease. Chicago Univ. Press.

(2) Cole, L. W. 1911. The relation of strength of stimulus to rate of learning in the chick. *Jour. Animal Behav.*, i.

(3) Dodson, J. D. 1917. Relative values of reward and punishment in habit formation. *Psychobiology*, i.

(4) Donaldson, H. H. 1924. The Rat. Philadelphia.

(5) Hoge, M. A., and R. J. Stocking. 1912. A note on the relative value of punishment and reward as motives. *Jour. Animal Behav.*, ii.

(6) Jenkins, Warner and Warden. 1926. Standard Apparatus for the Study of Animal Motivation. *Jour. Comp. Psych.*, 6: 361-82.

(7) Moss, F. A. 1924. A study of animal drives. *Jour. Exp. Psych.*, vii.

(8) Nicholls, E. E. 1922. A study of the spontaneous activity of the guinea pig. *Jour. Comp. Psych.*, ii.

(9) Richter, Curt P. 1922. A behavioristic study of the rat. *Comp. Psych. Monographs*, i, No. 2.

(10) Simmons, Rietta. 1924. The relative effectiveness of certain incentives in animal learning. *Comp. Psych. Monographs*, ii, No. 7.

(11) Slonaker, J. R. 1907. The normal activity of the white rat at different ages. *Jour. Comp. Neur. and Psych.*, xvii, No. 4.

(12) Szymanski, J. S. 1918. Abhandlungen zum Aufbau der Lehre von den Handlungen der Tiere. *Pflüger's Archiv. f. d. ges. Physiol.*, clxx.

(13) Szymanski, J. S. 1918-1919. Die Abhängigkeit der Lerngeschwindigkeit von der Antriebsstärke. *Pflüger's Arch. f. d. ges. Physiol.*, clxxii.

(14) Szymanski, J. S. 1918. Versuche über die Wirkung der Faktoren, die als Antriebs zum Erlernen einer Handlung dienen können. *Pflüger's Arch. f. d. ges. Physiol.*, clxxi.

(15) Tsai, Chiao. 1925. The relation of the sex and hunger motives in the albino rat. *Jour. Comp. Psych.*, No. 5.

(16) Yerkes, R. M., and J. D. Dodson. 1908. Relation of strength of stimulus to rapidity of habit formations. *Jour. Comp. Neur. and Psych.*, xviii.

2. THE EFFECT OF SHORT INTERVALS OF DELAY IN FEEDING UPON SPEED OF MAZE LEARNING [1]

C. J. WARDEN AND E. L. HAMILTON

In most experimental situations in which animals are employed it is necessary to arouse the test animal to the desired state of general activity by the use of an incentive stimulus of some sort. The incentive stimulus, if it is to be adequate, must be an object or situation which normally evokes in the organism positive responses of the locomotor type with respect to the stimulus in question. For obvious reasons food has been used most extensively in thus bringing about a more or less general stimulation of the organism, although other kinds of stimuli, such as sex objects for example, have been employed on occasion. In any case the same general principle applies. In order to reach the incentive stimulus the animal performs, or attempts to perform such feat as the experimental situation may require.

As a rule the animal is allowed immediate access to the incentive stimulus after performance so that the latter response is continuous with the response to the incentive stimulus. And it is, of course, quite proper that such procedure should be followed in the ordinary laboratory situation, since rewarding the animal immediately after performance most probably favors its general emotional adjustment to the artificial laboratory conditions. One interesting aspect of the general problem of motivation, however, is that concerning the effect upon behavior of separating the test response from the response to the incentive stimulus by intervals of time of various length. What will be the effect of thus delaying the latter response upon the learning of the former? Does the fixation of a pattern response, such as a maze habit, proceed more effectively and speedily when the reward immediately follows the performance?

[1] Published by C. J. Warden and E. L. Haas (now Mrs. E. L. Hamilton) in *The Journal of Comparative Psychology*, 1927, 7: 107-16 and reprinted from that periodical.

What, in fact, are the results of a progressive series of intervals of delay upon the rate of learning? The present study is a preliminary report of a systematic investigation of this general problem.

The present experiment was suggested by the work of Watson (1) who studied the effect of a 30-second interval of delay in feeding upon the rate of learning the sawdust type of problem box in the white rat. He used 6 animals in both test and control group and employed food as the incentive stimulus. The experimental conditions were kept constant and identical for both groups except for the fact that the animals of the test group were prevented from obtaining food after each daily performance by covering the food dish with a perforated lid which could be conveniently raised by the experimenter at the close of the 30-second interval of delay. In both groups, moreover, the animals were allowed to feed for 5 seconds from the dish and then transferred to their cages where the daily ration was provided. Watson found that the animals were very active during the period of delay. The time curves for the two groups are closely similar, leading Watson to conclude that the delay of 30 seconds did not alter the speed of forming the habit.

The work of Watson has been extended in two directions in the present study. In the first place a different type of performance has been selected, a simple maze replacing the problem box. This made necessary a different kind of apparatus for restraining the animal from securing the food during the period of delay, a detailed description of which will be given later. In the second place much longer intervals of delay were employed, i.e., a 1-minute and a 5-minute interval.. A change in procedure was also introduced. Instead of feeding the animals for only 5 seconds in the apparatus, and then giving them their main ration in their living cage, we allowed our animals to feed for 5 minutes in the apparatus after daily performance before removing them to a feeding cage close by, where they were allowed to feed for an additional 5-minutes, and then removed to their living cage. Any possible influence of the incentive stimulus, or of the response to it, upon the previous performance of running through the maze was thus given the best possible conditions in which to operate. In most cases active feeding had ceased before the animals were removed from the food box

The design of the maze employed is shown in figure 1. It proved to be a relatively easy pattern in spite of the fact that it contains 8 culs-de-sac. It was constructed after the usual manner of wood, with 4-inch runways, and was provided with a sliding door in front of the food box which prevented the animals from re-entering the maze after having reached the food box. The food dish consisted of the metal bell from an alarm clock, approximately 6 cm. in diameter, screwed down to the floor at the center of the food box.

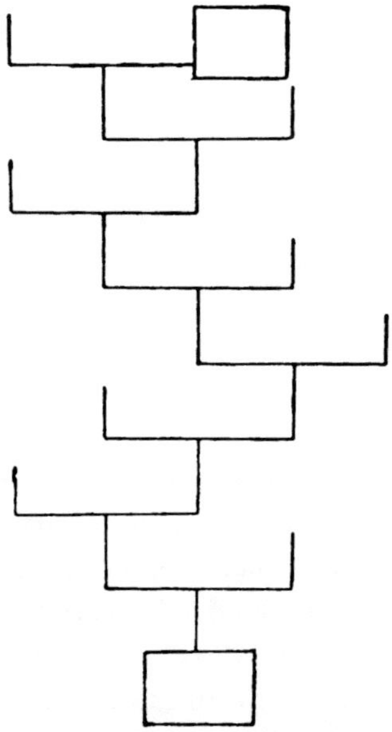

FIG. 1. MAZE PATTERN

The device utilized for restraining the animal from obtaining the food during the period of delay is illustrated in figure 2. The wooden screen was formed by extending the sides of the food box upward for a distance of 18 inches. This screen served the purpose of isolating the animal during the period of delay from possible distracting stimuli from without, to a large extent. It also eliminated the experimenter from the visual field of the animal quite effectually while permitting easy observation of the animal. The screen was provided with a door, *C*, for the removal of the animal from the food box after daily practice and feeding.

The restraining device proper consisted of a heavy metal funnel 9 cm. in diameter, lined inside with thin felt, which could be dropped down over the food dish at will. The funnel was heavily weighted at the top with lead so that it could not be removed from the dish by the animals, when once in place. It was operated by means of a wire cable as indicated in the diagram, the cable being coated with vaseline to prevent the rats from attempting to climb

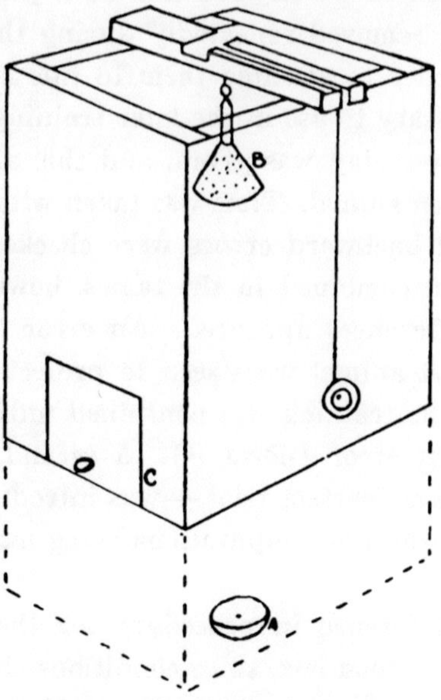

FIG. 2. RESTRAINING DEVICE

A, bell-shaped food dish; *B*, perforated metal funnel, attached to picture wire; *C*, hinged door, through which the animals were removed from the food box; *D*, food box (shown in broken lines).

it. Perforations in the funnel permitted normal olfactory stimulation during the period of delay. The operation of the device was practically noiseless.

A total of 43 white rats of approximately 80-gram weight were used. The control group included 15 animals and the two test groups 14 each. The diet consisted of whole wheat bread and milk and a sample of the regular diet was used as the incentive stimulus. The general illumination of the room was kept as constant as possible and was identical for all groups. Seven rats of each group were tested daily at 3 p.m., and the balance at 7 p.m., thus equalizing the time of day factor for the groups. The animals were allowed to feed, on each of 3 days previous to the training period, for 5 minutes in both entrance box and food box and then for a like period in the adjacent feeding cage. During this preliminary contact with the apparatus, the funnel was suspended about 6 inches above the food dish in which food had been placed. The animals were inserted and removed repeatedly during this period through the door, *C,* in order to acustom them to this method of removal which it was necessary to use in the later training series.

Only one trial per day was given, and this always at the same time of day for each animal. Time was taken with a stopwatch, and both forward and backward errors were checked. The two types of error have been combined in the tables, however, since no significant group differences appeared. An error was checked when the ears of the test animal were seen to project into the entrance of a cul-de-sac. The training was continued until 4 out of 5 trials were made without error (norm B). A second norm of mastery (norm A) — the first perfect trial — was introduced as a check on the more severe norm, the computations being made of course from the same records.

The essential difference in procedure for the different groups centered around the food box, since conditions elsewhere were kept uniform throughout. In the case of the control group the funnel was raised exposing the food at the instant the door closed behind the animal. After 5 minutes of feeding the rat was removed through the door, *C,* and placed in a small cage, with food, for an additional 5 minutes and then retured to its living cage. The procedure in the case of the two test groups was precisely the same, except for the fact that the funnel was raised only after the interval of delay, in each case, was over.

The results are presented in tables 1, 2, and 3 for both norms of mastery and for all criteria of performance (trials, errors, time). The statistical treatment of the average (norm B) appears in the last four columns of each table, this average representing probably on the whole the best group index. The median for the same norm agrees in general with the average, and the scores of norm A are also in essential agreement with those of norm B. In fact these four indices of central tendency exhibit a remarkable uniformity throughout, and the data for trials, errors and the time all point in the same direction.

Let us compare, first, the control and the 5-minute groups. If we take the scores at their face value it is evident that we must conclude that a 5-minute interval between the pattern response to be learned and the feeding activity has no disadvantageous effect upon the fixation of the maze habit. In fact, the values for both trials and errors are — in most instances — slightly lower than those of the control group, although this difference in favor of the 5-minute group is in no case significant. The 5-minute interval of delay then did not increase the number of trials required to master the maze, nor did it increase the number of errors made during the period of mastery.

The average total time spent within the maze was, however, somewhat greater for the 5-minute group, as will appear from an inspection of table 3. The difference is not large but is consistent regardless of norm of index of central tendency employed. Inasmuch as the number of trials, and of errors, was practically identical for the two groups, this difference in time would seem to reduce itself to speed of locomotion (average speed) in running through the passage ways. The animals of the 5-minute group evidently ran a little less speedily, or else made more frequent or longer stops, on the whole. Whether this difference in time per trial was due to the influence of the interval of delay cannot be certainly known from the data at hand. It may have been due instead to an accumulation of rats in this group that naturally run at a slower than average pace. There is a large individual difference in natural speed of running in the white rat, as anyone knows who has attempted to make use of a temporal norm of mastery in connection with either maze or problem box. Even if the slowing up in speed of locomotion were clearly due to a difference in motivation arising from the interval of delay it would still be difficult to connect this in any important way with the process of fixation. There is little reason to suppose that this slight difference

TABLE 1

Showing mean number of trials required in mastery of maze

GROUP	NORM A		NORM B					
	Median	Average	Median	Average	S. D.	S. D. (average)	D / S. D. (D)	V
Control	7.0	6.9	10.0	10.6	2.2	0.6	C. @ O.M. 2.00	20.6
One-minute	7.5	7.3	12.0	12.4	2.8	0.7	O.M. @ F.M. 2.40	22.6
Five-minute	5.5	5.6	10.0	10.2	2.2	0.6	C. @ F.M. 0.04	21.6

TABLE 2

Showing mean number of errors during mastery of maze

GROUP	NORM A		NORM B					
	Median	Average	Median	Average	S. D.	S. D. (average)	D / S. D. (D)	V
Control	22.0	23.8	22.0	24.7	6.8	1.8	C. @ O.M. 2.3	27.5
One-minute	29.5	29.6	29.5	31.2	8.4	2.2	O.M. @ F.M. 2.9	26.9
Five-minute	22.0	21.7	24.5	24.0	4.7	1.3	C. @ F.M. 0.3	19.6

TABLE 3

Showing mean number of seconds required in mastery of maze

GROUP	NORM A		NORM B					
	Median	Average	Median	Average	S. D.	S. D. (average)	D / S. D. (D)	V
Control	208.0	195.9	237.0	232.5	34.8	9.0	C. @ O.M. 2.8	15.0
One-minute	229.5	233.5	263.5	294.9	75.3	20.1	O.M. @ F.M. 0.7	25.5
Five-minute	211.0	222.1	263.5	275.6	61.5	16.4	C. @ F.M. 2.3	22.3

in speed of locomotion would bear any essential relation to rate of fixing the maze habit.

The results of the 1-minute group do not yield themselves to so obvious and simple an interpretation. The data are consistent throughout in showing a slower rate of learning for this group in comparison with either of the other two, regardless of norm of mastery, or index of central tendency employed. Moreover when the average for norm B is considered the difference is large enough to be fairly reliable in most cases, and always so as between the 1-minute group and the control. The acceptance of this result at its face value would lead to the conclusion that although a 5-minute interval of delay is not disadvantageous, a shorter one such as one minute may be so. Such a conclusion, while not impossible, does not appear to fit the logic of the situation very well. Nor is it supported by any fact of observation; there was no noticeable difference either qualitative or quantitative in the behavior of the 1- and 5-minute groups during the period of delay. They appear equally active in both cases.

The extreme precautions taken to insure uniformity of experimental conditions practically eliminate the possibility of any constant error in the results. Nor do the data afford any clue concerning a possible accumulation of a variable error, save, perhaps, in the matter of sampling. As already noted the 1-minute group showed, on the whole, a greater tendency to variability than the other two groups. Moreover, a single animal in this group extends the upper limit of the range in trials (norm B) from 16 to 20, as indicated in table 4. These facts suggest that perhaps the 1-minute group is less homogeneous than the other two, and hence not as fair a sample. A standard error of sampling formula can hardly be applied to such small groups. However, a rough check on this point seemed to be furnished by the following simple procedure. Each of the three groups was divided into two sub-groups of equal size by chance, and averages for these sub-groups computed for all criteria. Fifteen numbers were drawn from a hat and placed alternately into X or Y pile, each number representing a given individual animal. A large difference in the averages of the two sub-groups should mean lack of homogeneity in the major group, although

TABLE 4

Distribution table covering number of trials required for complete mastery (norm B)

GROUP	TRIALS													
	7	8	9	10	11	12	13	14	15	16	17	18	19	20
Control	1	2	2	3	1	4	0	1	1	0	0	0	0	0
One-minute	0	0	1	3	1	5	0	2	0	1	0	0	0	1
Five-minute	2	1	1	5	2	1	0	2	0	0	0	0	0	0

TABLE 5

Averages based upon a chance division of each group into sub-groups X and Y

GROUP	TRIALS		ERRORS		SECONDS	
	X	Y	X	Y	X	Y
Control	10.1	11.1	24.0	25.3	215.4	252.0
One-minute	10.9	14.0	29.6	33.9	250.9	339.0
Five-minute	10.0	10.4	23.9	24.1	250.1	301.0

such a simple method gives no more than a rough indication. An inspection of table 5 will show that the difference between the averages of the sub-groups, arranged on a chance basis, is very slight for both the control and 5-minute group, but is quite considerable for the 1-minute group. This is additional evidence that the 1-minute group is less homogeneous than the other two groups, and raises the question of the validity of the various indices of this group.

It seems best, therefore, to leave the matter of the effect of the 1-minute interval open for further investigation. This of course does not disturb in any way the clear cut evidence that a 5-minute interval of delay between the act to be fixated and the feeding sponse had no measurable effect upon the rate of fixation. This latter result must be considered as so much evidence against the law of effect, insofar as the latter insists that the value of the "satisfying state" is a function of its "nearness in time" to the act, or series of acts to be fixated.

Bibliography

(1) Watson, J. B. The effect of delayed feeding upon learning. *Psychobiology*, i, 51-59.

3. THE EFFECT OF DELAYED INCENTIVE ON THE HUNGER DRIVE IN THE WHITE RAT[1] (EXPERIMENT 2: THE LEARNING METHOD)

E. L. Hamilton

I. Introduction

While the Columbia Obstruction Method doubtless offers the most direct and satisfactory method of measuring animal drives, the marked influence of the incentive-drive situation in animal learning suggested the use of this more indirect method as a check on our results reported in Part II, 2.

Watson (14) was perhaps the first to attack the problem of determining definitely the effect of delayed incentive on rate of learning. The apparatus employed was a modification of the sawdust box provided with a device for restraining the animal from obtaining the food until an interval of time had passed. The food was placed in a cylindrical container 5 cm. in diameter and 3 cm. high, supplied with a lid perforated with several mm. holes to allow for olfactory stimulation. A small vertical rod was screwed into the center of the lid and passed through a hole in the wire mesh of the problem box, passing out through the top. By means of this rod the experimenter could allow the animal to get food after a given interval.

He used 6 male rats approximately 60 days of age and 6 male rats about 100 days of age. The twelve animals were divided into two groups, six of which had to learn the problem by the usual method of immediate feeding and six others by the method of delayed feeding. The animals were distributed at random into the two groups, with the only precaution that each group contained three young animals and three older ones.

[1] Reprinted with modifications from *Genetic Psychology Monographs*, 1929, 5, No. 2, 137-66.

As preliminary training, all twelve animals were allowed to get their food in the box a number of times before the training series. At all times during the tests the lid to the food box was left on in the case of both groups. The lid was lifted by the time the animals reached the food box for the control group. The animals of either group were allowed to eat for five seconds on each trial after entering the box and then were lifted out and taken back to their living cages. Only one trial per day was given. The same method was adopted for the delayed incentive group except that in this case the lid to the food box was held down for 30 seconds. Watson found that the animals were very active during the period of delay. Table 1 shows the average time per trial for the two groups. Examination of the data indicates that a delay of 30 seconds between the solving of the problem and obtaining the incentive (food) has no effect upon speed of forming this habit. Such decided changes in procedure as the increase of trials to two trials per day and an interchange of incentive conditions between the immediate and delayed groups produced no noticeable effect upon the results.

Simmons (6), in her study of the effectiveness of various incentives in aimal learning, made a brief test of the effect of delayed feeding on the learning of a complex maze. She used five male and five female albino rats between three and four months of age. The rats were selected by chance so that the two groups represented random samples with the exception that each group had approximately an equal number of males and females. As preliminary training, the rats were fed in a special feeding cage, for 7 to 10 days before the experiment was begun, at the same time of day at which they were later fed during the experiment proper. Two groups were run, a standard and a delayed group. The former were fed, during the training period, immediately after practice each day, as is usual in ordinary laboratory procedure. The delayed group were fed in the feeding cage one and one-half to two hours after the last run. Work was done in the late afternoon by electric light. Two trials per day were given to each group, and the maze was considered mastered at that trial after which nine out of ten successive runs were made without error. The results appear in table 2. The immediate feeding group had records slightly su-

TABLE 1
Table from Watson showing average time per trial

Trials	Average delayed feeding Seconds	Average immediate feeding Seconds
1	385	923
2	58	111
3	84	89
4	30	24
5	18	36
6	17	23
7	13	9
8	12	13
9	9	8
10	9	15
11	13	9
12	8	13
13	8	9
14	9	7
15	11	24
16	7	8
17	6	9
18	8	6
19	10	5
20	9	5
21	8	6
22	6	6
23	8	7
24	6	7
25	6	8
26	6	7
27	7	6
28*	5	4
29	5	5
30	5	5
31	4	5
32	4	4
33	3	4
34†	6.3	3.6
35	4.0	4.3
36	4.6	4.3
37	6.0	5.3
38	3.3	3.3
39	3.3	4.3
40	5.0	5.3
41	3.3	4.3

* At this trial the two-trials-per-day schedule was introduced.

† At this trial the delayed group was fed immediately and the control group delayed 30 seconds. Only three animals (60 days of age) of each group were included.

perior in respect to number of trials, but inferior in respect to time and errors. Simmons does not consider these differences large nor, in view of the large mean deviations and lack of agreement between the three criteria, very significant. Also the difference in error score between the groups was confined largely to the first five trials and the records were almost coincident thereafter. She concludes that there is no significant difference between the records obtained under the two conditions.

The more recent study of delayed incentive on rate of learning by Warden and Haas [Hamilton] (8) will be found in Appendix 2 of this volume. Larger groups were used and a better control of conditions was secured. The data appeared to warrant the conclusion that delays of one and five minutes had no appreciable effect on the rate of learning.

TABLE 2

Table from Simmons showing averages with initial trial omitted

CONDITIONS	TRIALS	ERRORS	TIME
Immediate feeding	27.8	127.7	1761.44
Delayed feeding	33.3	93.3	1105.10

It will be seen from the above review that very little work has been done on this topic. Such results as have been obtained must be considered inconclusive. Watson's study was limited to a single delay interval of 30 seconds, and Simmons' to one of approximately 2 hours. The need for a more systematic investigation of the effect of delayed incentive on rate of learning is clearly indicated.

A. Method and procedure

Apparatus. The apparatus used with the learning method of studying delay was the standard unit maze designed by Warner and Warden (13). Figure 4 shows the types of unit of which the maze is constructed. The material used in building the maze was 18-gauge, galvanized, Armco sheet metal covered with aluminum paint to insure uniformity of appearance. As is seen in the figure, the units consist of walls and a roof, with no floor. They were set up on a platform 7x8 feet in size, which had been covered, after very smooth finishing, with battleship linoleum (cemented down). This platform served as a base or floor. The angle of junction of cul-de-sac to pathway is the same as the angle at the end of the cul-de-sac, so that the animal can get no cues from looking down into the latter as to whether it is a blind alley or not. The roof of both pathway unit and cul-de-sac is made of very heavy wire mesh neatly riveted into the frame, while the two ends of the cul-de-sac are also made of wire mesh in order to secure lighting conditions in the alley similar to those of the pathway.

Several problems arose in connection with the apparatus. The first concerned the pattern to be used in the present study. It was

FIG. 4. Photograph showing types of unit of the Warner-Warden Standard unit maze

(For Figures 1, 2 and 3 of this monograph see Part II, 2)

desired that the pattern be sufficiently difficult to bring out individual and group differences and yet simple enough to be learned in a reasonable time. The pattern finally chosen (Figure 5) met these requirements very well. The conditions of the study made the use of a screen and peep-hole impossible for practical purposes, since the noiseless operation of the doors was especially important. Great care was needed to prevent the animal from becoming conditioned against the restraining compartment, and one of the most common causes of fear in experiments with rats is the matter of occasionally pinching their tails in the doors of an apparatus. Automatically controlled doors were impractical and doors operated by a system of levers and strings proved to be too slow for the rat's quick movements. In many animal mazes the entrance and food boxes are a considerable distance from each other, necessitating the use of two experimenters, one to introduce the animal and the other to remove it, or else the experimenter must change his position during a given trial. In the pattern used, the entrance and food boxes were in close proximity to each other, making the change of position of the experimenter unnecessary. The experimenter took a position midway between the two boxes and remained quietly throughout a given trial operating manually the doors for both the entrance and the food boxes.

The fact that the pathways were somewhat deeper than those of

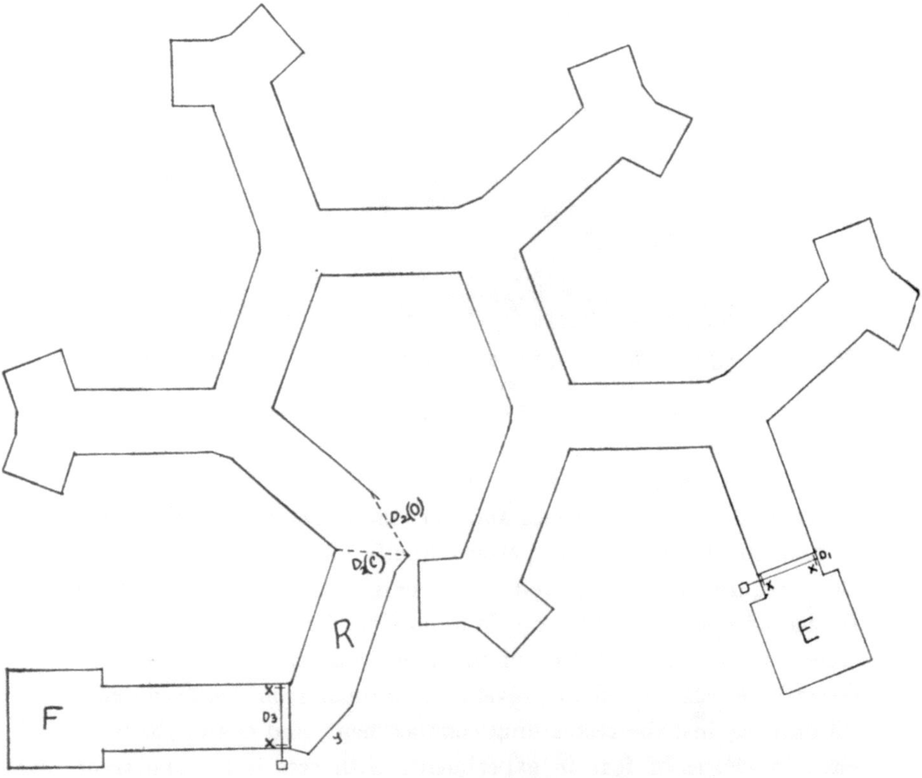

FIG. 5. MAZE PATTERN

E, entrance compartment; *F*, food box; *R*, restraining compartment; d_1, door separating entrance compartment from maze; d_2, door separating restraining compartment from maze; *o* and *c* represent door (d_2) in open and closed positions, respectively; d_3, door separating restraining compartment from food box; *xx*, rubber washers on brass rod to which door is fastened; *y*, piece of thin metal used to close opening in *R*.

most rat mazes (5 inches in the clear), together with the fact that a wire mesh was used as a covering for the maze, aided in the elimination of the experimenter from the visual field of the animal. A stop-watch of the lever type was used in the experiment and proved to be practically noiseless.

Several slight modifications were made in the apparatus as described by Warner and Warden (13). In previous work with this maze in the Columbia laboratory it was found that the animals hesitated as if frightened upon coming from the covered pathways into the food box, which was not covered with wire mesh. To prevent this behavior, a piece of mesh, of the same size as that forming the roof of the pathway and cul-de-sac units, was fastened over the long, narrow portion of the food box, so that the latter would more nearly resemble the other parts of the maze. A thin metal door (Figure 5, d_1, d_3) of the swinging type replaced the heavier metal sliding door at both the entrance and food compartments. The door, in each case, was fastened to a brass rod which turned in holes in the sides of the unit. Rubber washers (xx) prevented any noise which might result from the rod turning on the metal, and felt fastened to the bottom of the doors made their operation practically noiseless.

The problem of restraining the animal from the food during the delay period was met by using a pathway unit as a restraining compartment (R). This was placed between the last pathway unit of the pattern and the food box. Since this involved joining a pathway unit with another pathway unit instead of a cul-de-sac, as in the ordinary maze pattern, two openings were left in the restraining compartment. The opening nearest the food box was closed up by a thin piece of metal (y) especially cut to fit it. The other opening was used to good advantage. A thin metal door (d_2) of the swinging type with vertical pivot was made and served two purposes. When it was open it formed the wall of the maze between the last pathway unit and the restraining compartment; when closed, it served to prevent the animal from retracing its steps into the maze, thus forming an effectvie restraining compartment. A wire hook attached to the door provided a secure fastening. The door (d_3) to the food box was held closed by a brass stop while the animal was in the restraining compartment.

A piece of plate glass was used, to cover that part of the food box in which the food was placed, to prevent the animals from climbing out during the allotted time for feeding. The experimenter soon developed a noiseless method for replacing and removing this glass.

Animals. The subjects for this experiment were 105 male albino rats of approximately 100 grams in weight and 2 months of age at the beginning of the experiment, obtained from the Albino Supply, Inc., of Philadelphia. Upon arrival at our laboratory, the animals were placed in large cages, 30x15x15 inches, about 12 to 15 animals in a cage, and kept for two weeks before they were used. This two-week period previous to testing enabled them to become accustomed to laboratory conditions and to the experimenter, who handled them daily in feeding. All the animals remained in normal health throughout the experiment.

The animals were fed whole-wheat bread and milk throughout the experiment, with the addition of a weekly ration of greens. This diet was chosen instead of McCollum's Standard Diet because of the high incentive value of bread and milk (6). The animals were given their ration after the daily trial in the maze, the same food being used both for incentive and for regular feedings. Fresh water was supplied daily from inverted bottles with nozzles to prevent fouling. The animals were supplied with water even during the 24-hour starvation period. Before the starvation period, which marked the beginning of the experiment, fresh sawdust was supplied to insure complete absence of any food in the cage.

Five groups of animals were used. Table 10 shows the number of animals in each group. In addition to the 0-delay, or control group, four groups were used with the following intervals of delay: 1 minute, 3 minutes, 5 minutes, and 7 minutes.

The starvation period was 24 hours for all groups. The number of trials per day was constant for the groups, only one trial per day being given to each animal.

Procedure. The sex drive was kept constant for all groups by using males. It seemed advisable to eliminate all oestrous cycle complications by this restriction in the interests of a more adequate control of extraneous drive conditions. They were segregated as to sex during the two weeks previous to the beginning of the experiment and throughout the period of learning. The factor of time of day was controlled by running all the animals between the hours of 9 p.m. and 4 a.m. Since the laboratory was practically deserted at this time, noise distractions were reduced to a minimum. The room in which the experiment was conducted was lighted indirectly by electric light, the same number of lamps being lighted each night. The illumination was thus uniform in all parts of the room. The maze was kept in a constant position in the room, throughout the experiment and care was taken not to shift the pattern to new positions on the platform. The temperature was kept fairly constant by means of a thermostat.

TABLE 10
Grouping of animals

LENGTH OF DELAY PERIOD	NO. OF ANIMALS
0 (Control)	16
1 Minute	22
3 Minutes	22
5 Minutes	20
7 Minutes	25

The preliminary training consisted in feeding each animal singly in the entrance and food compartments 1 minute each and in a small feeding cage nearby for 10 minutes thereafter. A glass plate was kept on the food box to prevent the animal from climbing out during the 1-minute feeding period. This preliminary feeding was given the animals for two days. The animals were starved for 24 hours before each feeding period, as during the training series. The first trial, which occurred on the third day, was included in the preliminary training. The arguments for this procedure, which is the rule in the Columbia laboratory, have been developed by Cook (2).

The procedure for the 0-delay, or control group, was that usually employed in maze experiments, except that the animal was fed for a period of 1 minute in the food box and for 9 minutes in an adjacent feeding cage after the daily trial. The door between the maze proper and the restraining compartment (Figure 5, d_2) was open when the rat was placed in the maze. The stop-watch was started at the same time the animal was inserted in the maze. When the animal's tail cleared the door (d_2) at the restraining compartment, the door was closed and the watch stopped. The door (d_3) between the restraining compartment and the food box was raised immediately, allowing the animal to pass to the food box, and then closed. d_3 was raised, since this procedure was necessary in the case of the delay groups. The watch was started again, when d_3 was closed, to time the 1-minute feeding period in the food box. After the animal had fed for 1 minute, it was removed to the adjacent feeding cage and allowed to remain for 9 minutes.

The procedure for the delay groups was the same as that for the control group, with the following exception: Upon entering the restraining compartment, the animal was delayed for a period of time (1 minute, 3 minutes, 5 minutes, or 7 minutes) by closing the door (d_2) behind it. After the period of delay was over, the animal was permitted to enter the food box through d_3, which was raised, as in the control group.

Data were taken in terms of trials, errors, and time in seconds. An error was checked when the animal's nose projected past the entrance to a cul-de-sac. This criterion of error was chosen because the entrance of the nose in a cul-de-sac was more easily observed, and hence could be more accurately checked than other possible movements of the animal. Backward errors were checked for each unit of the pathway. Notes were made on the general activity of the animals in the delay compartment during the period of delay.

A change in the procedure was introduced at the 100th trial in the case of those animals which had not learned after running 99 trials and which had shown no indication of learning for 50 trials. On the 100th trial and thereafter, these animals were allowed immediate access to the food box, as in the case of the control group. This change was introduced to ascertain whether these animals were lacking in ability to learn the maze or whether failure to learn up to that time was due to the factor of delay.

The animals were run until they had attained the norm of mastery of four perfect trials out of five (Norm B), except the nine animals which had not reached the norm at the 99th trial. The data were also computed on the basis of the norm of two perfect trials out of three (Norm A), and under this norm all of the animals had learned by the 99th trial. In computation of scores the last four trials involved in the norm of mastery were omitted.

B. Results

The main results are based on the scores obtained by using the norm of two perfect trials out of three (Norm A) for the reason that only under this norm did all animals complete the learning process. The scores obtained under Norm B will be treated in a later section.

Norm A. The results are presented in Tables 11, 12, 13, 14, 15, 16, and 17, and in Figures 6, 7, and 8. Table 11 gives the frequency distributions covering trials, errors, and time for the various groups. Beginning at the fourth line, it reads as follows: one animal in the 0-delay group, one in the 1-minute group, and one in the 3-minute group, none in the 5-minute, and one in the 7-minute group learned in four trials. Two animals in the 0-delay group and none in any of the other groups made between 13 and 16 errors in learning. Four animals in the 0-delay group, one in the 1-minute group, none in the 3-minute group, one in the 5-minute group, and none in the 7-minute group required from 91 to 120 seconds to learn, etc. Table 12 gives the frequency distributions covering trials, errors, and time for the various groups, using larger intervals (1-5 for trials, 1-15 for errors, 1-150 for seconds). Table 13 is a cumulative frequency table showing percentage of group having learned by the end of each successive five trials. Figure 6 shows the same thing in graphic form. Table 14 presents the measures of central tendency and measures of variability for each period of delay. Table 15 gives the measure of reliability of the difference between the averages of various groups for trials, errors, and time. Table 16 gives the average errors per trial for each five trials. Figure 7 shows the same thing graphically. Table 17 presents the average time per trial for each five trials. Figure 8 gives the same data in graphic form.

When we examine the averages for the groups based on the scores for trials (table 14), we note that the averages and medians of the delayed incentive groups are considerably higher than those

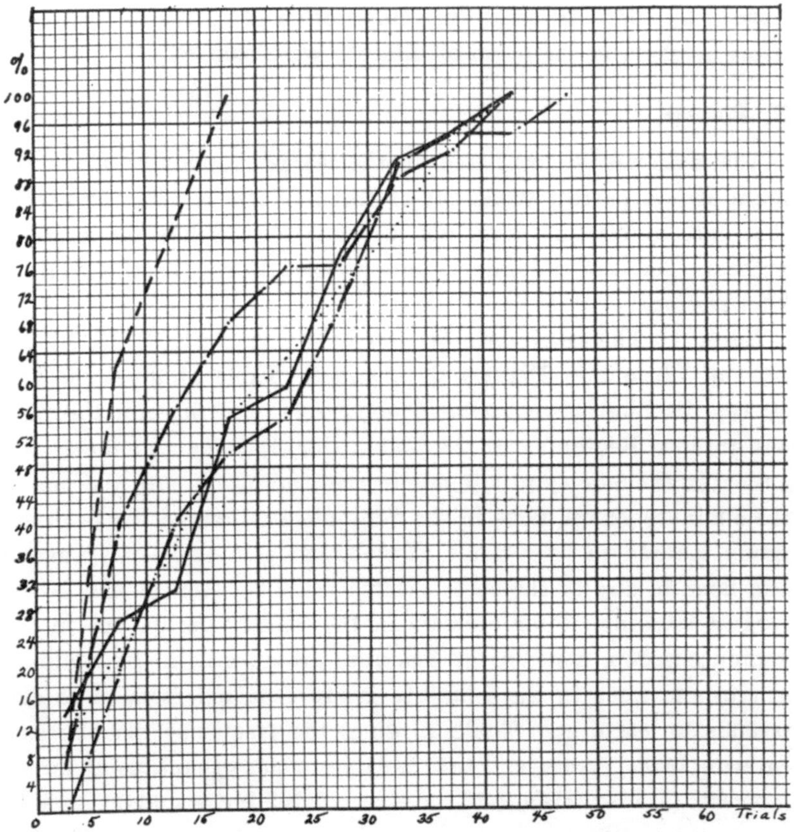

FIG. 6. CUMULATIVE FREQUENCY CURVES SHOWING PERCENTAGE OF GROUP HAVING LEARNED BY THE END OF EACH SUCCESSIVE FIVE TRIALS (NORM A)

Control group (dash-line); 1-minute group (solid line); 3-minute group (dot line); 5-mnute group (dot-dot-dash line); 7-minute group (dot-dash line).

of the control group. It will be seen from the table that the effect of a 1-minute delay is practically to double the number of trials required to learn the maze. The difference between the averages of the 1-minute group and the control group is highly reliable (3.82), as is shown in Table 15. The scores on errors and time show very much the same tendency.

The scores on trials and errors for the 3-minute, 5-minute, and 7-minute groups are only slightly different from those of the 1-minute group (Table 14), and in no case is the difference very re-

TABLE 11

*Frequency distributions covering trials, errors, and time for the various groups (Norm A)**

No. of trials	Groups 0	1 Min.	3 Min.	5 Min.	7 Min.	No. of errors	Groups 0	1 Min.	3 Min.	5 Min.	7 Min.	No. of seconds	Groups 0	1 Min.	3 Min.	5 Min.	7 Min.
1						1-4						1-30					
2						8						60		1			1
3		1				12					1	90	6	1	1		1
4	1	1	1		1	16	2					120	4	1		1	
5		1	1		1	20	1	1			2	150	1	4	2	2	3
6	3			3	1	24	3	1	1	1	3	180	1	1		2	2
7	2		1		3	28	2	1	2		1	210	2	1	3		2
8		1	1		1	32	1	1	1	2	1	240	1	3	3		2
9	2	2	1		3	36	1			1	2	270	1	1	1	1	
10	2			1		40	2	3	1	1		300		1			2
11			1		2	44			2		1	330			2	1	2
12	3		2			48			2	1	2	360		2		2	2
13				2	1	52		2			1	390		1	2	1	1
14		1		1	1	56	1	1	1	1	2	420		2		3	
15				1		60		1				450			1		1
16		2	1	1		64	2	3	2			480			1		
17	1	1	1		1	68					2	510		1			
18	1	1	1	1	1	72	1	1		1		540			1		
19	1		1		1	76			1	1		570				1	
20		1				80		1		1		600					
21		1	1			84		1		3	1	630			2	1	
22					1	88			1		1	660		1	1		
23			1		1	92				1	1	690				2	
24				1		96			2			720					
25						100		1	2			750			1	1	
26		1		1		104		1	1			780		1			1
27			1	1		108						810					1

TABLE 11 — *Continued*

No. of trials	Groups 0	1 Min.	3 Min.	5 Min.	7 Min.	No. of errors	Groups 0	1 Min.	3 Min.	5 Min.	7 Min.	No. of seconds	Groups 0	1 Min.	3 Min.	5 Min.	7 Min.
28				1		112			1	3		840				1	2
29		2				116		1		1		870					
30		1	1			120					1	900					1
31		1	1		1	124		1	2	1		930					
32		1	1		1	128						960					
33		1		1		132						990				1	
34				1	1	136		1			1	1020					
35				2		140						1050					
36		1	1		1	144						1080			1		
37			2	1		148						1110					
38						152						1140					
39						156				1	1	1170					
40						160						1200					
41					1	164						1230					
42		1				168						1260					
43						172					1	1290					1
44					1	176						1320					
45			1			180						1350					
46						184						1380					
47				1		188						1410					
48						192						1440					
49						196						1470					
50						200						1500					

*Two perfect trials out of three.

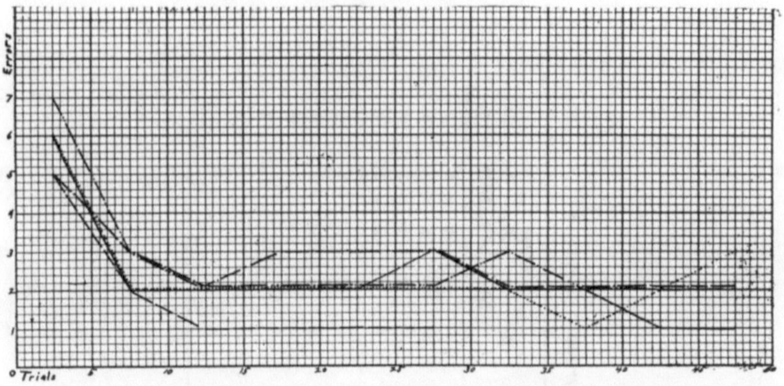

FIG. 7. AVERAGE ERRORS PER TRIAL FOR EACH FIVE TRIALS

Control group (dash line); 1-minute group (solid line); 3-minute group (dot line); 5-minute group (dot-dot-dash line); and 7-minute group (dot-dash line).

liable (the range in reliability index is .12 to 1.44), as will be seen from Table 15. This means that there is little difference between the effects of 1-minute, 3-minute, 5-minute and 7-minute delays when trial and error scores are considered.

The error curves (Figure 7) indicate that the animals in the delayed incentive groups did not differ markedly from each other in errors per trial. They consistently made more errors than the con-

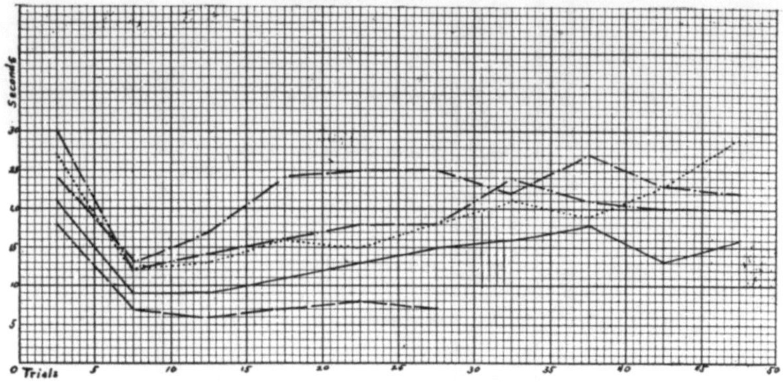

FIG. 8. AVERAGE TIME PER TRIAL FOR EACH FIVE TRIALS

Control group (dash line); 1-minute group (solid line), 3-minute group (dot line); 5-minute group (dot-dot-dash line); and 7-mnute group (dot-dash line).

TABLE 12

*Frequency distributions covering trials, errors, and time for the various groups (Norm A)**

No. of trials	Groups 0	1 Min.	3 Min.	5 Min.	7 Min.	No. of errors	Groups 0	1 Min.	3 Min.	5 Min.	7 Min.	No. of seconds	Groups 0	1 Min.	3 Min.	5 Min.	7 Min.
1-5	1	3	2	0	2	1-15	2	0	0	0	1	1-150	11	7	3	3	5
10	9	3	3	4	8	30	7	4	4	1	7	300	5	7	8	3	8
15	3	1	3	4	4	45	3	3	5	5	5	450		5	4	7	6
20	3	5	4	2	3	60	1	4	1	1	3	600		1	2	1	0
25		1	2	1	2	75	3	4	3	2	2	750		1	4	4	0
30		4	2	3	0	90		2	1	5	2	900		1	0	1	5
35		3	2	4	3	105		2	5	0	1	1050			0	1	0
40		1	3	1	1	120		1	1	4	1	1200			1		0
45		1	1	0	2	135		2	2	1	1	1350					1
50				1		150				0	0						
						165				1	1						
						180					1						

*Two perfect trials out of three.

trol-group animals. Thus, these results corroborate our conclusions based on scores for trials and for errors.

In the case of the time scores, however, the averages tend to increase with increase in length of delay period. The differences between the averages also tend to be slightly more reliable (the range in reliability is .12 to 2.26), as is shown in Table 15. In general, time scores are not considered as valid an index of learning in maze

TABLE 13

*Cumulative frequency table showing percentage of group having learned by the end of each successive five trials (Norm A)**

NO. OF TRIALS	GROUPS				
	Control	1 Minute	3 Minutes	5 Minutes	7 Minutes
1-5	6.3	13.6	9.1	0.0	8.0
10	62.5	27.3	22.7	20.0	40.0
15	81.3	31.8	36.4	40.0	56.0
20	100.0	54.5	54.5	50.0	68.0
25		59.1	63.6	55.0	76.0
30		77.3	72.7	70.0	76.0
35		90.9	81.8	90.0	88.0
40		95.5	95.5	95.0	92.0
45		100.0	100.0	95.0	100.0
50				100.0	

* Two perfect trials out of three.

experiments as trial and error scores, and hence more weight will be given to the latter. The inconsistency of the time scores with the general tendency of similarity among the delay groups seems to indicate that the longer delays caused the animals to run more leisurely through the maze. The lengthened time score arising in connection with delay is supported by behavior observations made by the experimenter. The animals appeared to be less motivated and ran slowly through the maze in a hesitant manner, while the control-group animals ran more quickly and directly through the maze to the food box. The time curves (Figure 8) show this same tendency — increase in time per trial with increase in length of delay period. The 7-minute group curve strikingly illustrates this tendency, rising much more suddenly than the other curves and remaining at a higher level. Thus the ainmals in the delay groups took a longer time per trial with increase of length of delay period.

The notes made on the behavior of the animals in the delay com-

TABLE 14

*Results for each period of delay (Norm A)**

Period of delay	Trials					Errors					Time				
	Median	Range	Aver.	S.D.	V.	Median	Range	Aver.	S.D.	V.	Median	Range	Aver.	S.D.	V.
0	9.5	4-19	10.3	4.40	42.72	29.0	13- 71	34.8	17.53	50.37	109.0	75- 258	131.0	54.44	41.56
1 Minute	19.0	3-42	20.3	11.14	54.88	60.5	17-135	64.3	32.12	49.95	237.0	49- 772	278.3	177.95	63.94
3 Minute	18.5	4-45	20.7	11.76	56.81	62.0	21-124	68.5	32.61	47.61	318.0	83-1061	380.4	234.99	61.77
5 Minute	21.0	6-47	22.2	11.77	53.02	80.5	31-154	77.1	35.22	45.68	381.0	106- 988	428.5	244.41	57.04
7 Minute	13.0	4-44	17.5	11.94	68.23	45.0	9-171	60.2	43.36	72.03	293.0	57-1288	390.0	310.02	79.49

*Two perfect trials out of three.

TABLE 15

*Reliability of the difference between the averages of the various groups (Norm A)**

Groups	Trials				Errors				Time			
	Difference	S.D. of the difference	Difference / S.D. of the difference	Chances in 100 of a true difference greater than 0	Difference	S.D. of the difference	Difference / S.D. of the difference	Chances in 100 of a true difference greater than 0	Difference	S.D. of the difference	Difference / S.D. of the difference	Chances in 100 of a true difference greater than 0
0 and 1 Min.	10.0	2.62	3.82	100	29.5	8.13	3.63	100	147.3	40.31	3.65	100
0 and 3 Min.	10.4	2.74	3.80	100	33.7	8.22	4.10	100	249.4	51.92	4.80	100
0 and 5 Min.	11.9	2.85	4.18	100	42.3	9.02	4.69	100	297.5	56.35	5.28	100
0 and 7 Min.	7.2	2.63	2.74	99.7	25.4	9.71	2.62	94	259.0	63.47	4.08	100
1 Min. and 3 Min.	.4	3.45	.12	55	4.2	9.76	.43	66	102.1	62.84	1.62	94
1 Min. and 5 Min.	1.9	3.46	.55	71	12.8	10.44	1.23	89	150.2	66.55	2.26	98.9
1 Min. and 7 Min.	2.8	3.37	.83	80	4.1	11.05	.37	64	111.7	72.69	1.54	93
3 Min. and 5 Min.	1.5	3.64	.41	65	8.6	10.52	.82	79	48.1	74.16	.65	74
3 Min. and 7 Min.	3.2	3.47	.92	82	8.3	11.11	.75	77	9.6	79.71	.12	54
5 Min. and 7 Min.	4.7	3.55	1.32	90	16.9	11.72	1.44	93	38.5	82.67	.47	67

*Two perfect trials out of three.

TABLE 16*

Average errors per trial for each five trials

No. of trials	Control	1 Min.	Groups 3 Min.	5 Min.	7 Min.
1-5	5	6	6	7	5
10	2	2	2	3	3
15	1	2	2	2	2
20	1	2	2	2	3
25	1	2	2	2	3
30	1	3	2	2	3
35		2	2	3	2
40		2	1	2	2
45		1	2	2	2
50		1	3	2	2

* In most learning experiments it is not possible to obtain an adequate picture of the average time per trial or of the average errors per trial for any group, due to the fact that individual animals reach the norm of mastery after different numbers of trials and are dropped out of the experiment. Thus the latter part of the curve must be based on smaller and smaller numbers of animals. In our experiment, however, we continued to run the animals in an effort to have them reach the higher norm (Norm B.) We, therefore, had data on a large number of animals beyond the point where they had learned under the lower norm (Norm A), and in constructing our tables (16, 17) and curves (Figures 7, 8) for time and errors per trial, we used this available data so as to increase in number of animals making up the average at a given point. This method probably gives a closer approximation to a true curve than possible statistical methods. The exact number of animals upon which the average at any point is based may be ascertained by reference to Figure 11.

TABLE 17

Average time per trial for each five trials

No.of trials	Control	1 Min.	Groups 3 Min.	5 Min.	7 Min.
1-5	18	21	27	30	24
10	7	9	12	12	13
15	6	9	13	14	17
20	7	11	16	16	24
25	8	13	15	18	25
30	7	15	18	18	25
35		16	21	24	22
40		18	19	21	27
45		13	23	20	23
50		16	29	20	22

partment reveal no noticeable differences between the groups. All animals were equally active, running back and forth, biting and clawing at the doors, etc. In no case did an animal remain quiescent or merely wait for the door to be opened. When released from the compartment, the animals in all cases ran directly to the food

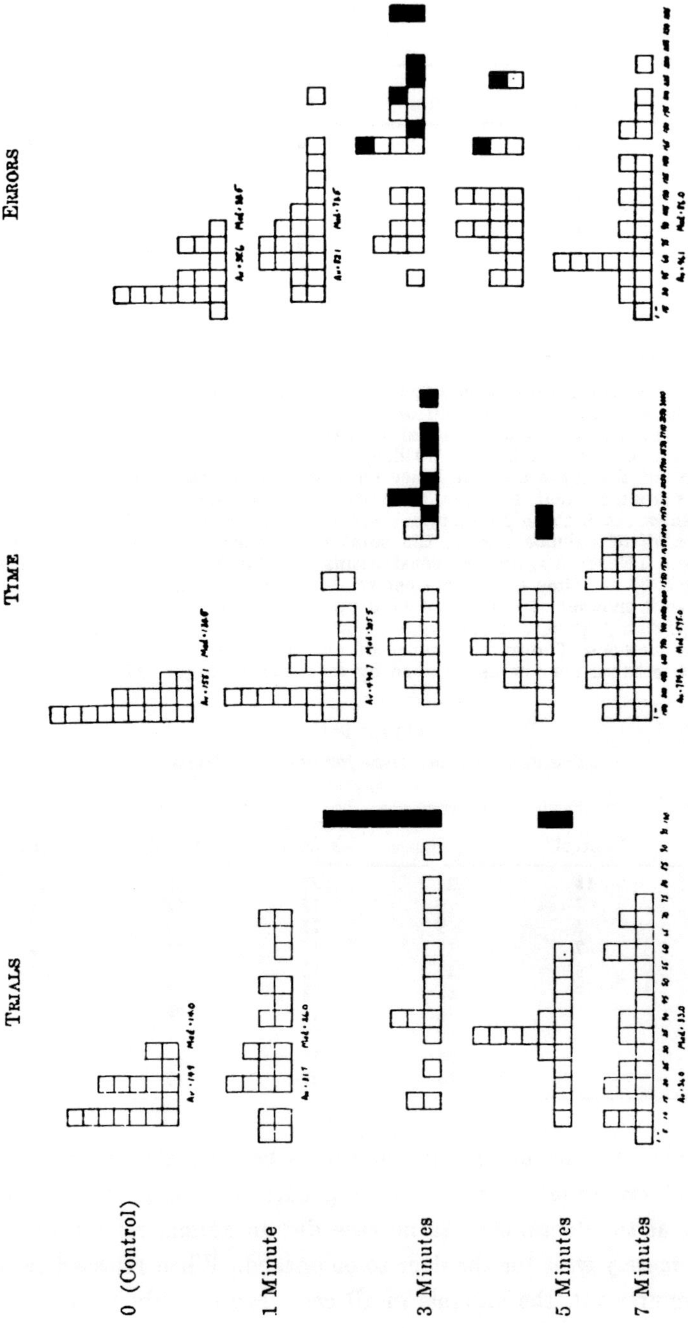

Fig. 9. Distribution of Animals According to Trials, Errors, and Time Taken to Learn (Norm B)
Each square represents one animal

and ate for the full minute allowed them in the food box, except on the first few trials during which they exhibited the usual exploratory activity which is natural for the rat in a relatively new situation.

Norm B. An attempt was made to run all of the animals until they attained the norm of four perfect trials out of five. Since 9 animals had not reached the norm at the 99th trial, and had shown no indication of learning for 50 trials, the change in conditions from delayed to immediate incentive was introduced. The effect of this change will be discussed in this section.

The results are presented in Tables 18, 19, and 20 and in Figures 9, 10, 11, and 12.

Figure 9 presents the distribution of animals according to trials, errors, and time taken to learn, each square representing one animal. The animals which had not learned by the 99th trial are indicated in solid black.

Table 18 is a cumulative frequency table showing percentage of group having learned by the end of each successive five trials. Figure 10 presents the data in graphic form.

Table 19 shows the effect of change in procedure from delayed to immediate incentive conditions, in terms of average time per trial, on the 9 animals which had not reached Norm B by the 99th trial. Figure 11 shows the same thing in graphic form.

Table 20 shows the effect of change in procedure from delayed to immediate incentive conditions, in terms of average errors per trial, on the 9 animals which had not reached Norm B by the 99th trial. Figure 12 shows the same data graphically.

We note from Figure 9 that 9 of the animals failed to learn, under this norm, in 99 trials. Obviously true averages and medians could not be obtained for the groups in which these animals occurred (3-minute and 5-minute groups). The averages and medians of the remaining groups are indicated in Figure 9. It will be seen from the distributions according to trials, errors, and time (Figure 9) that the delay groups all learned more slowly than the control group, showing the same general tendency as under Norm A. It appeared to be of some interest to see what would happen, under conditions of immediate feeding, to the 9 animals which had

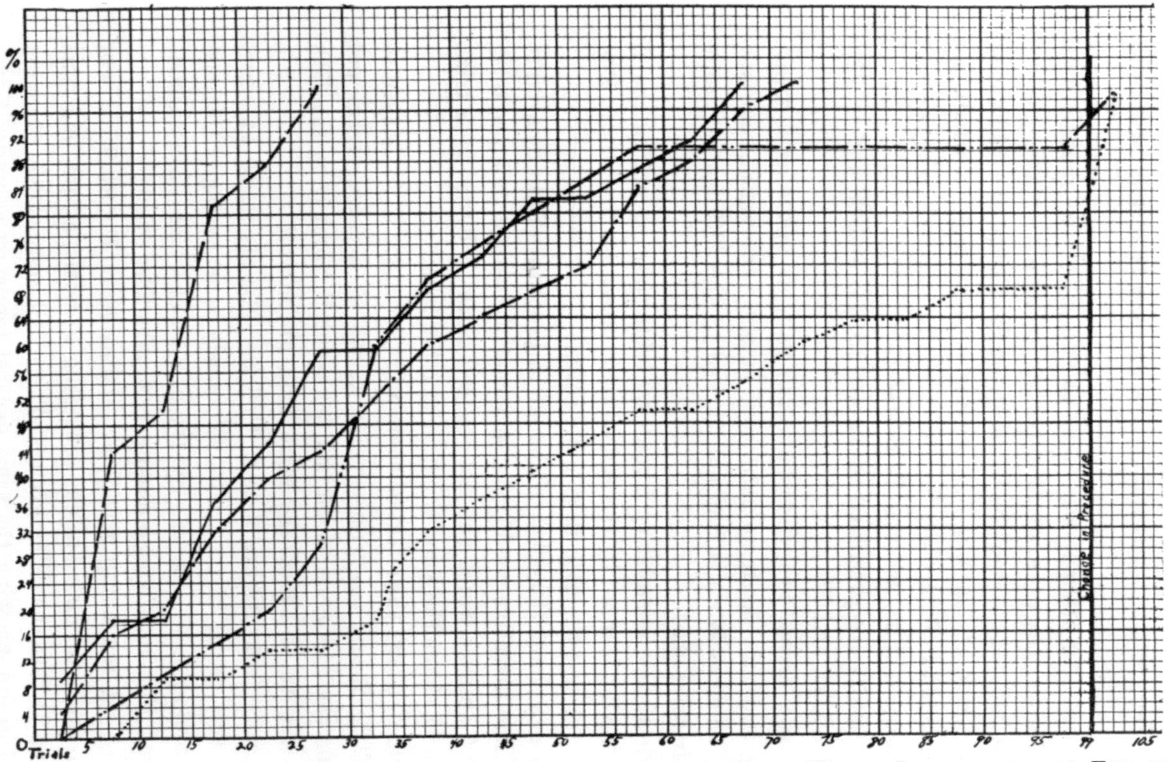

Fig. 10. Cumulative Frequency Curves Showing Percentage of Group Having Learned by the End of Each Successive Five Trials (Norm B)

Control group (dash line); 1-minute group (solid line); 3-minute group (dot line); 5 minute group (dot-dot-dash line); 7-minute group (dot-dash line).

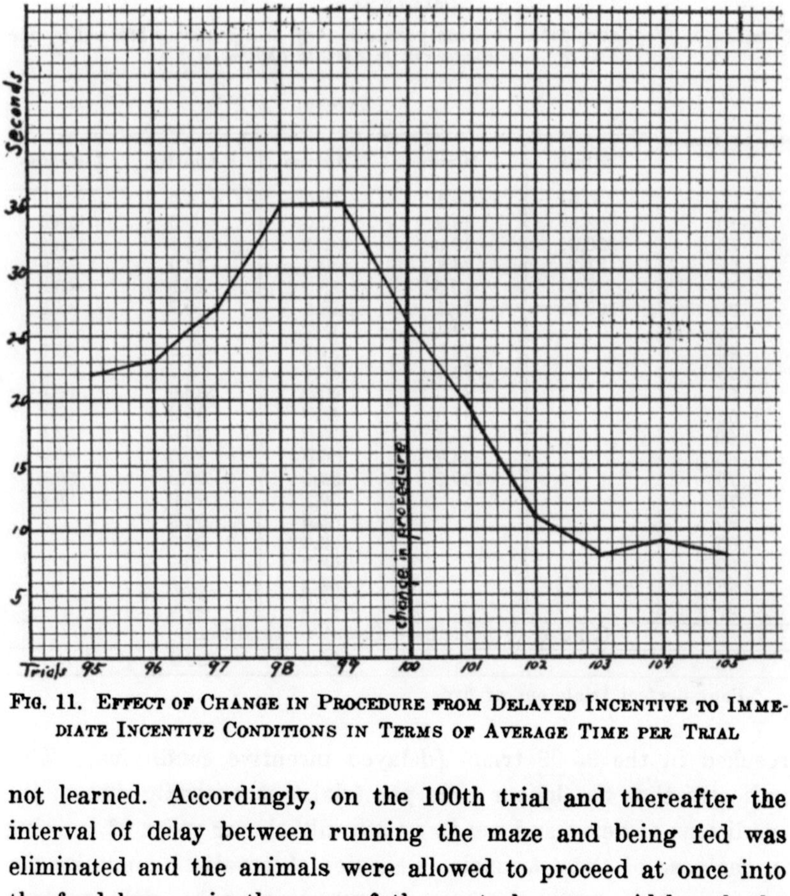

FIG. 11. EFFECT OF CHANGE IN PROCEDURE FROM DELAYED INCENTIVE TO IMMEDIATE INCENTIVE CONDITIONS IN TERMS OF AVERAGE TIME PER TRIAL

not learned. Accordingly, on the 100th trial and thereafter the interval of delay between running the maze and being fed was eliminated and the animals were allowed to proceed at once into the food box, as in the case of the control group. Although the animals had shown no progress in learning for 50 trials past, they speedily began to learn under the new conditions. As will be seen from Figure 10, by the 105th trial all 9 animals had reached the degree of mastery required under Norm B. The change in condition thus brought about almost immediate learning. This indicates that the animals were not lacking in ability to learn the maze and hence failure to learn up to the 99th trial was probably due to the factor of delay.

Figure 11 shows that the average time per trial decreased very markedly after the change in procedure was introduced. On the 101st trial the curve descended to a point lower than had been

TABLE 18

*Cumulative frequency table showing percentage of group having learned by the end of each successive five trials (Norm B)**

NO. OF TRIALS	GROUPS				
	Control	1 Minute	3 Minutes	5 Minutes	7 Minutes
1-5	0.0	9.1	0.0	0.0	4.0
10	43.8	18.2	0.0	5.0	16.0
15	50.0	18.2	9.1	10.0	20.0
20	81.3	36.4	9.1	15.0	32.0
25	87.5	45.5	13.6	20.0	40.0
30	100.0	59.1	13.6	30.0	44.0
35		59.1	18.2	60.0	52.0
40		68.2	31.8	70.0	60.0
45		72.7	36.4	75.0	64.0
50		81.8	40.9	80.0	68.0
55		81.8	45.5	85.0	72.0
60		86.4	50.0	90.0	84.0
65		90.9	50.0	90.0	88.0
70		100.0	54.5	90.0	96.0
75			59.1	90.0	100.0
80			63.6	90.0	
85			63.6	90.0	
90			68.2	90.0	
95			68.2	90.0	
99			68.2	90.0	
Change to immediate incentive conditions					
105			100.0	100.0	

* Four perfect trials out of five.

reached in the 95-99 trials (delayed incentive conditions). This indicates that the longer time per trial was probably due to the conditions of delay and not to a naturally slower speed of running in the case of these animals. Behavior observations made by the experimenter support these conclusions. During the trials in which the incentive was delayed, the animals appeared to be less motivated, running slowly through the maze in a somewhat hesitant manner as above mentioned. Under the immediate incentive conditions, however, they ran quickly and directly through the maze to the food box, resembling in this respect the control-group animals in their behavior.

The error curve (Figure 12) shows the same general tendency as the time curve. The curve descended on the 102nd trial to the zero line. The general level is considerably lower than that of the 95-99 trials (delayed incentive conditions). This supports our conclusion that failure to learn up to the 99th trial was probably due

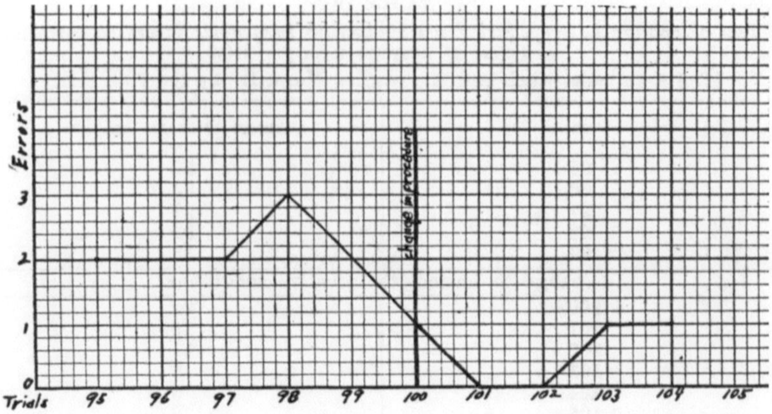

FIG. 12. EFFECT OF CHANGE IN PROCEDURE FROM DELAYED TO IMMEDIATE INCENTIVE CONDITIONS IN TERMS OF AVERAGE ERRORS PER TRIAL

to the conditions of delay rather than to inability to learn. The nature of the errors made previous to the 100th trial also offers corroborative evidence. The animals persisted in making one or two errors, going into the same cul-de-sac each time, which would suggest that the motivation was not strong enough to bring about perfect learning. In several cases, animals would run three perfect trials out of four and then make errors on the following trials so that the norm was not reached for many trials later. Evidently the delayed-incentive conditions worked against consistency in errorless runs since all animals reached Norm A by the 50th trial, while under the higher norm 9 had not learned by the 99th trial. That is, by reducing the motivation, delay operated to lower the quality or degree of perfection to which the habit was developed.

In relating our results to those of previous investigators, we find that few studies are comparable. Simmons (6) used a 2-hour period of delay, so naturally no comparison with the present work, which was restricted to much shorter periods, is possible. Watson (14) found that a delay as short as 30 seconds had no effect upon speed of learning. However, the interval he used was only half as long as our shortest, so the results are not comparable.

In a former study made by the writer (8), intervals of delay of 1 minute and 5 minutes were found to have no effect on speed of

TABLE 19

Effect of change in procedure from delayed to immediate incentive conditions in terms of average time per trial

(The 9 animals which had not reached Norm B* by the 99th trial)

TRIAL	AVERAGE TIME IN SECONDS
95	22
96	23
97	27
98	35
99	35
Change to immediate incentive conditions	
100	26
101	19
102	11
103	8
104	9
105	8

* Four perfect trials out of five.

TABLE 20

Effect of change in procedure from delayed to immediate incentive conditions in terms of average errors per trial

(The 9 animals which had not reached Norm B* by the 99th trial)

TRIAL	AVERAGE ERRORS
95	2
96	2
97	2
98	3
99	2
Change to immediate incentive conditions	
100	1
101	1
102	0
103	0
104	1
105	1

* Four perfect trials out of five.

learning. In the general tendency shown, this finding would agree with Watson's conclusion, but is not supported by the present experiment in which all periods of delay employed (including a 1-minute and 5-minute period) decreased the speed of learning. However, the previous study differed from the present one in one important respect: the animal was delayed in the same compartment with the food. That is, in the former experiment the animal was directly stimulated by the presence of the unobtainable food, whereas in the present case the direct stimulation of the food could occur only after the animal was released from the delay compartment into the food box. In the present experiment the matter of place of delay was much better controlled — the delay compartment was entirely separated from the food compartment. Evidently the situation of being delayed in the same place with the food, even though the latter be inaccessible, has little or no effect on the fixation process. While such a situation involves temporal delay, the pattern of the performance of running the maze is not definitely broken since the animal is not delayed until it comes upon the spot where the food is located. Our present technique involved not merely temporal delay but temporal delay at a spatial position which interrupted the general performance pattern of reaching the food.

Perhaps the conditions in our present experiment may be said to be analogous to the situation of children, at a birthday party, who are not allowed to enter immediately the room where the birthday cake is with its candles and decorations, but are made to wait in an anteroom. The former study might then be analogous to the situation in which the same children are not detained in an anteroom but are allowed immediate access to the refreshment room. However, they are prevented from touching or eating anything for an interval of time. The latter situation has the added element of direct stimulation of the desired objects, while the former involves mere knowledge about them. Watson, who also did not use the anteroom technique, but delayed the animals in the same compartment with the food, obtained similar results with those of the previous study of the writer, but different from those in the present experiment.

The results here obtained seem to show clearly that delay decreased the drive as measured by rate of fixation. In the case of all of the delay groups, the rate of learning was reliably slower than that for the control group, in terms of trials and errors. The fact that running time was increased somewhat, with longer delays, indicates a motivation difference which, however, did not affect measurably the rate of fixation. Obviously, trials and errors are rather gross units of measurement of efficiency and if finer units could have been used, a decreased efficiency reflecting the slowing down in running time might have been found. On the other hand, there is no need to suppose that speed of maze running is perfectly, or even highly, correlated with rate of fixation.

There is some indication that delay affected the quality of the learning, i.e., the habit was not formed as perfectly under the delay conditions, as seen by a comparison of results under the two norms of mastery. Although the animals in all groups had attained the lower norm (Norm A) [2] by the 50th trial, 9 of them had failed to learn under the higher norm (Norm B)[3] by the 99th trial. Delay had apparently decreased the drive in these animals and there was insufficient motivation to bring about further perfection of the habit. This interpretation is borne out by the fact that the 9 animals immediately developed the habit to the higher level (Norm B)[4] when the delay was eliminated.

It would seem to be impossible to relate our findings in any very important way to the law of effect as formulated by Thorndike. Our results agree with an increasingly large body of data showing the advantageous effects of motivated over unmotivated learning situations. If the terms "satisfying state" and "annoying state," as employed in the statement of the law of effect (7, p. 4), be taken in their usual psychological connotation as indicative of the subjective states of the organism, then naturally our findings could not be stated in terms of this law since we have no index of the subjective states of our animals. It might be argued that they "enjoyed" the delay in some anticipatory way with as much cogency as that they were "annoyed" by it. The drive and incentive values obtained serve as objective indices of certain tendencies of the organism under the various conditions of motivation employed — and that is all that can be said. The argument covering this point has been made by Carr (1) and other writers.

As a corollary to the law of effect, Thorndike develops the principle that the strength of the bond is determined by the temporal closeness of the response and the satisfying or annoying state which follows. If we assume, as Thorndike seems to, that the incentive (food) produces a satisfying state, then our results would suggest

[2] Two perfect trials out of three.

[3] Four perfect trials out of five.

that very definite restriction must be placed upon the value of the temporal factor in this type of learning. For, while any interval, within the limits investigated, cuts down very markedly the efficiency of the learning process, the longer intervals were hardly more effective than the shorter. Our results do not afford us a basis for conjecture as to whether delay operated as a positive "punishment" or merely as a distraction from the incentive situation in a purely negative sense, nor do they offer any clues as to the physiological (or neurological) processes underlying the behavior called out.

C. Summary of results

Norm A[4]

(1) An interval of delay as short as 1 minute between the running of the maze and the feeding response markedly decreased the rate of learning, in terms of the scores for trials, errors, and time. (The increase in trials taken to learn was 97 per cent; in errors, 85 per cent; in time, 112 per cent.)

(2) Longer delay periods of 3 minutes, 5 minutes, and 7 minutes did not further decrease the speed of learning, as indicated by the scores on trials and errors. The time scores, however, were considerably lengthened as the interval of delay increased in length, indicating a slower rate of progress through the maze.

Norm B[5]

(1) When a high norm of mastery (four perfect trials out of five) was used, 9 of the 105 animals failed to learn by the 99th trial, although all had learned by the 50th trial under Norm A.

(2) Change of conditions from delayed to immediate incentive had a markedly favorable effect on the performance of the abovementioned 9 animals. Time and error scores decreased immediately and all of the animals learned within 5 trials after the change in procedure, although they had made little or no progress for approximately the last 50 trials under the delayed incentive conditions.

[4] Two perfect trials out of three.

[5] Four perfect trials out of five.

D. Comparison of obstruction and learning methods

In a comparison of the results obtained in the two experiments it is essential to have clearly in mind the method and procedure employed in each experiment. The most important differences in method have been arranged in tabular form as follows:

(1) The length of the test period was arbitrarily limited to 20 minutes in the obstruction experiment, whereas time per trial was determined by the animal and varied from time to time in the case of the maze experiment.

(2) Only a single test period was given each animal in the Obstruction Method, hence only a short time required per animal. No animal in the longest delay group (3-minute) required a longer time than one hour for the complete test. In the maze study, however, one trial a day was given until the maze had been learned, and thus the experiment extended over a much longer period of time per animal. An animal in the 3-minute group might require as long as three hours of testing time.

(3) The period of starvation was 48 hours in the obstruction study. This period was used in order to get the maximal state of the hunger drive which should best bring out the effects of delay. In the maze experiment a period of 24 hours was used, since one trial per day was given.

(4) The number of intervals of delay was determined by the number of crossings possible within the 20-minute period in the obstruction study. None of our animals reached the theoretical limit which has been shown to be about 44 crossings (9). On the other hand, the number of delay periods per trial in the learning experiment naturally was limited to one, since only a single daily trial was given.

(5) In reacting to the incentive situation in the obstruction experiment, the act of securing the incentive did not need to be learned. That is, a more direct measure of drive was obtained, largely uncomplicated by the chance factors that always operate in a learning situation of the maze type.

(6) The obstruction study involved "punishment" (shock), whereas the maze involved no punishment, but was a type of free activity which seems to be natural to the rat.

(7) Fewer chance factors enter into the obstruction test, since such factors as illumination, sex drive, emotional disturbance, etc., could be more carefully controlled. The fact that only a single test period involved but one incentive situation, while the maze involved many incentive situations distributed over a much longer period of time, also made possible better control of conditions in the obstruction experiment than in the maze experiment.

It is apparent from the differing nature of the methods employed in this study that only general comparisons of the effect of delay

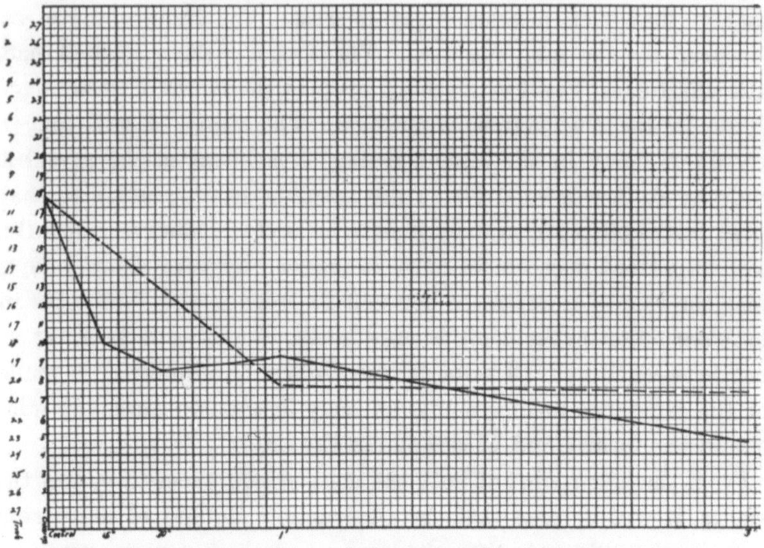

FIG. 13. MAIN RESULTS IN OBSTRUCTION AND MAZE EXPERIMENTS.

The solid line represents the average number of crossings for the combined (male and female) groups (Obstruction Method). The dash line represents the average number of trials required to learn (Maze Experiment).

can be made. In order to compare the general trend of the results in the experiments, curves were constructed on the basis of the main results in each experiment as shown in Figure 13. Since no comparison is possible beyond the 3-minute interval, the curve for the maze was not extended beyond this point. It will be seen from the curves that a delay of 1 minute markedly decreased the drive and to much the same extent in both experiments. We do not know what effect shorter delays such as 15 seconds and 30 seconds might have had in the maze-learning experiment. During the progress of the maze experiment, no further decrease in the drive was obtained with the longer delays of 3 minutes and 5 minutes, so it appeared to be of more interest to use a longer interval of 7 minutes in an attempt to determine whether the plateau effect was a real tendency or not. It would have been interesting also to study the effect of intervals shorter than 1 minute, but this was impossible without undue extension of the experiment.

In general, the same tendency is apparent in both experiments; namely, that even a short delay decreases the hunger drive, as measured by the methods used. The longer delays studied had little effect except that of a second drop in the curve of some magnitude in the obstruction experiment. The study of still longer periods of delay in the maze-learning experiment might reveal a tendency to decreased drive in the learning scores similar to that noted in the obstruction experiment. It is clear, however, that the most important decrease in drive is induced by very short delays, since the rate of decrease as between 15 seconds and 3 minutes in the obstruction experiment is only about 50 per cent as great as that between 0 and 15 seconds, even when no allowance is made for the difference in increments of delay in the two cases ($2\frac{3}{4}$ minutes in the former and 15 seconds in the latter).

From the standpoint of both the method and procedure, and the results, it will be seen that the Obstruction Method of measuring drive is more adequate than the learning method in many respects. As above noted, the Obstruction Method is superior in that it furnishes a more *direct* measure of drive, uncomplicated by the factor of fixation; the technique is better standardized and controlled than the maze method and hence fewer chance factors enter into the drive index. Furthermore, the time consumed in obtaining an index of drive is much less than half as long as that required in the direct maze-learning method.

Bibliography

(1) Carr, H. 1914. Principles of selection in animal learning. *Psychol. Rev.*, 21, 157-65.

(2) Cook, S. A. 1928. The effect of various temporal arrangements of practice on the mastery of an animal maze of moderate complexity. *Arch. Psychol.*, 15, No. 98, 33 pp.
(3) Jenkins, M. 1928. The effect of segregation on the sex behavior of the white rat as measured by the obstruction method. *Genet. Psychol. Monographs*, 3, 455-571.
(4) Jenkins, T. N., L. H. Warner, and C. J. Warden. 1926. Standard apparatus for the study of animal motivation. *J. Comp. Psychol.*, 6, 361-82.
(5) Long, J. A., and H. M. Evans. 1922. The oestrous cycle in the rat and its related phenomena. *Mem. Univ. Calif.*, 6, 148 pp.
(6) Simmons, R. 1924. The relative effectiveness of certain incentives in animal learning. *Comp. Psychol. Monographs*, 2, No. 7, 79 pp.
(7) Thorndike, E. L. 1913. Educational psychology, II. New York, Teachers College, 452 pp.
(8) Warden, C. J., and E. L. Haas. 1927. The effect of short intervals of delay in feeding upon speed of maze learning. *J. Comp. Psychol.*, 7, 107-15.
(9) Warden, C. J., and H. W. Nissen. 1928. An experimental analysis of the obstruction method of measuring animal drives. *J. Comp. Psychol.*, 8, 325-42.
(10) Warner, L. H. 1928. A study of the hunger drive in the white rat by means of the obstruction method. *J. Comp. Psychol.*, 8, 273-99.
(11) ———1927. A study of sex behavior in the white rat by means of the obstruction method. *Comp. Psychol. Monographs*, 4, No. 22, 68 pp.
(12) ———1928. A study of the thirst drive in the white rat by means of the obstruction method. *J. Genet. Psychol.*, 35, 178-92.
(13) Warner, L. H., and C. J. Warden. 1927. The development of a standardized animal maze. *Arch. Psychol.*, 15, No. 92, 64 pp.
(14) Watson, J. B. 1917. The effect of delayed feeding upon learning. *Psychobiol.*, 1, 51-59.

4. THE RELATIVE VALUE OF REWARD AND PUNISHMENT IN THE FORMATION OF A VISUAL DISCRIMINATION HABIT IN THE WHITE RAT [1]

C. J. Warden and Mercy Aylesworth [2]

One of the major problems within the general field of animal motivation is that concerning the relative incentive value of reward and punishment. The terms reward and punishment as here used carry no subjective connotation whatsoever, but are used merely as convenient classificatory terms for stimuli that normally evoke positive and negative reactions respectively when employed as incentives in experimental situations. The chief interest in this topic centers around the relation of this factor to speed or efficiency of habit formation, and more intensive work has been done on the discrimination type of habit than upon either problem box or maze.

The present study was suggested by the investigation of Hoge and Stocking (1) who found some evidence that punishment has a higher incentive value, either alone or in combination with reward, than reward alone in the building up of a simple visual discrimination habit in the white rat. Their conclusion to this effect was based upon very slight data, only two animals being tested in each group. In one case correct responses were rewarded by a morsel of food; in a second, correct responses went unrewarded while wrong responses were followed by a shock, while in the third instance a correct response was followed by food and an incorrect response by a shock. They found that both rats in the reward-punishment group perfected the habit — one in 490 trials and the other in 550 trials, while one rat in the punishment group learned in 550 trials and the other failed to learn in 620 trials, and neither

[1] Reprinted from *The Journal of Comparative Psychology*, 1927, 7: 117-27.

[2] The problem was suggested and the data gathered by the junior author; the senior author collaborated in technical details and in writing this report.

of the animals in the reward group had learned after 590 trials. The problem involved a discrimination between a 2 c.p. and 16 c.p. light, the norm of mastery being 15 correct responses on each of two successive days, and the apparatus a modification of the Yerkes discrimination box. The shock used was furnished by a Porter inductorium and applied in the usual manner.

The purpose of the present investigation was to repeat the work of Hoge and Stocking, using larger groups and modifying the technique in certain details. The main improvement in method, perhaps, consisted in a more standardized shock than that used by Hoge and Stocking, the mechanism for the control of which being identical with that employed in connection with the Obstruction box (2) of this laboratory. In the present work a current of 0.11 m a. with a resistance of 1000 volts was used, and readings were taken daily before beginning the taking of data. The resulting shock was strong enough to cause the animals to lift their feet energetically and sometimes squeal a little in scampering off the grills, without causing them to be too inactive for our purposes after getting off. We used a group of 10 white rats approximately 3 months of age in connection with each of the three incentive conditions previously studied by Hoge and Stocking, i.e., reward, reward-punishment, and punishment.

The apparatus employed was a modification of the Yerkes-Watson design (see their description, Behavior Monographs No. 2) and need not be figured in detail. The source of light was a 75-watt Mazda Westinghouse bulb (120 volts) set 10 inches behind the plate at an angle of 45 degrees, in such wise that the center of the bulb coincided with the center of the circular opening (6 cm.) of the plate. The distance from center to center of the openings on either side of the stimulus box was 5¼ inches, and that from the center of the opening to the floor of the control box was 4 inches. A bulb was placed on either side of the box so that a shift from light to dark could be made without disturbing the shutter, by the simple expedient of pressing an electric button conveniently placed on the outside of the stimulus box. The current was supplied to the two bulbs through a common intake. The ground plan of the control section of the apparatus in which the animal was placed for testing is shown in figure 1. The box is 37 inches long, 29 inches wide, and 11½ inches deep (inside dimensions) and the partitions as figured are drawn to scale. The electric grills were equipped with independent switches so that either could be turned off at will by the experimenter, stationed directly in front of the entrance compartment. Since a dark room was not available, the control box was entirely screened in by a hood of black cloth extending 24 inches above the top of the box. Inside the hood and in the exact center

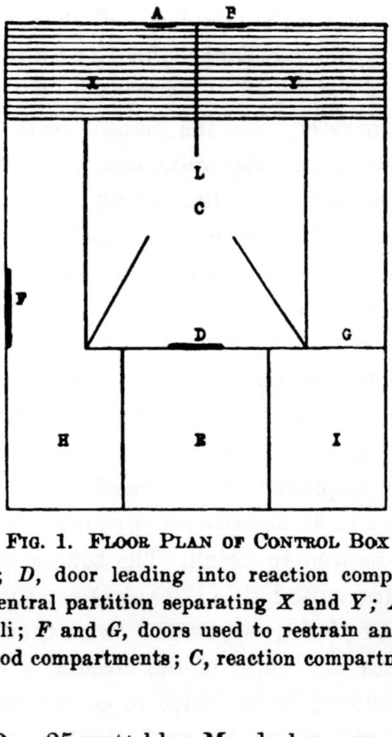

Fig. 1. Floor Plan of Control Box

E, entrance box; *D*, door leading into reaction compartment; *X* and *Y*, electric plates; *L*, central partition separating *X* and *Y*; *A* and *B*, position of lights used as stimuli; *F* and *G*, doors used to restrain animal (*F* open and *G* closed); *H* and *I*, food compartments; *C*, reaction compartment.

above the door, *D*, a 25-watt blue Mazda lamp was placed and kept burning constantly during the experiment. Small automobile mirrors were fastened to the top of the box above the grills, *X* and *Y*. The animals were visible in the dim light at all times under this arrangement and were observed through a peep hole in the hood above the center of *E*.

The problem was to learn to discriminate between the stimulus patches outlined by the circular openings in the plates, when one was illuminated from behind and the other not. The animals were trained to the dark patch inasmuch as 8 of the 10 in each group showed either a preference for the illuminated patch or no preference at all, the other two in each case showing only a slight preference for the dark patch. The preference series consisted of 10 trials per day for two days with food in both compartments (*I* and *H*), regardless of later incentive conditions. As the data show, the discrimination was much less difficult than that required in the Hoge and Stocking experiment.

The procedure was kept as uniform for the three groups as possible. The characteristic response of the animal under the different conditions of motivation made necessary minor variations in procedure as will appear from the description that follows. When the response was correct, in the reward group, a nibble of milk-soaked bread was allowed in the appropriate goal box; when incorrect the animal was lifted from the goal box, which contained no food, by the experimenter. Food was placed in neither goal compartment in the punishment series; when a correct response was scored the animal merely escaped the shock and was removed from the goal compartment as above. An incorrect response resulted in a shock and, as a rule, the animal retreated from the grill and remained in the reaction compartment, *C*, since there was no incentive for it to continue on across the grill into the goal compartment for that side. Accordingly the animals of this group were removed from the reaction compartment after such retreat from the grill, or from the appropriate goal compartment if they actually crossed the grill instead of retreating. Contact with the grill beneath the light, resulting in a shock, was thus the basis of scoring errors for this group regardless of the nature of the more or less insignificant response that followed this experience. The same procedure was followed likewise when an animal of the reward-punishment group made an error; it was removed from either position, the goal or reaction compartment, depending upon whether it retreated from, or crossed over the grill. In certain instances in both the reward and reward-punishment groups the animals would dash around to the correct goal compartment after making a wrong response before the restraining door (*F* or *G*) could be closed. This occurred in approximately 5 per cent of the trials of the former, and 2 per cent of the trials of the latter group.

The original plan was to give 15 trials per day and the order of shifting the lights was arranged for such a schedule. This order which was repeated over and over, without regard to the number of trials given per day, was as follows for the dark stimulus patch: left, right, right, left, right, left, right, right, left, left, left, right, right, left, right. Five trials per day were given for the first 7 days, ten trials per day thereafter until the 27th day, after which the 15 trial per day schedule was followed in the case of the reward group, which showed little or no evidence of learning at this stage. The reward-punishment group had completed the problem by the 15th day; the punishment group were so nearly finished on the 27th day that they were continued on the 10 trial per day schedule. The change to the 15 trial schedule on the 27th day for the reward group was made in the hope that they might master the problem within the limit of time available for the study.

During the first day 10 minutes were allowed for each trial provided the animal did not make a "choice" before. From the second day onward this time was reduced to 5 minutes. Animals of the reward group usually reacted within a few seconds. However, the use of the shock in the other two groups induced hesitation and inactivity to such an extent that many of the animals did not react within the 5-minute period. In the reward-punishment group the animals had to be removed without having reacted in 59 out of 760 trials (7.8 per cent) while in the punishment group the corresponding value was 425 out of 1500 trials (28.3 per cent). The effectiveness of the food in inducing an increased tendency to react promptly in the former group is thus clearly indicated. Failure to react within the 5-minute period was checked as an error, although the data on this point are given separately from the errors, in table 1, under the heading "no reaction."

The positive value of the reward incentive stimulus (food) in bringing about activity, and especially the definite act of going promptly to the goal was also very evident. When reward alone was used the complete reaction of going to one or the other goal compartments occurred in practically every case. Naturally there was

TABLE 1

Showing the percentage of right, dwrong, and "no reaction" responses for each series of trials

DAYS	REWARD		PUNISHMENT			REWARD-PUNISHMENT		
	Right	Wrong	Right	Wrong	"No reaction"	Right	Wrong	"No reaction"
1	30	70	42	44	14	34	62	4
2	28	72	10	10	80	40	26	34
3	36	64	24	14	62	42	26	32
4	32	68	44	18	38	62	18	20
5	38	62	46	18	36	68	20	12
6	32	68	56	14	30	64	30	6
7	44	56	52	8	40	84	14	2
8	42	58	57	11	32	84	15	1
9	45	55	51	7	42	80	20	
10	47	53	55	10	35	95	5	
11	43	57	59	6	35	86	12	2
12	47	53	56	4	40	87	10	3
13	48	52	54	6	40	90		10
14	45	55	71	9	20	100		
15	54	46	67	12	21	100		
16	47	53	80	10	10			
17	48	52	86	6	8			
18	52	48	75	13	12			
19	55	45	60	12	28			
20	57	43	85	12	3			
21	63	37	83	7	10			
22	58	42	80	15	5			
23	52	48	80	10	10			
24	49	51	70	20	10			
25	58	42	100					
26	51	49	70	30				
27	55	45	50	40	10			
28	57	43	50	30	20			
29	57	43	80	20				
30	60	40	90	10				
31	59	41	70	30				
32	55	45	80	10	10			
33	57	43	100					
34	56	44	100					
35	61	39						
36	57	43						
37	62	38						
38	55	45						
39	61	39						
40	65	35						
41	62	38						

The term "no reaction" means that the animal did not make a "choice" within the five-minute test period. Such failure to react did not occur in the case of the reword group.

little or no incentive for the animals of the punishment group to go to the goal compartments. In 376 trials out of 1500 (25 per cent) they were removed without having reached the goal — they reacted to the grill-stimulus patch situation and stopped at that, although they had been fed during the 20 trials of the preference series in whichever goal compartment they happened to enter. This type of response occurred in only 62 out of 760 trials (8 per cent) in the case of the reward-punishment group which had, of course, a more definite incentive to keep running till they reached the goal. When both "no reaction" and failures to go on to the goal after reacting are combined, we find that in 63.3 per cent of the trials the punishment group did not actually reach the goal while the corresponding value for the punishment-reward group is 15.8 per cent.

Bread and milk constituted the main article of diet both before and during the experiment. The animals of each group were allowed to feed for 10 to 15 minutes (or until they were satiated) one hour after the daily schedule of trials for the group had been completed. This interval had the effect of separating the ordinary taking of nourishment from the incentive situation presented in the apparatus. The groups were tested at the same time of day in so far as the conditions of the experiment permitted, the work beginning at 11 a.m. each day. Each trial was timed with a stop watch but these data have not been included in this report since they appear to add nothing of significance.

The results are given in table 1 and the graph of figure 2 in terms of the percent of right, wrong, and "no reaction" responses per day throughout the 41 day training period. At the end of this period the reward group were scoring only about 60 per cent right responses and had made little gain in the preceding 300 trials (20th day onward). The punishment group made a perfect record on the 33rd day and the reward-punishment group on the 13th day, these differences being so great that no question of interpretation arises. The reward-punishment motivation is by far the best; reward alone is markedly inferior to punishment alone, just how much so we cannot tell since the former group did not perfect the habit at all. The rise in the error curve of the punishment group on the second and third days was due to the large number of "no reaction" responses occurring on those days. The marked rise in this curve on days 26-28 and on several days thereafter was due in large part to the erratic behavior of rat no. 4. This rat after making almost perfect records for several days suddenly took to jumping over the grill on the right hand side, without regard to the stimuli, and this habit had to be broken up before further progress could be made.

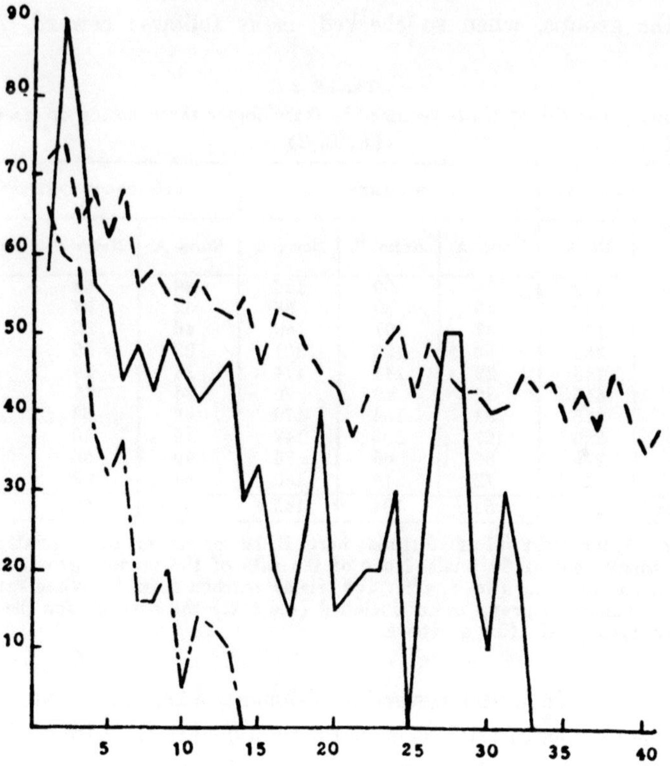

FIG. 2. CURVE SHOWING ERROR ELIMINATION FOR THE THREE GROUPS

Percentage of wrong (including "no reaction") responses are plotted on the ordinate against days of the training period on the abscissa.

— — — reward. ———— punishment. — · — · — reward-punishment.

The data on number of trials required to form the habit are given in table 2 for each animal of the three groups as well as the averages for the groups. The records were checked against three norms of mastery, i.e., 9 correct out of 10, 18 correct out of 20, and

27 correct out of 30 responses. None of the animals of the reward group reached the highest norm and only one (no. 8) reached the norm of 18 correct responses out of 20 consecutive trials, doing so only after 265 trials. The data for the lowest norm only, therefore, are included in the table for the reward group. A comparison of the averages for the three groups is thus possible only on the basis of the data of norm A. The average number of trials to learn for the groups, when so checked, is as follows: reward, 293.5,

TABLE 2
Showing number of trials required to learn under three norms of mastery (A, B, C)

RAT	REWARD	PUNISHMENT			REWARD-PUNISHMENT		
	Norm	Norm A	Norm B	Norm C	Norm A	Norm B	Norm C
1	439	36	89	126	39	64	72
2	242	50	60	68	32	37	47
3	173	47	137	145	46	52	62
4	362	60	125	301	23	99	108
5	145	22	141	174	13	49	57
6	446	32	42	70	24	75	84
7	276	81	134	170	39	56	59
8	255	127	136	147	19	40	45
9	279	35	66	75	49	56	60
10	318	72	114	183	44	69	79
Average	293.5	56.2	104.4	145.9	32.8	59.7	67.3

Norm A, 9 correct out of 10 trials; norm B, 18 correct out of 20 trials; norm C, 27 correct out of 30 trials. None of the rats of the reward group reached norm C and only one (no. 8, after 265 trials) reached norm B. When rat no. 4 of the punishment group is not included (see text) the average for the group is 128.6 instead of 145.9 as above.

punishment, 56.2, and reward-punishment, 32.8. It is noteworthy that the number of trials is more than doubled in the case of the two latter groups when the norm is raised to 27 correct responses out of 30 trials (norm C).

It is evident that norm A is not high enough to insure dependable results in work of this sort. It does not rule out the possibility of apparent success in discrimination which is, in fact, only apparent and due to chance factors. For, as appears from table 1, the reward group as a whole never averaged on any given day higher than 65 per cent correct responses, and yet each of them made a record of 9 correct out of 10 trials at least once during the training period. That this record was due to chance factors mainly is shown by the fact that only one animal of the group (rat number 8) was able to reach norm B and none norm C, although in many cases the requirement of norm A had been met before one-half of the training series (223 trials) was completed. The use of a high norm of mastery appears to be more necessary in discrimination work than in habit formation of the maze type (3). The order of shifting the lights, and the greater likelihood of the setting up of position habits in a simple situation of this sort, probably operate as important factors in this connection.

The general principle suggested by the work of Hoge and Stocking seems to be fully established by the present work, for this type of problem. This conclusion should, however, be limited to simple and easy discrimination, inasmuch as several investigators (Yerkes and Dodson, Dodson, Cole, etc.) have shown that in more difficult discrimination too much punishment (shock) must not be employed if the best results would be obtained. Before an unqualified generalization can be made concerning the relative incentive value of reward, punishment, and reward-punishment these must be studied in relation to a series of discriminations ranging between wide extremes of difficultness.

A criticism of the control box used by ourselves and others, in so far as its use in studying the relative value of incentives is concerned, should be made by way of offering a suggestion regarding further work on this topic. The shock is usually administered by means of a grill placed directly in front of the stimulus patches while the food or reward is placed at some distance from the patches and can be reached only by making a turn after crossing the grill. Doubtless the value of the punishment incentive is greatly increased by the fact that it tends to stop the animal directly in front of the discrimination stimuli while the reward incentive (food) does not. In order to obtain strictly comparable data the reward and punishment factors should be administered at points equidistant from the stimulus patches. That is, either the food (or other reward object) should be located in the front end of the control box near the grill, or perhaps better, the grill should be placed in, or near the goal compartments rather than directly in front of the stimulus patches.

So far as we know, this matter of the identity of location of incentive stimuli has not been adequately controlled in any work so far done on the relative value of reward and punishment in this type of habit formation.

Bibliography

(1) Hoge, U. A., and R. J. Stocking. 1912. A note on the relative value of reward and punishment as motives. *Jour. Anim. Behav.*, 2, 43-50.

(2) Jenkins, T. N., L. H. Warner, and C. J. Warden. 1926. Standard apparatus for the study of animal motivation. *Jour. Comp. Phychol.*

(3) Warden, C. J. 1926. A comparison of different norms of mastery in animal maze learning. *Jour. Comp. Psychol.*, 159-79.

INDEX

INDEX

COLUMBIA UNIVERSITY PRESS
Columbia University
New York

FOREIGN AGENT
OXFORD UNIVERSITY PRESS
Humphrey Milford
Amen House, London, E. C.

COLUMBIA UNIVERSITY PRESS
Columbia University
New York

FOREIGN AGENT
OXFORD UNIVERSITY PRESS
Humphrey Milford
Amen House, London, E. C.

Bei Fragen zur Produktsicherheit wenden Sie sich bitte an:
If you have any questions regarding product safety,
please contact:

Walter de Gruyter GmbH
Genthiner Straße 13
10785 Berlin
productsafety@degruyterbrill.com